Fundamental Research in
Homogeneous
Catalysis
Volume 2

A Continuation Order Plan is available for this series. A continuation order will bring delivery of each new volume immediately upon publication. Volumes are billed only upon actual shipment. For further information please contact the publisher.

Fundamental Research in
Homogeneous
Catalysis

Volume 2

Edited by

Yoshio Ishii
Chubu Institute of Technology
Aichi, Japan

and

Minoru Tsutsui
Texas A & M University
College Station, Texas

Plenum Press · New York and London

Library of Congress Cataloging in Publication Data

International Workshop on Fundamental Research in Homogeneous Catalysis, 1st,
  Santa Flavia, Italy, 1976; 2d, Shiga, Japan, 1977.
  Fundamental research in homogeneous catalysis.

  Vol. 2 edited by Y. Ishii and M. Tsutsui.
  Sponsored by the Toray Science Foundation and others.
  Includes bibliographies and index.
  1. Catalysis—Congresses. I. Tsutsui, Minoru, 1918-   II. Ugo, Renato. III.
Ishii, Yoshio, 1914-   IV. Toray Science Foundation. V. Title.
QD505.I6 1976   541'.395   77-13024
ISBN-13: 978-1-4615-7043-1   e-ISBN-13: 978-1-4615-7041-7
DOI: 10.1007/ 978-1-4615-7041-7

Proceedings of the Second International Workshop on
Fundamental Research in Homogeneous Catalysis
held at Shiga-ku, Japan, September, 1977

© 1978 Plenum Press, New York
Softcover reprint of the hardcover 1st edition 1978
A Division of Plenum Publishing Corporation
227 West 17th Street, New York, N.Y. 10011

Preface

The upsurge in development of homogeneous catalysis in both academic and industrial areas has prompted us to organize the Second International Workshop at Shiga, Japan, 1977.

The objective of this workshop was similar to that of the first: to identify opportunities for the solution of energy problems and industrial production problems by homogeneous catalysis.

This workshop on homogeneous catalysis was sponsored by the Toray Science Foundation, the Yoshida Foundation for Science and Technology, the Yamada Science Foundation, and the National Science Foundation.

The Robert A. Welch Foundation Grant A-420 partially supported the time spent by M. Tsutsui for the organization of the workshop and the editorial work of these proceedings.

Organizing Committee, April, 1978
Jack Halpern (U.S.A.)
Yoshio Ishii (Japan)
Jiro Tsuji (Japan)
Minoru Tsutsui (U.S.A.)
Akio Yamamoto (Japan)

Contents

METAL CLUSTERS IN CATALYSIS XVII[1]

CATALYTIC HYDROCARBON REACTIONS

E. L. Muetterties

Cornell University

Ithaca, New York 14853

INTRODUCTION

Catalytic scissions of carbon-hydrogen and carbon-carbon bonds[2,3] are processes of substantive scientific and technological interest; for example, they are critical elements in petroleum refining operations. Both reactions are catalyzed with relative facility by solid state, particularly metallic surfaces[2,3], but are catalyzed by coordination complexes only in rather special cases[4]. With discrete coordination or organometallic complexes, catalytic carbon-hydrogen bond cleavage is well established in the reversible interaction of hydridometal complexes with olefins and dienes[4-7] and for reversible reactions of aromatic hydrocarbons with a variety of transition metal complexes like $(\eta^5-C_5H_5)_2TaH_3$ and platinum (II) halides[8,9]. In addition, this type of cleavage reaction, typically monitored by H-D exchange, is observed for the hydrogen atoms in ligands like alkyl- and arylphosphines, aryl phosphites, and alkylcyclopentadienyl, with the ligand <u>bound to a transition metal</u>[4,8,9]. Catalysis of H-D exchange in aliphatic hydrocarbons presently has no precedent in coordination catalysis with but one dramatic exception[10-12], D^+ exchanges with the hydrogen atoms in methane and other saturated alkanes with a $PtCl_4^{2-}$ catalyst dissolved in acetic acid and promoted by a noncomplexing mineral acid. Also demonstrated as catalysts, although of lower activity than platinum(II), are $RhCl_3 \cdot 3H_2O$[13] and Na_3IrCl_6[14].

For C-C bond cleavage, the most remarkable reaction catalyzed by coordination complexes is the olefin metathesis reaction, equation 1, which appears to proceed by a carbene chain mechanism[15,18].

$$RCH=CHR' \rightleftharpoons RCH=CHR + R'CH=CHR' \qquad \underset{\sim}{1}$$

This reaction is taken to equilibrium at 20° within minutes by a reaction mixture of $WOCl_4$ and an alkylating reagent like $C_2H_5AlCl_2$[19].
Some of the metathesis catalysts, e.g., $C_6H_5WCl_3$[20a], catalyze at
-78° the complete conversion of cyclopropane to ethylene but not
propylene. Interestingly, for the _equilibrium_ case of 2 cyclopropane
$\rightleftharpoons$ 3 ethylene at -78°, the cyclopropane is substantially favored;
ethylene becomes favored only at higher temperatures[20b]. Despite the
fact that the conversion of cyclopropane to propylene is highly favorable, $\Delta G^\circ_{300^\circ} = -9.98$ kcal/mole, coordination catalysis of the
conversion was unknown prior to our studies with the exception[21] of
$[Rh(CO)_2Cl]_2$ which slowly isomerizes some substituted cyclopropanes
but not cyclopropane itself, at 60°. Platinum(II) complexes[22] react
with cyclopropane and cyclopropane derivatives to give the platinacy-
clobutane ring, and $[Rh(CO)_2Cl]_2$ primarily forms[21], the rhodacyclo-
pentanone ring system. On the other hand, metal complex catalysis
of carbon-carbon bond cleavage in highly strained ring systems is a
well documented reaction. For example, rhodium(I) complexes catalyze
in a two-step sequence conversion of cubane to _syn_-tricyclooctadiene,
tricyclooctane to 1,5-cyclooctadiene, and bicyclohexane to cyclohexene[23]. Nickel(0) complexes behave in an analogous fashion[24] to
those of rhodium whereas silver(I) catalyzes a different reaction sequence via a heterolytic cleavage of the C-C bonds. In all these
polycyclic hydrocarbons, cyclobutane rings represent the only reactive sites, but no catalysis of cyclobutane conversion to butene has
been reported with a coordination or organometallic complex; even
the very reactive metathesis catalysts do not interact with cyclobutane. There is one report[25] of a cyclobutane scission but this
was for _cis-anti-cis_ 1,2,3,4-tetraphenylcyclobutane which yielded
not a cyclobutene but two molecules of _trans_-stilbene on reaction
with $PtCl_4^{2-}$ at 100° in CH_3COOH. The unique reactivity of the highly
strained cyclobutane rings in molecules like cubane and tricyclooctane can be traced to the relative high electron density that lies
off the C-C bond axis; electrophilic attack by a low valent metal
complex is greatly facilitated as compared to cyclobutane itself.
Consistent with these experimental observations for four-membered
ring systems, bicyclic systems that have highly strained cyclopropane rings are readily and catalytically rearranged by low valent
transition metal complexes, e.g., the conversion of quadricyclane
to norbornadiene[26]. Catalytic cleavage of aliphatic alkanes which
contain no strained ring structural element is unknown in transition
metal coordination chemistry.

The present state-of-the-art for coordination chemistry and for
coordination catalysis clearly provides a mechanistic background
that identifies certain necessary conditions for the scission of
carbon-carbon and carbon-hydrogen bonds by coordination catalysts.
For example, the carbon-hydrogen bond cleavage reaction may occur
by direct oxidative addition of the carbon-hydrogen bond of an

uncomplexed hydrocarbon to a metal atom in a coordinately unsaturated complex. This type of reaction is known for aromatic hydrocarbons as in the presumed but eminently plausible mechanistic sequence[26], 2, for H-D exchange in aromatic hydrocarbons catalyzed by

$$(C_5H_5)_2TaH_3 \rightleftharpoons H_2 + (C_5H_5)_2TaH \xrightarrow{D_2} (C_5H_5)_2TaHD_2$$

$$(C_5H_5)_2TaHD_2 \rightleftharpoons HD + (C_5H_5)_2TaD \xrightarrow{C_6H_6} (C_5H_5)_2TaDH(C_6H_5)$$

$$(C_5H_5)_2TaDH(C_6H_5) \rightleftharpoons (C_5H_5)_2TaH + C_6H_5D$$

2

coordination complexes.

Of course, the arene could π complex with the metal center before the C-H addition occurs but this would not be a mandatory step
in the reaction sequence. Nearly all other known carbon-hydrogen
bond additions across metal centers in coordination complexes involve a carbon-hydrogen bond in a ligand bound to the metal center.
The exemplar is the β-hydrogen abstraction in transition metal alkyl
complexes whereby a metal-olefin and metal-hydrogen bond are formed.
Is there then a plausible scheme whereby a saturated hydrocarbon
might react directly with a metal center in a coordination complex?
This would be the reverse of the well known reductive elimination
of an alkane from a hydridoalkylmetal complex in which the hydride
and alkyl ligands are vicinal. The basic reaction is thermodynamically feasible. In fact, oxidative addition of the carbon-hydrogen
bond in saturated hydrocarbons is a common reaction at some clean
metal surfaces, e.g., iron and nickel. Although it is quite improb-
able that a coordination complex of a zerovalent or low valent metal
could approach the reactivity of a clean metal surface, the demonstrated[10-12] $PtCl_4^{2-}$ catalyzed H-D exchange between D^+ and CH_4 shows
the feasibility of catalytic hydrocarbon reactions with coordination
complexes. A high degree of coordination unsaturation in the complex obviously would be the direction to go for emulation of a metal
surface — but if the remaining ligands have carbon-hydrogen bonds
an internal C-H oxidative addition should largely prevail as previously noted for low valent metal phosphine and phosphite complexes,
e.g., as in equation 3. Acceptable ligands then would be carbon

$$Fe[P(CH_3)_3]_4 \rightleftharpoons [(CH_3)_3P]_3Fe\begin{matrix} H & CH_2 \\ | & / | \\ & \\ P(CH_3)_2 \end{matrix}$$

3

monoxide and either phosphorus trifluoride or perfluoroalkylphos-
phines (at least for the more noble transition metals where metal-
fluorine bond energies are relatively low) and halide or sulfide

ions. The rate of H-D exchange in $PtCl_4^{2-}$ catalyzed exchange in aliphatic and aromatic hydrocarbons is a linear function of the ionization potential of the alkane provided that the alkane has either a straight chain or a cyclic structure[4]. It has been suggested that electron transfer from the alkane to the platinum atom is a key, early step in the exchange sequence. If this surmise is accurate, an electrophilic metal center is a critical feature. Accordingly, strong σ donor ligands such as alkylphosphines or alkyl phosphites are not desirable. Halides and strong π acceptor ligands like PF_3 or $(CF_3)_3P$ would seem to be preferred at least in the context of this postulated mechanistic sequence. There is experimental evidence that dimeric species[4] are mechanistically important in the Pt(II), Rh(III) and Ir(III) halide — acetic acid catalyzed reactions. Possibly, species such as "Rh_2^{4+}" are key reaction intermediates; however, we[32] have found that Co(II) is also an active catalyst for these H-D exchange reactions in aliphatic hydrocarbons, and analogous dinuclear cobalt species are unknown. Copper(II) and vanadium(III) halides or acetates were inactive for these exchange reactions[32].

One approach to C-C bond activation, although highly limited in scope, is to employ a "saturated" hydrocarbon that could initially complex with the metal center of a coordinately unsaturated complex. Cyclopropane is the simplest example; since in a formal molecular orbital purview cyclopropane resembles an olefin. Thus, a simple probe of this type of hydrocarbon transformation would be the cyclopropane ⇀propylene reaction. A more sensitive but less demanding probe would be the quadricyclane ⟶ norbornadiene reaction. Either of these reactions would accurately probe the proclivity of the metal center toward an oxidative C-C bond addition.

We describe here some of our attempts to effect catalytic reactions of saturated hydrocarbon reactions with coordination or organometallic catalysts. These exploratory attempts have largely focused on the reactions of thermally stable osmium and iridium carbonyl clusters, $Os_3(CO)_{12}$ and $Ir_4(CO)_{12}$, metal acetate clusters, and the naked bismuth clusters, Bi_x^{y+}. Two classes of saturated hydrocarbons were studied, acyclic and cyclic alkanes. H-D exchange between methyl protons in methyl substituted benzenes and D_2 is also reported for two complexes, $\eta^3-C_3H_5Co[P(OCH_3)_3]_3$ and $\eta^6-C_6(CH_3)_6Ru-\eta^4-C_6(CH_3)_6$, that are catalysts of the arene hydrogenation reaction.

EXPERIMENTAL

All hydrocarbons were reagent grade and were vacuum distilled from either CaH_2 or $LiAlH_4$ prior to use. Reactions were effected either in closed-system, sealed glass tubes or in "open" inert-atmosphere reaction systems. All analyses were based on gas chromatography, nuclear magnetic resonance, and gas chromatography-mass spectroscopy. The coordination complexes were prepared by standard literature procedures.

RESULTS AND DISCUSSION

We have found no evidence of stoichiometric or catalytic reactions of saturated hydrocarbons up to 180° in the presence of the metal carbonyl clusters, $Ir_4(CO)_{12}$, $Os_3(CO)_{12}$ and $H_2Os_3(CO)_{10}$. The latter cluster was especially selected because it is thought to be a thermal precursor to the unsaturated $Os_3(CO)_{10}$ cluster. Model saturated alkanes utilized in this study included n-hexane, nujol, cyclohexane, cyclopentanes, and various alkylcycloalkanes.[28,29] Closed and open reaction systems yielded negative results up to $180^\circ C$. Thermal instability precluded an investigation of these clusters at temperatures substantially above $160-180^\circ$. The metal carbonyl cluster approach to homogeneous catalytic reactions of aliphatic hydrocarbons does not appear to be promising although $Ru_3(CO)_{12}$ and $Os_3(CO)_{12}$ are known to effect stoichiometric cleavages of C-H and C-C bonds in hydrocarbons.

The naked metal cluster ions such as the bismuth set of Bi_4^{2-}, Bi_5^{3+}, Bi_8^{2+} and Bi_9^{5+} present an uncharted, potential access to homogeneous catalytic reactions of saturated hydrocarbons. Presently, a major tactical problem is the identification of a solvent for the naked cluster salts that is inactive toward hydrocarbons. Molten $NaAlCl_4$ is a solvent for the $AlCl_4^-$ salts of the bismuth cations but, unfortunately, the solvent retains much of the activity of $AlCl_3$ itself for catalysis of hydrocarbon rearrangements. Experimentally, there was no large difference between $NaAlCl_4$ alone and $NaAlCl_4$ plus catalytic amounts or $[Bi_5^{3+}][AlCl_4^-]_3$ in the catalytic rearrangement of n-hexane.[30] A meaningful test of the catalytic properties of these bismuth clusters cannot be achieved until a solvent, inert to alkanes, can be devised for these salts.

Facile catalytic conversion of cyclopropane to propylene is unknown although substituted cyclopropanes have been catalytically converted to the olefin with $[Rh(CO)_2Cl]_2$.[21] No other reports of rapid catalytic conversion of cyclopropane to propylene have been found although highly strained cyclopropane derivatives like quadricyclane[26] are rapidly isomerized through an effective C-C bond cleavage by a number of zero- or low-valent coordination complexes. We found cyclopropane essentially nonreactive toward $Ir_4(CO)_{12}$ and $Ru_3(CO)_{12}$ to the effective decomposition temperatures of the clusters; a characteristic product of the ruthenium cluster reaction at temperatures above 100° was $Ru_6C(CO)_{17}$ when an alkane was the solvent and $Ru_6C(CO)_{14}$ -arene when an arene was the solvent. No cyclopropane reaction was observed in the presence of $Os_3(CO)_{12}$ when an arene was the solvent; $H_2Os_3(CO)_9$ -benzyne type complexes seemed to be the predominant products. With alkane solvents, the osmium

cluster catalyzed the conversion of cyclopropane to propylene with
the formation of an Os_3 cluster derivative(s) that has not been
fully characterized to date. The rate of the catalytic reaction was
low. Complete conversion to propylene was achieved only after about
twenty days at $180°$, the conversion after 1 day was only about 4%.[31]

A key mechanistic point to be resolved in cyclopropane rearrangements is whether the reaction is purely intramolecular or occurs
with the interjection of a metal hydride species. Two contrasting
reaction mechanisms outlined in sequences 4 and 5 could be distin-
guished in experiments with isotopically labelling of the reagents

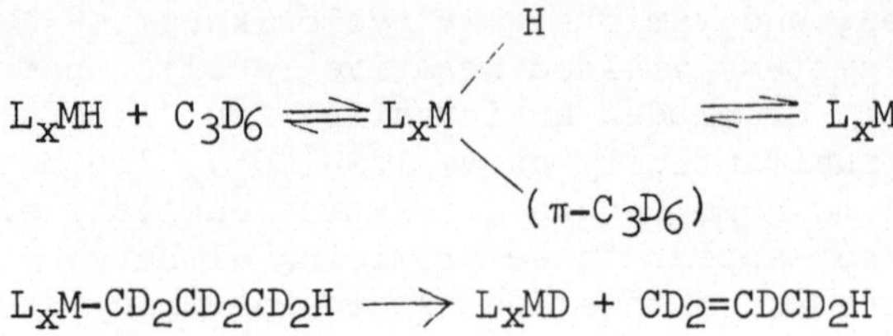

$$L_xMH + C_3D_6 \rightleftharpoons L_xM\overset{H}{\diagup} \rightleftharpoons L_xM-CD_2CD_2CD_2H$$

$$(\pi-C_3D_6)$$

$$L_xM-CD_2CD_2CD_2H \longrightarrow L_xMD + CD_2=CDCD_2H$$

4

$$L_xM + C_3D_6 \rightleftharpoons L_xM(\pi-C_3D_6)$$

$$L_xM(\pi-C_3D_6) \rightleftharpoons L_xM\overset{D}{\underset{C_3D_5}{\diagup}}$$

$$L_xM\overset{D}{\underset{C_3D_5}{\diagup}} \longrightarrow L_xM + CD_2=CDCD_3$$

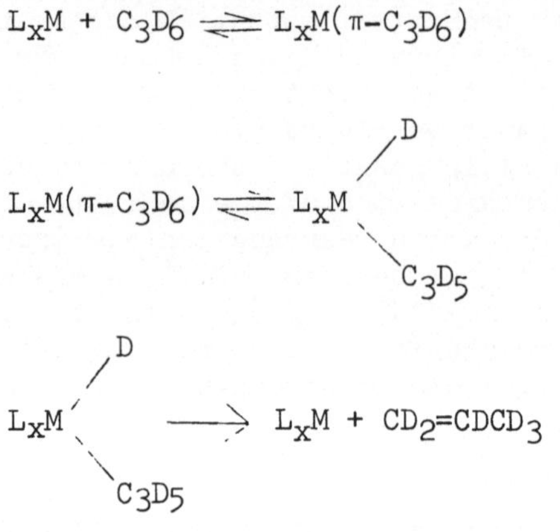

5

including the cyclopropane although degenerate olefin isomerization
in a system like 4 could greatly complicate the interpretation.

In the remarkable $PtCl_4^{2-}$ catalyzed H–D exchange between D^+
and CH_4[10], the initial interaction between the platinum atom and
the methane probably comprises an electrophilic attack by the platinum on the methane. Incisively, the statistics of the exchange
indicate a multiple rather than single exchange per sequence. On
this basis, it has been suggested[9,11] that carbene-like intermediates
may be involved. This reaction should not be unique to platinum(II)
and, accordingly, we have sought evidence of comparable exchange

reactions with Co(I), Rh(I) and Ru(0) complexes. We have found that
simple cobalt(II) complexes in CH_3COOD media will catalyze H-D exchange in alkylarenes. Another interesting finding has been the H-D
exchange between D_2 and the hydrogen atoms of the methyl substituents in toluene and xylenes. This novel exchange reaction has been
effected with $\eta^3-C_3H_5Co[P(OCH_3)_3]_3$ at 25° and with $\eta^6-C_6(CH_3)_6Ru-\eta^4-$
$C_6(CH_3)_6$ at 90°.[22,23] This exchange appears to be totally selective
to the CH_3 hydrogen atoms with the cobalt(I) catalyst and is evident
only in the hydrogenation products, the methylcyclohexanes. In the
D_2-toluene and D_2-xylene reactions, the recovered arenes showed no
deuterium incorporation, an observation consistent with earlier studies with the cobalt catalyst that indicated that an arene once coordinated does not dissociate until fully hydrogenated. In the case
of the ruthenium(0) catalyst, aromatic C-H hydrogen atom exchange
does occur but at a substantially lower rate than for the methyl
group hydrogen atoms. This enhanced methyl group H-D exchange was
evident in the product cyclohexanes and cyclohexenes as well as in
the recovered, "unreacted" arenes. The ruthenium complex did not
catalyze H-D exchange between D_2 and cyclohexane at 90°. We tenta-
tively suggest that a key set of reactions in the exchange process
comprises initial π arene complex formation followed by an oxidative
addition of the alkyl C-H bond to the metal to give σ and π benzyl-
metal hydride complexes[31,32], 6.

6

ACKNOWLEDGEMENT

This research was supported by the National Science Foundation.

REFERENCES AND NOTES

1. E. L. Muetterties, Pure and Applied Chemistry 00, 0000 (1978).
2. J. E. Germain, Ed., Catalytic Conversion of Hydrocarbons, Academic Press, New York 1969.
3. J. R. Anderson, Ed., Chemisorption and Reactions on Metallic Films, Academic Press, New York, 1971, Vols. 1 and 2.
4. For a recent interview of this subject see D. E. Webster, Adv. Organometallic Chem., 15, 147 (1977).
5. G. N. Schrauzer, Transition Metal Hydrides, Dekker, New York, 1971.
6. R. F. Heck, Organotransition Metal Chemistry, Academic Press, New York, 1974.
7. C. A. Tolman, Transition Metal Hydrides, E. L. Muetterties, Ed., Dekker, New York, 1971.
8. G. W. Parshall, Accounts Chem. Res., 3, 139 (1970).
9. G. W. Parshall, Accounts Chem. Res., 8, 113 (1975).
10. N. F. Gol'dschleger, M. B. Tyabin, A. E. Shilov, and A. A. Shteinman, Zh. Fiz. Khim., 43, 2174 (1969).
11. M. B. Tyabin, A. E. Shilov, and A. A. Shteinman, Dokl. Akad. Nauk, 198, 381 (1971)
12. R. J. Hodges, D. W. Webster, and P. B. Wells, J. Chem. Soc. A, 3230 (1971).
13. M. R. Blake, J. L. Garnett, I. K. Gregor, W. Hannan, K. Hoa and M. A. Long, J. Chem. Soc., Chem. Comm., 930 (1975).
14. J. L. Garnett, M. A. Long, and K. B. Peterson, Aust. J. Chem., 27, 1823 (1974).
15. N. Calderon, Accounts Chem. Res., 5, 127 (1972).
16. J. L. Herrison and Y. Chauvin, Makromol. Chem., 141, 161 (1970).
17. T. J. Katz and J. McGinnis, J. Amer. Chem. Soc., 97, 1592 (1975).
18. E. L. Muetterties, Inorg. Chem., 14, 951 (1975).
19. M. T. Mocella, R. Rovner, and E. L. Muetterties, J. Amer. Chem. Soc., 98, 4689 (1976).
20a. P. G. Gassman and T. H. Johnson, J. Amer. Chem. Soc., 98, 6057 (1976),
20b. F. D. Mango, J. Amer. Chem. Soc., 99, 6117 (1977).
21. F. J. McQuillin and K. G. Powell, J. Chem. Soc., Dalton, 2129 (1972).
22. F. J. McQuillan and K. G. Powell, J. Chem. Soc., Dalton, 2123 (1972) and references therein.
23. For a discussion of these reactions see J. Halpern, In Organic Synthesis via Metal Carbonyls, I. Wender and P. Pino Eds., Wiley, New York, 1977. Vol. 3, p. 705.
24. H. Takaya, M. Yamakawa, and R. Noyori, Chem. Lett. 781 (1973).
25. I. J. Harvie and F. J. McQuillan, J. Chem. Soc., Comm., 806

(1974).

26. L. Cassar and J. Halpern, Chem. Comm., 1082 (1970).
27. E. K. Barefield, G. W. Parshall, and F. N. Tebbe, J. Amer. Chem. Soc., 92, 5234 (1970).
28. K. G. Caulton, M. G. Thomas, B. A. Sosinsky and E. L. Muetterties, Metal Clusters in Catalysis (IX). Hydrocarbon Reactions, Proc. Nat. Acad. Sci., 73, 4274 (1976).
29. M. J. D'Aniello, Jr. and E. L. Muetterties, unpublished work.
30. C. G. Demitrus and E. L. Muetterties, unpublished work.
31. J. W. Johnson and E. L. Muetterties, to be published in J. Amer. Chem. Soc.
32. E. L. Muetterties and J. R. Bleecke, to be published.

CLUSTER CATALYSIS. HOMOGENEOUS CATALYTIC OXIDATION
OF CARBON MONOXIDE AND OF KETONES WITH MOLECULAR
OXYGEN USING COMPLEXES OF RHODIUM(0), IRIDIUM(I),
AND PLATINUM(0)

D. M. Roundhill

Department of Chemistry
Washington State University
Pullman, Washington 99164

Transition metal compounds have been widely used as reagents
and catalysts in the oxidation of organic compounds.[1] One approach
is to use a complex having the metal ion in a high oxidation state,
and to use the compound as a stoichiometric oxidant. A second
method is to use molecular oxygen as oxidant and to initiate an
autoxidation reaction by using a metal ion which will readily undergo a one electron oxidation. Among the former reagents, two commonly used compounds are the potassium salts of the dichromate and
permanganate ions. These compounds will oxidize a wide variety of
organic functional groups while the metal ion is reduced to a
lower oxidation state. Since it is difficult to re-oxidize these
metals back to their highest oxidation state these compounds are
always used as oxidants in a stoichiometric reaction. Potentially
it is advantageous to use molecular oxygen as oxidant because of its
availability and because there is no large concentration of metal
containing waste products remaining after reaction. The problems
to be overcome when using molecular oxygen as oxidant are ones regarding reactivity and selectivity. Molecular oxygen is kinetically inert to most organic compounds, and when activated to react
as an oxidant the problem then becomes one of diverting a free
radical autoxidation reaction toward a specific function, and
avoiding complete combustion of the organic compound to carbon
dioxide and water. Two metal ions frequently used in conjunction
with molecular oxygen for oxidation reactions are the ferrous and
cobaltous ions. These ions will undergo an electron transfer
reaction with oxygen leading to the formation of the superoxide
ion, which is involved in the propagation steps of

$$M(II) + O_2 \rightleftharpoons M(III) + O_2^-$$

a radical chain process. The use of these simple salts allows
one to control reactivity and selectivity by temperature, pressure,
and solvent variation, but little can be done to change the
chemistry about the metal center.

The problem of synthesizing metal complexes for specific use
as homogeneous oxidation catalysts with molecular oxygen as oxidant is a difficult one. The possibility of building a complex
which will activate molecular oxygen to react as an oxidant by a
non-radical process is indeed an attractive one, but except for a
few isolated claims[2] there remains little direct evidence that
this goal can be achieved. A considerable number of complexes are
known which will form dioxygen complexes[3] but use of these compounds
in conjunction with molecular oxygen as oxidant invariably leads to
a free radical autoxidation reaction with simultaneous oxidation
of the ligands about the metal center. It may be possible to synthesize ligands which will survive the conditions of oxidation
reactions but this goal has not yet been attained.

In this article we will focus discussion on our results
obtained using the hexametallic cluster compound $Rh_6(CO)_{16}$ as an
oxidation catalyst for carbon monoxide and aliphatic ketones using
molecular oxygen as oxidant. Since transition metal carbonyl
cluster compounds have a high stability to decomposition they may
find a role in catalytic oxidations where the high energy intermediates formed during reaction will decompose most potential
catalysts to an insoluble mixture of metal oxides. Muetterties in
his work on cluster catalysis is using the approach that new and
labile clusters will be required before a wide range of reactions
can be catalyzed by this range of compounds,[4] nevertheless since
we are focussing our efforts on catalytic oxidations we believe
that we can use the robust carbonyl clusters to good effect. We
therefore approach the problem with some optimism that the cluster
compound will maintain some structural integrity throughout the
reaction cycle, and that by maintaining a high pressure of carbon
monoxide on the system we can ensure the presence of carbonyl containing species during oxidation, and thereby prevent destruction
of the compound to metal oxide. Finally we can envisage the
possibility of controlling the nature of the species in solution
by variations in carbon monoxide pressure, and also in taking
advantage of the free radical autoxidation of carbon monoxide to
carbon dioxide to initiate desirable, and terminate undesirable,
chain reactions.

In a preliminary communication we have reported that the
compound $Rh_6(CO)_{16}$ is an effective homogeneous catalyst for the
oxidation, with molecular oxygen, both of carbon monoxide to carbon dioxide, and of aliphatic ketones to carboxylic acid.[5] Little
mechanistic information was presented in that article regarding
the initial steps of the oxidation reaction. For comparative purposes we have now extended the work to investigate other low-
valent transition metal complexes which are known to form dioxygen

adducts.   The compounds used are chlorocarbonylbis(triphenylphosphine)iridium(I) and tris(triphenylphosphine)platinum(0).   The
choice was made because these and similar low-valent complexes of
the platinum metal group have been found to be effective for the
oxidation of alkenes,[6] tertiary phosphines,[7] and isocyanides.[8]
More recently these compounds have also been used for the homogen-
eously catalyzed oxidation of benzaldehyde.   Hojo, et al., reported
the reaction and proposed the involvement of a molecular mechanism
rather than a free radical chain process,[9] but more recently
Sakamoto et al., have presented persuasive evidence[10] that the reaction using typical compound $PdO_2(PPh_3)_2$ proceeds via a free
radical autoxidation mechanism.   The oxidation of ketones is
much more difficult to achieve than is the oxidation of benzaldehyde, and there is only a brief literature on the former subject.
The purpose of this article is to present and discuss our recent
results on the transition metal complex catalyzed autoxidation of
ketones, and to comment on the possible roles and significance
both of the cluster compound $Rh_6(CO)_{16}$, and of the simultaneous
oxidation of carbon monoxide to carbon dioxide.   Only a limited
amount of experimental detail is included in this article because
of our recent publication of a paper on the subject.[11]

In the preliminary communication[5] we reported that the
oxidation of carbon monoxide to carbon dioxide occurs in the presence of $Rh_6(CO)_{16}$ as catalyst at temperatures of 80° or greater,
when a stoichiometric ratio of carbon monoxide is present at a
pressure of 15-20 atm.   A variety of ketones have been used as
solvent and substrate for this oxidation and in each case a catalytic oxidation to carboxylic acid occurs.   From the four separate
ketones, acetone, 2-butanone, 3-pentanone, and cyclohexanone, the
respective carboxylic acids formed are acetic acid from the first
two, an equimolar mixture of acetic and propionic acids from the
third, and adipic acid from the latter ketone.   Acid identification
has been carried out by gas chromatographic techniques for the
monobasic acids, and in the case of adipic acid the compound has
been isolated pure and characterized both by its melting point and
the mass spectra of its methyl and ethyl esters.   These latter
compounds can be formed by effecting the oxidation of cyclohexanone
in the presence of the appropriate alcohol.   The alcohol suffers no
apparent oxidation and the respective esters can be distilled from
the reaction mixture after depletion of the oxygen.

$$CH_3COCH_3 + O_2 \longrightarrow CH_3CO_2H$$

$$CH_3CH_2COCH_3 + O_2 \longrightarrow 2\ CH_3CO_2H$$

$$CH_3CH_2COCH_2CH_3 + O_2 \longrightarrow CH_3CO_2H + CH_3CH_2CO_2H$$

$$+ O_2 \longrightarrow HO_2C(CH_2)_4CO_2H$$

The identity of the products obtained in these oxidations shows
that the ketone suffers carbon-carbon bond cleavage at the keto
carbon with oxidation of both the alkyl and acyl groups to carboxyl
functions. These results also show that oxidation of a methylene
group adjacent to a carbonyl occurs preferentially over that of a
methyl group in a similar structural position. This selectivity
is particularly apparent in the case of 2-butanone where the product is acetic acid without any simultaneous formation of propionic
acid. For acetone an analogous oxidation pathway will lead to the
formation of an equimolar mixture of acetic and formic acids.
Under the prevailing experimental conditions we have only been able
to identify the former carboxylic acid. This means therefore that
we cannot be definitive as to whether formic acid is not formed
during the reaction, or whether the compound is absent because of
subsequent decomposition under the prevailing reaction conditions.

The oxidation reaction using $IrCl(CO)(PPh_3)_2$ or $Pt(PPh_3)_3$ as
initiators is carried out in the absence of carbon monoxide. The
quantitative data for all oxidations are shown in Table 1.
Carboxylic acids are formed with all the metal complexes but there
are some important differences which will be discussed later. It
should be noted, though, that the compounds $IrCl(CO)(PPh_3)_2$ and
$Pt(PPh_3)_3$ cannot be considered as catalysts for this oxidation
reaction since the ligands are also oxidized and the complex cannot be recovered at the end of the reaction. The solution does
nevertheless remain homogeneous and there is no formation of metal
or metal oxide at the end of the reaction. For the cluster compound $Rh_6(CO)_{16}$ the situation is somewhat different. This compound
has a very low solubility in the ketone at room temperature, but
after completion of the reaction a portion of the material has
dissolved. This filtrate is catalytically active for carbon
monoxide and ketone oxidation, and remains homogeneous at the end
of a series of catalytic cycles. It appears therefore that there
is a homogeneous component to this catalytic oxidation reaction,
but since a portion of the compound is likely present as solid
throughout the reaction, we cannot discount the possibility that
there is also a component to the oxidation where the cluster compound functions as a heterogeneous catalyst. It must be emphasized that we do not claim that the compound $Rh_6(CO)_{16}$ remains
unchanged throughout the catalytic oxidation. Indeed we envisage
a situation occurring in solution where the cluster compound undergoes one or more reversible changes during the reaction leading to
the carboxylato bridged carbonyl dimer $[Rh(OCOR)(CO)_2]_2$. At the
beginning of the reaction the concentration of carboxylic acid in
the solvent is very low[12] but increases as the reaction proceeds
to completion. It has been found that refluxing the compound
$Rh_6(CO)_{16}$ with carboxylic acid converts it into the bridged
carboxylato carbonyl dimer $[Rh(OCOR)(CO)_2]_2$.[13] Since under the
conditions of our catalytic reaction the concentration of carbon
monoxide is decreased at the same time as the carboxylic acid

TABLE 1

Ketone (10 ml)	Time (h)	Temperature (°)	Compound (μ mol)	Pressure (lb in^{-2})	Carboxylic acid (mequiv)
Acetone	20	100	None	CO(330)/O$_2$(170)	0.1
Acetone	29	100	Rh$_6$(CO)$_{16}$ (41)	"	0.8
2-Butanone	48	77	None	O$_2$(500)	0.5
2-Butanone	22	77	IrCl(CO)(PPh$_3$)$_2$ (47)	"	11.0
3-Pentanone	19	80	None	"	0.5
3-Pentanone	22	80	IrCl(CO)(PPh$_3$)$_2$ (48)	"	30
3-Pentanone	49	80	IrCl(CO)(PPh$_3$)$_2$ (40)	"	2.6
			α-Naphthol (2340)		
Cyclohexanone	19	80	None	"	0.6
Cyclohexanone	21	80	IrCl(CO)(PPh$_3$)$_2$ (49)	"	16.3
Cyclohexanone	22.5	77	Pt[O$_3$C(Me)$_2$](PPh$_3$)$_2$ (40)	"	3.0
Acetophenone	26	100	None	"	0.1
Acetophenone	24	100	IrCl(CO)(PPh$_3$)$_2$ (38)	"	0.1
Phenylacetone	1	100	IrCl(CO)(PPh$_3$)$_2$ (41)	"	Trace (EXPLOSION)

concentration increases, it is apparent that the solution at the
end of the reaction must contain a considerable fraction of the
rhodium in the form of the carboxylato bridged carbonyl dimer. It
is pertinent to note that if the reaction vessel containing the

$$Rh_6(CO)_{16} \underset{CO + O_2}{\overset{RCO_2H}{\rightleftharpoons}} [Rh(OCOR)(CO)_2]_2$$

catalyst solution is further pressurized with carbon monoxide and
oxygen, the oxidation reaction continues with no apparent decrease
in rate. Furthermore in a separate experiment we have confirmed
that a synthetic sample of the complex $[Rh(OCOMe)(CO)_2]_2$ is an
effective catalyst for these oxidations and is converted to the
cluster compound $Rh_6(CO)_{16}$ under the high pressure of carbon monoxide and oxygen prevalent at the beginning of the oxidation. In
an elementary sense of the meaning of the word catalyst the cluster
compound can be accurately assigned this designation since it is
recovered unchanged at the end of the reaction when the initial
conditions are reapplied. Nevertheless we have noted earlier that
the compound does not maintain its structural integrity throughout
the whole duration of the reaction sequence. We consider the compounds $Rh_6(CO)_{16}$ and $[Rh(OCOR)(CO)_2]_2$ to be present at the two
extremes of the concentration ratios of $CO:RCO_2H$, and since the
conversion between hexametallic and bimetallic rhodium complexes
is unlikely to proceed in a single step it is possible that other
multimetallic carboxylato carbonyl compounds are present at
intermediate $CO:RCO_2H$ concentration ratios.

A number of experiments have been performed which verify that
the oxidation reactions can be considered to be free radical
autoxidation processes. Firtly we find that the oxidation, after
initiation, proceeds at a rate which is linearly dependent on time,
and independent of oxygen pressure (Fig. 1). We have also found
the oxidation to be strongly inhibited by the presence of α-
naphthol, a compound used for termination of free radical chains.
Thus when 3-pentanone (10 ml) is oxidized for 22 h. at 80° under
an initial oxygen pressure of 500 lb. in^{-2} in the presence of the
complex $IrCl(CO)(PPh_3)_2$ (37 mg), the quantity of carboxylic acid
produced is 30 mmol. When the same experiment is performed for
49 h. in the presence of $IrCl(CO)(PPh_3)_2$ (32 mg) and α-naphthol
(337 mg) the yield of carboxylic acid is only 2.6 mmol. This
rather dramatic decrease in product yield shows that the predominant pathway for carboxylic acid formation in this reaction is a
free radical route. Furthermore we conclude from our earlier work
on the reactivity of low-valent d[16] metal ions to phenols[14] that
it is unlikely that the metal complexes used in this work will be
decomposed by α-naphthol. It should also be noted that an explosion occurs on attempted oxidation of phenylacetone. We believe
that this condition results from the buildup of too high a

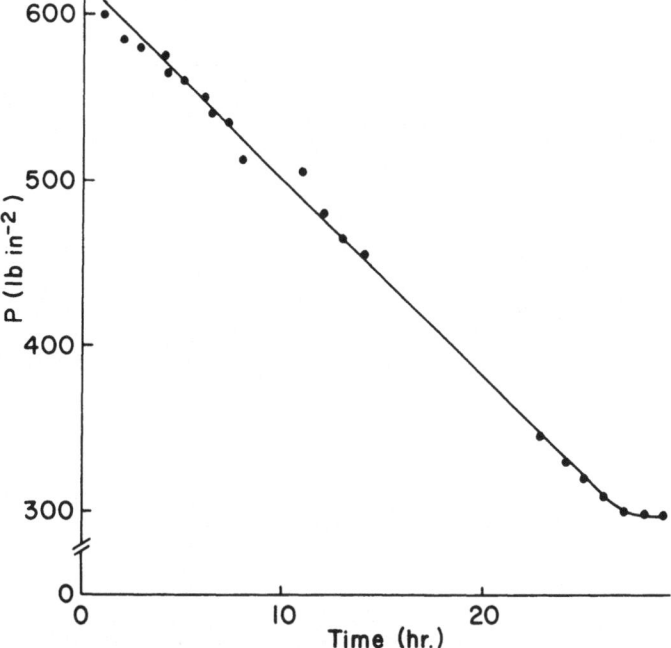

Fig. 1.   Rate plot for the oxidation of CO with $Rh_6(CO)_{16}$ (15 mg)
          in acetone (10 ml) at 100°C.   The initial pressures of
          CO and O_2 at 20°C are 330 and 170 lb in^{-2} respectively.

concentration of the resonance stabilized benzyl radical which results in an explosive chain reaction being propagated as the temperature of the reaction is raised to 80°; the minimum temperature for reaction. From these considerations we propose the following initiation and propagation steps are involved in the reaction.

Initiation:

$$MO_2 + R-\overset{\overset{H}{|}}{\underset{\underset{H}{|}}{C}}-\overset{\overset{}{}}{\underset{\underset{O}{\parallel}}{C}}-R' \longrightarrow MOOH + R-\overset{\overset{\bullet}{}}{\underset{\underset{H}{|}}{C}}-\overset{}{\underset{\underset{O}{\parallel}}{C}}-R'$$

$$MO_2 + H^+ \longrightarrow MOOH^+$$

$$MOOH \longrightarrow M + HO_2$$

$$MOOH \longrightarrow M\overset{\bullet}{O} + \overset{\bullet}{O}H$$

$$MOOH^+ \longrightarrow MO^+ + \overset{\bullet}{O}H$$

Propagation:

$$R-\overset{\overset{\bullet}{}}{\underset{\underset{H}{|}}{C}}-\overset{}{\underset{\underset{O}{\parallel}}{C}}-R' + O_2 \longrightarrow R-\overset{\overset{O-\overset{\bullet}{O}}{|}}{\underset{\underset{H}{|}}{C}}-\overset{}{\underset{\underset{O}{\parallel}}{C}}-R'$$

$$R-\overset{\overset{O-\overset{\bullet}{O}}{|}}{\underset{\underset{H}{|}}{C}}-\overset{}{\underset{\underset{O}{\parallel}}{C}}-R' + R-\overset{\overset{H}{|}}{\underset{\underset{H}{|}}{C}}-\overset{}{\underset{\underset{O}{\parallel}}{C}}-R' \longrightarrow R-\overset{\overset{O-OH}{|}}{\underset{\underset{H}{|}}{C}}-\overset{}{\underset{\underset{O}{\parallel}}{C}}-R' + R-\overset{\overset{\bullet}{}}{\underset{\underset{H}{|}}{C}}-\overset{}{\underset{\underset{O}{\parallel}}{C}}-R'$$

$$R-\overset{\overset{O-OH}{|}}{\underset{\underset{H}{|}}{C}}-\overset{}{\underset{\underset{O}{\parallel}}{C}}-R' \longrightarrow R-\overset{\overset{O}{\parallel}}{C}-\overset{}{\underset{\underset{O}{\parallel}}{C}}-R' + H_2O$$

$$2\ R-\overset{\overset{O}{\parallel}}{C}-\overset{}{\underset{\underset{O}{\parallel}}{C}}-R' + O_2 \longrightarrow R\overset{\overset{O}{\parallel}}{C}-\overset{\bullet}{O}_2 + R'-\overset{\overset{O}{\parallel}}{C}-\overset{\bullet}{O}_2 + R-\overset{\overset{O}{\parallel}}{\overset{\bullet}{C}} + R'-\overset{\overset{O}{\parallel}}{\overset{\bullet}{C}}$$

$$R-\overset{\bullet}{C}O_3\ (R'-\overset{\bullet}{C}O_3) + H_2O \longrightarrow RCO_3H\ (R'CO_3H)$$

$$R-\overset{\overset{}{}}{\underset{\underset{O}{\parallel}}{\overset{\bullet}{C}}}\ (R'-\overset{\overset{}{}}{\underset{\underset{O}{\parallel}}{\overset{\bullet}{C}}}) + \overset{\bullet}{O}H \longrightarrow R-\overset{\overset{}{}}{\underset{\underset{O}{\parallel}}{C}}-OH\ (R'-\overset{\overset{}{}}{\underset{\underset{O}{\parallel}}{C}}-OH)$$

Transition metal hydroperoxide complexes of platinum metals have been identified by earlier workers using EPR spectroscopy. From kinetic studies on the reactivity of the dioxygen complexes $Pt(PPh_3)_2O_2$ and $IrX(CO)L_2O_2$ to electrophiles[16,17] it is apparent that hydroperoxide complexes of these metals will result from protonation, and from calculations on the charge on the oxygen atoms in end-on bonded dioxygen complexes[18] it appears that hydrogen atom abstraction from the ketone is quite feasible. Subsequent to the formation of this hydroperoxide intermediate, homolytic fission of the O-O or M-O bond can readily occur to give hydroxyl or hydroperoxyl radicals for chain initiation. The reaction scheme outlines a series of propagation and termination steps which finally result in the formation of aliphatic peracids which will readily convert to the carboxylic acids RCO_2H and $R'CO_2H$ from the ketone RCH_2COR'. In this proposed reaction scheme we suggest that the $\alpha\beta$-diketone is formed as an intermediate during the reaction, and it is pertinent to note that introduction of such compounds into the catalytic oxidation reaction results in their rapid conversion to carboxylic acids. This result contrasts with the oxidation of 2-butanone by ferric complexes of 1,10-phenanthroline where the final product is 2,3-butanedione.[19] Finally in order to ascribe any catalytic activity to the metal complexes it is necessary to verify that homolysis and radical chain propagation cannot occur equally as well in the absence of a metal complex. The data in the table verifies that the yield of carboxylic acid is considerably reduced in the absence of metal complex. A previous report on the oxidation of ketones to carboxylic acids with molecular oxygen in the absence of metal complex has used 1% of di-t-butyl hydroperoxide as initiator for the free-radical reaction.[20] These authors did, however, use the significantly higher temperature of 140° to carry out the oxidation, but we have not yet determined whether it is possible to lower this temperature to 80°.

The finding of a free radical autoxidation mechanism implies that the oxidation could be catalyzed by simple metal salts or by the group of cobalt complexes which will form end-on bonded oxygen complexes.[21] The use of simple cobalt salts causes serious problems. Under reflux conditions no significant oxidation to carboxylic acid occurs, and under the high pressure conditions used throughout this study a violent explosion occurs. The oxidation reaction does proceed under high pressure conditions with the complexes $Co(salen)_2$, providing excess pyridine is present. Ligand oxidation also occurs and the reaction mixture is heterogeneous at the end. A similar situation exists when the manganese(II) complex with phthalocyanine[22] is used with excess pyridine; the resulting compound is manganese dioxide. We find therefore that such complexes can be effectively used for initiating these oxidations, but there is difficulty both in controlling the rate of reaction, and in preventing ligand oxidation with the resulting formation of heterogeneous products.

The observation that the compound $Rh_6(CO)_{16}$ is an effective catalyst also for the oxidation of carbon monoxide to carbon dioxide is deserving of attention. This reaction proceeds simultaneously with the oxidation of ketones to carboxylic acids. The presence of carbon monoxide is necessary during this latter oxidation in order to prevent formation of insoluble non-carbonyl containing rhodium products after all the ligating carbonyl groups have been oxidized to carbon dioxide. We have not determined the ultimate catalyst turnover number for this reaction but it is significant that after oxidation of over 12,000 moles of carbon monoxide per mole of $Rh_6(CO)_{16}$ have been oxidized the catalyst solution has not decreased in its activity to this reaction. Despite an earlier report on the stoichiometric conversion of the compound $Pt(PPh_3)_2O_2$ into $Pt(CO_3)(PPh_3)_2$ with CO,[23] we find that neither of the compounds $Pt(PPh_3)_3$ and $IrCl(CO)(PPh_3)_2$ are effective catalysts for the oxidation of CO to CO_2 with molecular oxygen at 100°C.

The heterogeneously catalyzed gas phase oxidation of carbon monoxide to carbon dioxide has been studied in detail using materials such as palladium[24] and nickel,[25] or oxides such as zinc[26] and titanium[27] oxide. The reactions proceed at high temperatures and the mechanism is considered to involve chemisorbed oxygen atoms. The activation energy for the dissociation of the oxygen molecule into atoms is estimated[28] to be 50 k.cal.mol.[-1] If the catalyzed

$$O_2 \longrightarrow 2\ O \qquad E_a = 50\ \text{k.cal.mol.}^{-1}$$

oxidation of CO by $Rh_6(CO)_{16}$ involves the formation of oxygen atoms it is likely that they can be free until termination by CO leads to the formation of CO_2. This diradical oxygen atom, however, can potentially be involved in propagation steps of other chain reactions, notably, in our case, ketone oxidation. Indeed we believe it may be highly significant that the cleavage or oxidation of the methyl group in acetone is only observed when the catalyst is $Rh_6(CO)_{16}$ in the presence of carbon monoxide. It is indeed reasonable to propose that the parallel reaction resulting in the oxidation of carbon monoxide to carbon dioxide influences the ketone oxidation reaction. Walling has discussed in some detail autoxidation conditions where the initiation, propagation, and termination steps of one reaction have a major effect on a second one,[29] and this possibility is an intriguing one for this catalysis with a cluster compound. Further experiments will obviously be necessary in order to verify whether we are observing any causality effects between the simultaneous oxidations, but we believe that such considerations should be fruitful in introducing selectivity into transition metal oxidations. In principle it is possible to induce the same effect with triphenylphosphine complexes where the oxidation to triphenylphosphine oxide will cross catalyze the oxidation of the second substrate, nevertheless in practice the production of large quantities of the non-volatile triphenylphosphine oxide would be prohibitively expensive and would lead to separation problems.

The precise function of $Rh_6(CO)_{16}$ as an initiator for a free-radical oxidation reaction is at present very unclear. At the beginning of this article we emphasized that transition metal catalyzed autoxidations can be initiated by either one-electron reductants, or by complexes of dioxygen which are converted into hydroperoxides. Despite a considerable quantity of research on transition metal cluster compounds[30] it is not possible to make a rational choice between even the two likely reaction mechanisms. There is a considerable accumulation of data on the electrochemistry of transition metal organometallic compounds,[31] however there is little information on the electrochemical potential for the oxidation of $Rh_6(CO)_{16}$. In the absence of such data it must be considered possible that the free-radical autoxidation process is

$$Rh_6(CO)_{16} + O_2 \rightleftharpoons Rh_6(CO)_{16}^+ + O_2^-$$

initiated by electron transfer to the oxygen molecule to give superoxide ion. Unfortunately there is also a paucity of data on the formation of dioxygen transition metal cluster compounds. Nevertheless other π-acceptor ligands such as 1,5-cyclooctadiene will undergo substitution reactions with $Rh_6(CO)_{16}$ to give complexes of stoichiometry $Rh_6(CO)_{14}(diene)$,[32] where the rhodium framework is retained. Such a reaction pathway via a dioxygen complex could therefore lead to hydroperoxide complex chain initiation.

Acknowledgments

Thanks are due to the National Science Foundation and the Petroleum Research Fund for support of this work.

References

1. (a) L. J. Chinn, "Selection of Oxidants in Synthesis," Marcel Dekker, New York, 1971; (b) "Oxidation," Vols. 1 and 2, R. L. Augustine, Ed., Marcel Dekker, New York, 1969; (c) "Oxidation in Organic Chemistry," Part A, K. B. Wiberg, Ed.; Part B, W. S. Trahanovsky, Ed.; Academic Press, New York, 1965; 1973; (d) L. Reich and S. S. Stivala, "Autoxidation of Hydrocarbons and Polyolefins," Marcel Dekker, New York, 1969; (e) R. A. Sheldon and J. K. Kochi, _Advan. Catal._, 25, 272 (1976).

2. R. P. Hanzlik and D. Williamson, _J. Am. Chem. Soc._, 98, 6570 (1976).

3. L. Vaska, _Acc. Chem. Res._, 9, 175 (1976).

4. E. L. Muetterties, _Bull. Soc. Chim. Belg._, 84, 959 (1975).

5. G. D. Mercer, J. S. Shu, T. B. Rauchfuss, and D. M. Roundhill, _J. Am. Chem. Soc._, 97, 1967 (1975).

6. (a) J. P. Collman, M. Kubota, and J. W. Hosking, J. Am. Chem.
    Soc., 89, 4809 (1967); (b) A. Fusi, R. Ugo, F. Fox, A. Pasini,
    and S. Cenini, J. Organometal. Chem., 26, 417 (1971).

7. (a) G. Wilke, H. Schott, and P. Heimbach, Angew. Chem. Int.
    Ed. Engl., 79, 62 (1967); (b) J. Halpern and A. L. Pickard,
    Inorg. Chem., 9, 2798 (1970).

8. S. Otsuka, A. Nakamura, and Y. Tatsuno, J. Am. Chem. Soc., 91,
    6994 (1969).

9. J. Hojo, S. Yuasa, N. Yamazoe, I. Mochida, and T. Seiyama,
    J. Catal., 36, 93 (1975).

10. H. Sakamoto, T. Funabiki, and K. Tarama, J. Catal., 48, 427
    (1977).

11. G. D. Mercer, W. B. Beaulieu, and D. M. Roundhill, J. Am. Chem.
    Soc., 99, 6551 (1977).

12. Although the ketones are purified by distillation prior to use
    it is difficult to be certain that all traces of carboxylic
    acid have been eliminated.

13. B. F. G. Johnson, J. Lewis, and P. W. Robinson, J. Chem. Soc.
    A, 1100 (1970).

14. P. Kong and D. M. Roundhill, Inorg. Chem., 11, 749 (1972).

15. B. R. James and E. I. Ochiai, Can. J. Chem., 49, 975 (1971).

16. G. M. Zanderighi, R. Ugo, A. Fusi, and Y. B. Taarit, Inorg.
    Nucl. Chem. Lett., 12, 729 (1976).

17. W. B. Beaulieu, G. D. Mercer, and D. M. Roundhill, J. Am.
    Chem. Soc., in press.

18. B. D. Olafson and W. A. Goddard, III, Proc. Natl. Acad. Sci.,
    74, 1315 (1977).

19. V. D. Komissarov and E. T. Denisov, Russ. J. Phys. Chem., 43,
    426 (1969).

20. J. Rouchaud and B. Lutete, Can. J. Chem. Eng., 47, 157 (1969).

21. F. Basolo, B. M. Hoffman, and J. A. Ibers, Acc. Chem. Res., 8,
    384 (1975).

22. J. A. Elvidge and A. B. P. Lever, Proc. Chem. Soc., 195 (1959).

23. P. J. Hayward, D. M. Blake, G. Wilkinson, and C. J. Nyman, J. Am. Chem. Soc., 92, 5873 (1970).

24. V. I. Tret'yakov, A. V. Sklyarov, and B. R. Shub, Kinet. Katal., 12, 996 (1971).

25. A. M. Horgan and D. A. King, Trans. Farad. Soc., 67, 2145 (1971).

26. K. Tanaka and G. Blyholder, J. Chem. Soc. D, 736 (1971).

27. Y. Onishi, Bull. Chem. Soc. Jap., 44, 1460 (1971).

28. S. Oki and Y. Kaneko, J. Res. Inst. Catal., Hokkaido Univ., 18, 93 (1970); Chem. Abstr., 74, 25505j (1971).

29. C. Walling, J. Am. Chem. Soc., 91, 7590 (1969).

30. (a) R. B. King, Progr. Inorg. Chem., 15, 287 (1972); (b) P. Chini, G. Longoni, and V. G. Albano, Adv. Organometal. Chem., 14, 285 (1976).

31. R. E. Dessy and L. A. Barnes, Acc. Chem. Res., 5, 415 (1972).

32. T. Kitamura and T. Joh, J. Organometal. Chem., 65, 235 (1974).

METAL CARBONYL CATALYSIS OF CARBON MONOXIDE REACTIONS

R. B. King, A. D. King, Jr., M. Z. Iqbal, C. C. Frazier, and R. M. Hanes

Department of Chemistry, University of Georgia
Athens, Georgia 30602, U. S. A.

An important raw material for modern industrial processes is carbon monoxide (CO) which can be manufactured readily from steam and coal by the water gas reaction. In order to minimize the energy requirements of processes using CO as a raw material it is important to develop homogeneous catalysts which allow reactions with CO to proceed under the mildest possible conditions. Such catalysts use transition metal complexes and involve metal carbonyls as intermediates. Understanding the formation and reactivity of metal carbonyl derivatives in the presence of excess CO is important for the development of new and improved homogeneous catalysts capable of facilitating industrially important reactions of CO.

Infrared spectroscopy in the 2200-1600 cm.$^{-1}$ ν(CO) region is a simple and important method for detecting and analyzing mixtures containing metal carbonyl derivatives. In cases where the metal carbonyl derivative in question is a known stable compound or a close structural analogue of a known stable compound, then empirical comparison of the observed ν(CO) spectra with that of the model compound can be used for identification of the unknown metal carbonyl derivative. In cases where an appropriate model compound for the unknown metal carbonyl derivative is not available, then the observed band pattern in the ν(CO) region can be compared with that calculated for various possible compounds using group theoretical methods.[1]

In view of the importance of infrared spectroscopy in the analysis of metal carbonyl mixtures we have constructed at the University of Georgia some equipment capable of determining infrared spectra in the ν(CO) region at elevated pressures (up to 500 atmospheres CO). Such high pressure infrared spectroscopy experiments were first performed by Whyman and his co-workers at Imperial Chemical Industries using a titanium cell with sodium chloride windows.[2] We have improved and simplified the design of such high pressure infrared spectroscopic equipment by using a stainless steel high pressure cell with Irtran-I windows. Stainless steel is less expensive and more readily machined than titanium. Irtran-1 is mechanically stronger than sodium chloride while retaining sufficient transparency in the important 2200-1600 cm.$^{-1}$ region. Furthermore Irtran-1 is not attacked by water thereby allowing study of aqueous solutions in the high pressure infrared cell. Details of the construction of our equipment are described elsewhere.[3]

This paper summarizes our high pressure infrared studies relevant to understanding the following two important catalytic reactions of CO:

(1) The hydroformylation of olefins to give aldehydes, i.e.

$$RCH{=}CH_2 + CO + H_2 \longrightarrow RCH_2CH_2C(O)H \qquad \text{or} \qquad (1a)$$

$$RCH{=}CH_2 + CO + H_2 \longrightarrow RCH[C(O)H]CH_3 \qquad (1b)$$

(2) The so-called "water-gas shift" reaction, i.e.

$$CO + H_2O \longrightarrow CO_2 + H_2 \qquad (2)$$

Consideration of the role of metal alkyls as intermediates in the metal-catalyzed hydroformylation reaction has led to an infrared spectroscopic study of reactions of appropriate metal alkyl model compounds with CO followed by H_2. The high pressure infrared spectroscopic studies of reactions of iron carbonyl anions with CO has led to a new understanding of the mechanism of the metal catalysis of the water gas shift reaction which in turn has led to the development of some new catalysts for this reaction. Further details of the work cited in this paper will be published in future journal articles.

MODEL SYSTEMS FOR HYDROFORMYLATION CATALYSTS

An example of an active catalyst for the hydroformylation of olefins to give aldehydes (equations 1a and 1b) is $HCo(CO)_4$. The mechanism of this reaction appears to involve metal alkyl intermediates as exemplified by the following cyclic sequence of eight reactions:[4]

$$HCo(CO)_4 \longrightarrow HCo(CO)_3 + CO \qquad (3a)$$

$$HCo(CO)_3 + RCH{=}CH_2 \longrightarrow HCo(CO)_3(CH_2{=}CHR) \qquad (3b)$$

$$HCo(CO)_3(CH_2{=}CHR) \longrightarrow RCH_2CH_2Co(CO)_3 \qquad (3c)$$

$$RCH_2CH_2Co(CO)_3 + CO \longrightarrow RCH_2CH_2Co(CO)_4 \qquad (3d)$$

$$RCH_2CH_2Co(CO)_4 \longrightarrow RCH_2CH_2C(O)Co(CO)_3 \qquad (3e)$$

$$RCH_2CH_2C(O)Co(CO)_3 + H_2 \longrightarrow RCH_2CH_2C(O)Co(CO)_3H_2 \qquad (3f)$$

$$RCH_2CH_2C(O)Co(CO)_3H_2 \longrightarrow RCH_2CH_2C(O)H + HCo(CO)_3 \qquad (3g)$$

$$HCo(CO)_3 + CO \longrightarrow HCo(CO)_4 \qquad\qquad (3h)$$

This cyclic sequence involves the alternating production of 16- and 18-electron intermediates.[4] An important intermediate in this cyclic sequence is a metal alkyl as exemplified by $RCH_2CH_2Co(CO)_4$ produced in step 3d. Key reactions of such metal alkyls in this type of catalytic sequence for olefin hydroformylation include CO insertion into the metal-alkyl bond to give the corresponding acyl-metal derivative (step 3e) and the subsequent hydrogenolysis of the acyl-metal bond to give the aldehyde with liberation of the metal carbonyl hydride (step 3g).

Alkylcobalt tetracarbonyls related to $RCH_2CH_2Co(CO)_4$ (see step 3d) are too unstable for study in our high pressure infrared spectroscopy cell. For example, $CH_3Co(CO)_4$ decomposes above $-35\,°C$.[5] We therefore chose to investigate the reactions of some more stable types of alkylmetal carbonyls with CO and H_2 at elevated temperatures and pressures in order to see whether we could find such systems with reactivities appropriate for hydroformylation catalysts. In this study we included metal alkyls of the types $RM(CO)_5$ (M= Mn and Re), $RMn(CO)_4L$ (L = tertiary phosphine or phosphite), $RM(CO)_3C_5R'_5$ (M = Mo, R' = H and CH_3, M = W, R' = H), and $RFe(CO)_2C_5H_5$. The main features of the reactivity of these metal alkyls towards carbon monoxide are summarized below. Further details will be published in an appropriate full paper.[3]

$CH_3Mn(CO)_5$. This was the only metal alkyl where both the carbonylation to the corresponding metal acyl and the subsequent hydrogenolysis of this metal acyl to the corresponding aldehyde (i.e. steps 3d through 3g in the $HCo(CO)_4$ hydroformylation mechanism) could be observed in our high pressure infrared spectroscopic studies.

A solution of $CH_3Mn(CO)_5$ in tetradecane was found to react with CO at 320 atmospheres pressure and $67\,°C$ to give $CH_3COMn(CO)_5$ identified by the appearance of its characteristic acetyl $\nu(CO)$ band[6] at 1661 cm.$^{-1}$. After forming a solution of $CH_3COMn(CO)_5$ in this manner the CO pressure was released and 313 atmospheres of H_2 were added. Upon heating at $95\,°C$ the 1661 cm.$^{-1}$ acetyl $\nu(CO)$ band decreased in intensity with the concurrent appearance of a new $\nu(CO)$ band at 1738 cm.$^{-1}$. This new $\nu(CO)$ frequency was assigned to acetaldehyde by comparison with the reported[7] $\nu(CO)$ frequency of 1740 cm.$^{-1}$ and by comparison with the infrared spectrum of an authentic sample of acetaldehyde in tetradecane (1738 cm.$^{-1}$). When the formation of acetaldehyde from $CH_3COMn(CO)_5$ and H_2 was complete, the manganese carbonyl derivative present in the solution appeared to be $Mn_2(CO)_{10}$ rather than $HMn(CO)_5$. We have thus obtained infrared spectroscopic evidence for the successive reaction of $CH_3Mn(CO)_5$ with CO and H_2 according to the following scheme:

$$CH_3Mn(CO)_5 + CO \longrightarrow CH_3COMn(CO)_5 \qquad\qquad (4a)$$

$$2\ CH_3COMn(CO)_5 + H_2 \longrightarrow Mn_2(CO)_{10} + 2\ CH_3CHO \qquad (4b)$$

$CH_3Mn(CO)_4P[N(CH_3)_2]_3$. Substitution of one of the carbonyl ligands in $CH_3Mn(CO)_5$ by a trivalent phosphorus ligand does not appear to prevent CO insertion to form the acetyl derivative $CH_3COMn(CO)_4L$ but appears to prevent hydrogenolysis of this acetyl-manganese carbonyl derivative to give acetaldehyde. Thus a tetradecane solution of cis-$CH_3Mn(CO)_4P[N(CH_3)_2]_3$[8] reacted with 315 atmospheres CO at 81 °C to give $CH_3COMn(CO)_4P[N(CH_3)_2]_3$ identified by its $v(CO)$ frequencies at 2015 (w), 2000 (vs), 1968 (w), and 1659 (m) cm.$^{-1}$ Release of the CO pressure followed by treatment of this $CH_3COMn(CO)_4P[N(CH_3)_2]_3$ with 300 atmospheres H_2 at 72 °C gave $Mn_2(CO)_{10}$. The 1600 to 1750 cm.$^{-1}$ region of these spectra was clear indicating no appreciable formation of acetaldehyde. Also no $v(CO)$ frequencies were observed that could be assigned to the known[8] bimetallic complex $\{[(CH_3)_2N]_3PMn(CO)_4\}_2$.

$CH_3Re(CO)_5$. A tetradecane solution of $CH_3Re(CO)_5$ was heated slowly under 320 atmospheres CO. No change in the infrared $v(CO)$ frequencies occurred up to 140 °C. However, upon further heating to 200 °C the $CH_3Re(CO)_5$ was partially converted to $Re_2(CO)_{10}$. Even at 200 °C and 320 atmospheres CO about half of the $CH_3Re(CO)_5$ remained unchanged indicating the stability under CO pressure of this rhenium derivative as compared to the corresponding manganese derivative $CH_3Mn(CO)_5$ discussed above.

The failure of $CH_3Re(CO)_5$ to form $CH_3CORe(CO)_5$ by treatment with CO under pressure at elevated temperatures is not a consequence of the instability of $CH_3CORe(CO)_5$, which has been prepared[9] as a stable compound by the reaction of $NaRe(CO)_5$ with acetyl chloride. Instead, rhenium-carbon bonds appear to be appreciably stronger and therefore less reactive than corresponding manganese-carbon bonds.

$CH_3Fe(CO)_2C_5H_5$. A tetradecane solution of $CH_3Fe(CO)_2C_5H_5$[10] reacted with 320 atmospheres CO at 97 °C to form $CH_3COFe(CO)_2C_5H_5$ identified by comparison of its $v(CO)$ frequencies with those of authentic $CH_3COFe(CO)_2C_5H_5$ prepared by the reaction of $NaFe(CO)_2C_5H_5$ with acetyl chloride.[11] The formation of $CH_3COFe(CO)_2C_5H_5$ from $CH_3Fe(CO)_2C_5H_5$ and CO is relatively easy and begins even at 26 °C. However, the reaction is slow at this temperature and goes to completion at 97 °C. During this reaction rather large amounts of $[C_5H_5Fe(CO)_2]_2$ were also formed at 97 °C and 325 atmospheres CO as indicated by its infrared spectrum and its isolation from the solution in the high pressure infrared cell after completion of the reaction.

Treatment of a tetradecane solution of $CH_3COFe(CO)_2C_5H_5$ with 313 atmospheres H_2 at 155 °C resulted in removal of the acetyl group to give $[C_5H_5Fe(CO)_2]_2$. No infrared bands were observed in the range 1600-1780 cm.$^{-1}$ indicating the absence of acetaldehyde.

$CH_3Mo(CO)_3C_5H_5$. A concentrated solution of $CH_3Mo(CO)_3C_5H_5$[10] in tetradecane was heated under 316 atmospheres CO. No change was observed up to 52 °C. At 55 °C a shoulder appeared at 1985 cm.$^{-1}$ which developed into a stronger band upon further heating to 153 °C along with the disappearance of the $v(CO)$ frequencies of $CH_3Mo(CO)_3C_5H_5$. This new $v(CO)$ frequency can be assigned to $Mo(CO)_6$ formed by cleavage of the cyclopentadienyl ring and methyl groups from $CH_3Mo(CO)_3C_5H_5$ with the CO under pressure. This reaction was irreversible as indicated by no change in the infrared $\nu(CO)$ frequencies after cooling the solution from 153 °C to room temperature. No evidence for the formation of the bimetallic complex $[C_5H_5Mo(CO)_3]_2$ was observed at any time during this reaction of $CH_3Mo(CO)_3C_5H_5$ with CO.

The reaction of $C_2H_5Mo(CO)_3C_5H_5$ with CO proceeded analogously to form $Mo(CO)_6$ without any evidence for $[C_5H_5Mo(CO)_3]_2$ as an intermediate. However, analogous reactions of $C_6H_5CH_2Mo(CO)_3C_5H_5$ and $CH_2=CHCH_2Mo(CO)_3C_5H_5$ with CO formed appreciable quantities of $[C_5H_5Mo(CO)_3]_2$ as an intermediate in the formation of $Mo(CO)_6$.

$CH_3Mo(CO)_3C_5(CH_3)_5$. A solution of $CH_3Mo(CO)_3C_5(CH_3)_5$[12] in tetradecane was heated under 300 atmospheres CO. The 1985 cm.$^{-1}$ band indicating the formation of $Mo(CO)_6$ began to appear at 82 °C. At 110 °C a new band at 1704 cm.$^{-1}$ also began to appear. At 137 °C the 1704 cm.$^{-1}$ band became very strong and stopped growing further. Upon cooling the solution the 1704 cm.$^{-1}$ band persisted. The CO was then vented from the system and 320 atmospheres H_2 added. Heating to 212 °C failed to affect the 1704 cm.$^{-1}$ band. The relative intensity of the 1704 cm.$^{-1}$ band seemed to be independent of the $\nu(CO)$ frequency at 1985 cm.$^{-1}$ assigned to $Mo(CO)_6$. Vapor phase chromatography of the solution obtained from the infrared cell after the conclusion of this experiment indicated the presence of acetylpentamethylcyclopentadiene.[13] Furthermore, an authentic sample of acetylpentamethylcyclopentadiene[13] exhibited a $\nu(CO)$ frequency at 1704 cm.$^{-1}$ in tetradecane solution.

These observations indicate that $CH_3Mo(CO)_3C_5(CH_3)_5$ reacts with CO under pressure according to the following equation:

$$CH_3Mo(CO)_3C_5(CH_3)_5 + 4 CO \longrightarrow Mo(CO)_6 + CH_3COC_5(CH_3)_5 \quad (5)$$

Equation 5 is exactly the reverse of the reaction for the preparation of $CH_3Mo(CO)_3C_5(CH_3)_5$ by heating $Mo(CO)_6$ with acetylpentamethylcyclopentadiene.[12]

METAL CARBONYL CATALYSIS OF THE WATER GAS SHIFT REACTION

The initial objective in the reactions of transition metal alkyls with CO described above was to examine factors affecting the ease of alkyl migration[14] to a metal carbonyl group with the net result of CO insertion to give the corresponding metal acyl derivative. We also investigated the reaction of $HFe(CO)_4^-$ with CO at elevated temperatures and pressures hoping to obtain a similar type of CO insertion to form the known[15,16] formyl derivative $HC(O)Fe(CO)_4^-$. A process of this type would be a key step in a viable mechanism for a homogeneous metal-carbonyl catalyzed Fischer-Tropsch type hydrogenation of CO to methanol or methane. Instead, however, we observed that CO at elevated temperatures and pressures displaces the hydride ligand from $HFe(CO)_4^-$ to give $Fe(CO)_5$.

In a typical experiment a solution of $Fe(CO)_5$ (0.15 g.) was dissolved in a nitrogen-saturated solution of 0.12 g. of NaOH in 1.4 ml. of water and 48 ml. of 1-butanol. The infrared spectrum of this solution indicated that the expected[17,18] formation of $HFe(CO)_4^-$ had taken place. This 1-butanol solution of $HFe(CO)_4^-$ was then treated with 330 atmospheres CO in the high pressure infrared cell. No change in the infrared spectrum of the solution was observed upon standing for several hours at room temperature under 330 atmospheres CO. However, upon gradual heating under CO pressure the characteristic strong 1885 cm.$^{-1}$ band of $HFe(CO)_4^-$ gradually disappeared with the concurrent appearance of a 1995 cm.$^{-1}$ band indicating the presence of regenerated $Fe(CO)_5$. In this experiment with a base/$Fe(CO)_5$ ratio of 3.9 the formation of $Fe(CO)_5$ from $HFe(CO)_4^-$ and 330 atmospheres CO was complete at 93-98 °C. Increasing the base concentration increased the temperature necessary to react all of the $HFe(CO)_4^-$ to $Fe(CO)_5$ at a given CO pressure. Furthermore, at temperatures above 125 °C under 330 atmospheres CO a new $v(CO)$ frequency began to appear at 2300 cm.$^{-1}$ which may be assigned to CO_2 produced in the water gas shift reaction (equation 2). Cooling these mixtures to room temperature in the high pressure infrared cell resulted in partial regeneration of $HFe(CO)_4^-$ from reaction of $Fe(CO)_5$ with residual base, even when the CO pressure was maintained. No evidence for the formation of the formyl derivative $HC(O)Fe(CO)_4^-$ was observed in this or any other similar experiments involving the carbonylation of $HFe(CO)_4^-$.

These infrared spectroscopic studies at high CO pressures indicate that CO can displace the hydride ligand from $HFe(CO)_4^-$ to form $Fe(CO)_5$. This provides direct experimental verification of the missing link (i.e. equation 6a) in the following cycle for the catalysis of the water gas shift reaction (equation 2) by iron carbonyls:

$$HFe(CO)_4^- + CO \longrightarrow Fe(CO)_5 + H^- \qquad\qquad (6a)$$

$$H^- + H_2O \longrightarrow OH^- + H_2 \qquad\qquad (6b)$$

$$OH^- + Fe(CO)_5 \longrightarrow Fe(CO)_4C(O)OH^- \qquad\qquad (6c)$$

$$Fe(CO)_4C(O)OH^- \longrightarrow HFe(CO)_4^- + CO_2 \qquad\qquad (6d)$$

Equation 6b corresponds to the well-known vigorous reactions of ionic hydrides with water to give hydrogen gas and hydroxide ion.[19] The nucleophilic attack of a metal-bonded carbonyl group with hydroxide ion to give a metal-bonded carboxyl group (equation 6c) is well-established in metal carbonyl cation chemistry.[20] Carboxyl groups directly bonded to transition metals through the carbon atom such as in the proposed intermediate $Fe(CO)_4C(O)OH^-$ are well-known[21,22] to undergo facile decarboxylation yielding the corresponding metal hydride as in equation 6d. The sum of equations 6a-6d is the water gas shift reaction (equation 2).

Quantitative measures of the catalytic activity of the above iron carbonyl system have been obtained. Thus, preliminary results show that heating a mixture of 0.3 ml. of $Fe(CO)_5$, 2.6 g. of NaOH, 10 ml. of water, and 190 ml. of 1-butanol in a 700 ml. autoclave under 28.2 atmospheres CO in the temperature range 170-180 °C resulted in the production of 140 moles of H_2 per mole of $Fe(CO)_5$ per day. Significant catalytic activity of this system was also observed at even lower temperatures ($\sim$ 130°). Control experiments showed that both the base and the $Fe(CO)_5$ are essential for the catalytic activity of this system.

Basic solutions of the metal hexacarbonyls $M(CO)_6$ (M = Cr, Mo, and W) appear to be even more active catalysts for the water gas shift reaction (equation 2). Some systems of particularly high catalytic activity have been obtained with $W(CO)_6$. Thus heating a mixture of 0.0442 g. of $W(CO)_6$, 20 ml. of 10 M aqueous KOH, and 100 ml. of methanol under 7.7 atmospheres CO at 170 °C resulted in the production of 920 moles of H_2 per mole of $W(CO)_6$ per day. This reaction was continued until 66% of the CO originally introduced had reacted to form CO_2 and H_2. This reaction was also shown to be catalytic in base.

In addition some studies have been performed to examine the effects of temperature, pressure, and $KOH/W(CO)_6$ ratio on the catalytic activity of the above system. In accord with expectation, increasing the temperature increased the daily H_2 production from 14 moles of H_2 per mole of $W(CO)_6$ per day at 110 °C to 920 moles of H_2 per mole of $W(CO)_6$ per day at 170 °C. Increasing the CO pressure from 4.4 atmospheres to 14.6 atmospheres also increased the rate of H_2 production from 200 to 420 moles of H_2 per mole of $W(CO)_6$ per day at 140 °C. Furthermore, increasing the $KOH/W(CO)_6$ ratio from 350 to 1200 increased the rate of H_2 production from 90 to 240 moles of H_2 per mole of $W(CO)_6$ at 140 °C and 7.8 atmospheres CO.

We have attempted to detect metal carbonyl intermediates in the water gas shift reaction (equation 2) catalyzed by basic solutions of the metal hexacarbonyls $M(CO)_6$ (M = Mo and W) using infrared spectroscopy at elevated pressures. However, the only $\nu(CO)$ bands that we have found in 1-butanol solutions of $M(CO)_6$ (M = Mo and W) containing large excesses of KOH at 67 atmospheres CO are those arising from the free metal hexacarbonyl and from the CO_2 produced in the water gas shift reaction. Apparently the catalytically active metal carbonyl intermediate involved in these reactions is produced in concentrations too small to be detected by infrared spectroscopy under these conditions. Attempts to obtain these infrared spectra at lower CO pressures (e.g. 4 to 20 atmospheres) more nearly approximating conditions under which preparative water gas shift reactions are normally carried out were unsuccessful because solid decomposition products formed, coating the windows thus preventing light transmission through the infrared cell.

ACKNOWLEDGMENT

We are indebted to the U. S. Energy Research and Development Admin-
istration for partial support of this work under Contract EY-76-S-09-0933.

BIBLIOGRAPHY

1. F. A. Cotton, "Chemical Applications of Group Theory," Wiley-Inter-
    science, New York, 1971.

2. W. Rigby, R. Whyman, and K. Wilding, J. Phys. E: Sci. Instrum., 3,
    572 (1970).

3. R. B. King, M. Z. Iqbal, C. C. Frazier, and A. D. King, Jr., sub-
    mitted for publication.

4. C. A. Tolman, Chem. Soc. Revs., 1, 337 (1972).

5. W. Hieber, O. Vohler, and G. Braun, Z. Naturforsch., 13b, 192 (1958).

6. K. Noack, J. Organometal. Chem., 12, 181 (1968).

7. Sadtler Standard Spectra, 2, 2538 (Index 1965).

8. R. B. King and T. F. Korenowski, J. Organometal. Chem., 17, 95
    (1969).

9. W. Beck, W. Hieber, and H. Tengler, Ber., 94, 862 (1961).

10. T. S. Piper and G. Wilkinson, J. Inorg. Nucl. Chem., 3, 104 (1956).

11. R. B. King, J. Am. Chem. Soc., 85, 1918 (1963).

12. R. B. King and A. Efraty, J. Am. Chem. Soc., 94, 3773 (1972).

13. R. B. King, W. M. Douglas, and A. Efraty, Org. Syn., 56, 1 (1977).

14. K. Noack and F. Calderazzo, J. Organometal. Chem., 10, 101 (1967).

15. J. P. Collman and S. R. Winter, J. Am. Chem. Soc., 95, 4089 (1973).

16. C. P. Casey and S. M. Neumann, J. Am. Chem. Soc., 98, 5395 (1976).

17. H. W. Sternberg, R. Markby, and I. Wender, J. Am. Chem. Soc., 79,
    6116 (1957).

18. J. R. Case and M. C. Whiting, J. Chem. Soc., 4632 (1960).

19. D. T. Hurd, "Chemistry of the Hydrides," Wiley, New York, 1952.

20. T. Kruck and M. Noack, Ber., 97, 1693 (1964).

21. W. Hieber and T. Kruck, Z. Naturforsch., 16b, 709 (1961).

22. H. C. Clark, K. R. Dixon, and W. J. Jacobs, Chem. Comm., 548 (1968).

19. D. Liebert, "Chemistry of the Hinidder, SWP Inc. New York, 197...

20. F. Buchanan, M. Street, Proc. V., 163, Rush.

21. W. Miller and J. Roth, Z., Naturforsch., 181, 779 1935.

22. G. C. Schur, R. C. Liben, and J. T. Sander, Springer Company, 198 (1982).

ACTIVATION OF SATURATED HYDROCARBONS BY RUTHENIUM (I) COMPLEXES?

Brian R. James and Daniel K.W. Wang

Department of Chemistry
The University of British Columbia
Vancouver, British Columbia, Canada V6T 1W5

INTRODUCTION

Some studies on the activation of dihydrogen by ruthenium(III) complexes of the type RuX_3L_2 (X = halide, L = tertiary-phosphine or -arsine ligand) led us to the synthesis of the hydridoruthenium(II) complexes $[HRuXL_2]_2$, which were well-characterized by elemental analysis, molecular weight data, i.r., and ^1H and ^{31}P n.m.r.[1,2] In attempts to substantiate any role for such bisphosphine species in homogeneous hydrogenations catalyzed by the trisphosphine system[3] $HRuCl(PPh_3)_3$, we decided to study such processes using the isolated hydridobisphosphine dimer. Some unexpected and novel kinetics reported here suggest the involvement of ruthenium(I) species, and a reverse step detected in the hydrogenation mechanism indicates cleavage of a C-H bond in the saturated hydrocarbon moiety by such species. A brief communication on this work has appeared.[2]

EXPERIMENTAL

Dried, degassed, reagent grade solvents were used throughout, and NN'-dimethylacetamide (DMA) was purified as described previously.[4] Complexes were made under inert atmospheres using schlenk-tube techniques. A complex $RuCl_3(PPh_3)_2(DMA)\cdot DMA$, containing one coordinated and one solvate DMA, was made by stirring $RuCl_3\cdot3H_2O$ (0.5 g) with a two-fold excess of PPh_3 in DMA (20 ml) for 24 h at 20°C. After washing with DMA and drying under vacuum, green crystals were isolated in 70% yield. Found: C(58.2), H(5.4), N(3.0); calc. for $RuCl_3P_2C_{44}H_{48}N_2O_2$: C(58.3), H(5.3), N(2.8); ν^{Nujol}: 1630 cm^{-1} (uncoordinated DMA), 1590 cm^{-1} (coordinated DMA) 335 and 319 cm^{-1} (Ru-Cl).

The hydride dimer $[HRuCl(PPh_3)_2 \cdot DMA]_2$ was produced as solvated red crystals by stirring $RuCl_3(PPh_3)_2DMA \cdot DMA$ (1 g) with 1 g of proton sponge [1,8-bis(dimethylamino)naphthalene] in 50 ml DMA under 1 atm H_2 for 48 h at 20°C.   The product was washed with DMA and hexane and vacuum-dried (50%). Found: C(64.8), H(5.2), N(1.2); calc. for $RuClP_2C_{40}H_{39}NO$: C(64.2), H(5.2), N(1.8); ν^{Nujol}: 2035 cm^{-1} (Ru-H), 1650 cm^{-1} (uncoordinated DMA); $\tau^{C_6D_6}_{TMS}$: 23.8 (Ru-H). The $^{31}P-\{H\}$ n.m.r. spectrum in toluene at -60°C shows an AB pattern consistent with a chloride-bridged structure with two square pyramids sharing a basal edge; the spectrum is very similar to that we have measured for the $[RuCl_2(PPh_3)_2]_2$ complex[1] and which was first reported by Hoffman and Caulton[5] during measurements on solutions containing $RuCl_2(PPh_3)_3$.

The procedure for following H_2-uptake at constant pressure has been described elsewhere.[4] The H_2 solubility in DMA at 35°C was found to be 1.8×10^{-3} M atm^{-1}, Henry's Law being obeyed up to 1 atm.

RESULTS

Solutions of $[HRuCl(PPh_3)_2 \cdot DMA]_2$, $\underline{1}$, in DMA ($\sim 10^{-3}$ M) were found to efficiently catalyze the complete hydrogenation of acrylamide (~ 0.4 M) to propionamide under mild conditions (1 atm H_2, 35°C).   The gas-uptake plots were S-shaped with maximum rates being established after about 10 min.   The catalyst solution is initially red (λ_{max} = 500 nm, ε = 1530 M^{-1} cm^{-1} with Beer's Law being obeyed from 5×10^{-4} to 10^{-2} M), but on addition of acrylamide with increasing concentration, the colour changed through orange to yellow.   A quantitative spectrophotometric study of this reaction, monitored at 500 nm, analyzed for a 1:1 olefin-hydride reaction and since the yellow solutions contain no metal-hydride n.m.r. signal alkyl formation is indicated, with Markovnikov addition being most likely (equation 1):

$$HRuCl(PPh_3)_2 + CH_2=CHCONH_2 \overset{K}{\rightleftharpoons} (PPh_3)_2ClRu-\overset{H}{\underset{CH_3}{\overset{|}{\underset{|}{C}}}}-CONH_2 \qquad (1)$$

The equilibria were established after 5-10 min with K = 150 M^{-1}.   The Beer's Law behavior and the analysis according to equation 1, together with the kinetic data below, strongly imply that $\underline{1}$ is present as solvated monomers in the coordinating DMA solvent.   As the acrylamide was hydrogenated, the reactant solution changed from yellow to finally red as $\underline{1}$ was regenerated.

The kinetics of the catalytic hydrogenation were determined at 35°C from the region of maximum rate which was readily measured.   The Ru dependence is second-order up to 4×10^{-3} M (Fig. 1), while the H_2 dependence is first-order up to ca. 100 mm Hg and then decreases and

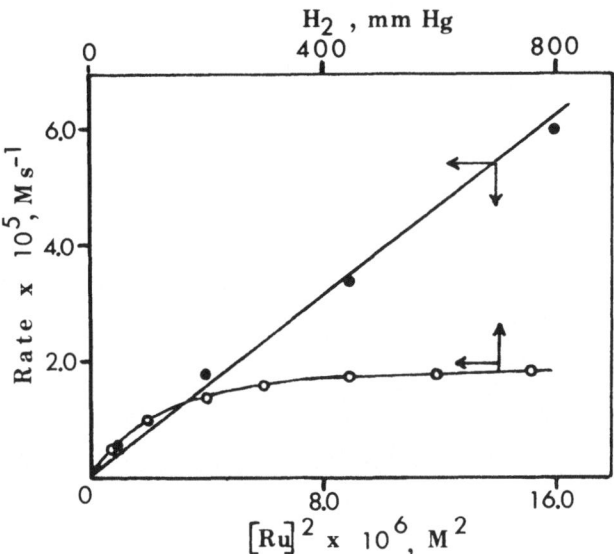

Fig. 1. (•) *Dependence of maximum rate on* [Ru]^2 *at 760 mm* H_2 *(1.8 × 10^{-3} M) and 0.4 M acrylamide.*
(o) *Dependence of maximum rate on* H_2 *pressure at 2 × 10^{-3} M Ru and 0.4 M acrylamide.*

approaches zero-order at pressures ⩾ 500 mm Hg (Fig. 1). An unusual inverse dependence on acrylamide from 0.01 to 0.8 M (Fig. 2), and an inverse dependence on added propionamide under first-order H_2 conditions (Fig. 3), are also observed. Addition of excess propionamide under conditions zero-order in H_2 had no effect on the rate.

The kinetics are readily accounted for by the mechanism outlined in equations 1-3.

$$RuCl(PPh_3)_2(alkyl) + HRuCl(PPh_3)_2 \underset{k_{-1}}{\overset{k_1}{\rightleftharpoons}} [RuCl(PPh_3)_2]_2 \qquad (2)$$
$$+ \text{ sat. product}$$

$$[RuCl(PPh_3)_2]_2 + H_2 \overset{k_2}{\rightarrow} 2 \; HRuCl(PPh_3)_2 \qquad (3)$$

Assuming a steady-state treatment for $[RuCl(PPh_3)_2]_2$ gives rate-law 4, which reasonably accommodates most of the experimental data.

$$-\frac{d[H_2]}{dt} = \frac{K \, k_1 k_2 [Ru_{total}]^2 [H_2][olefin]}{(1 + K[olefin])^2 (k_2[H_2] + k_{-1}[\text{sat. product}])} \qquad (4)$$

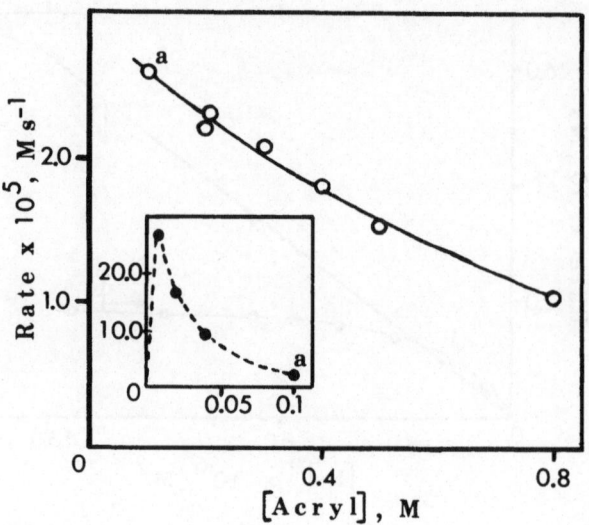

Fig. 2. *Dependence on [acrylamide] at 760 mm H_2 and 2×10^{-3} M Ru.*
The experimental point 'a' is the same on both scales. The
inset shows the predicted dependence at low [acrylamide].

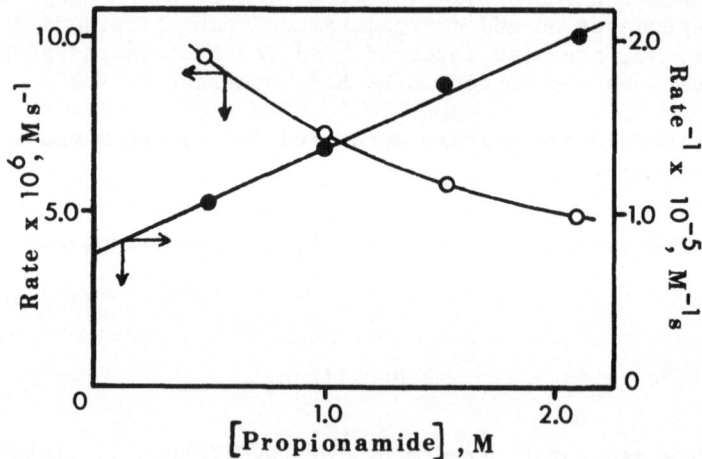

Fig. 3. *Dependence on [propionamide] at 100 mm H_2 (2.5×10^{-4} M)*
and 0.4 M acrylamide.

At high [acrylamide] and [H$_2$], K[olefin] $>>$ 1, the hydride is essentially fully converted to the alkyl, and equation 4 reduces to equation 5.

$$rate = k_1[Ru_{total}]^2/K[olefin] \qquad (5)$$

The slope of the straight line of Fig. 1, showing the Ru dependence, gives k_1/K = 1.6 s^{-1}, using for [olefin] a value of 0.38 M since the maximum rates were developed at such a concentration, i.e. at low conversion (see Discussion). Using the measured K value of 150 M^{-1} yields a k_1 value of 230 M^{-1}s^{-1} at 35°C. A test for the olefin dependence can also be made using equation 5, which holds at higher [olefin] and [H$_2$] (Fig. 4). The slope of the straight line drawn gives $k_1/K \sim 1.7$ s^{-1}.

Again at high [olefin], equation 4 can be rearranged to give equation 6.

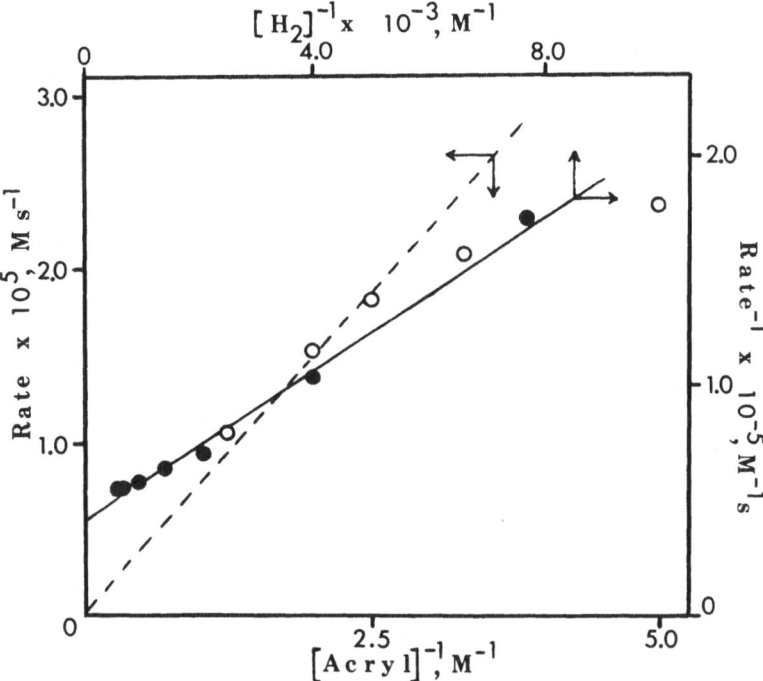

Fig. 4. *(o) Dependence on [acrylamide] as plotted according to equation 5.*
(•) Dependence on [H$_2$] as plotted according to equation 6.

$$\frac{1}{rate} = \frac{K[olefin]}{k_1[Ru_{total}]^2} + \frac{Kk_{-1}[olefin][sat.\ product]}{k_1 k_2 [Ru_{total}]^2} \cdot \frac{1}{[H_2]} \quad (6)$$

Figure 4 also shows a plot of $(rate)^{-1}$ vs. $[H_2]^{-1}$ at constant [olefin], $[Ru_{total}]$, and [sat. product]. In each case the maximum rate was attained at approximately the same conversion to propionamide, whose concentration is ca. 0.02 M. The intercept yields $k_1/K \sim$ 2.3 s^{-1} which is in reasonable agreement with the previously obtained values, considering the uncertainty in finding the intercept and possibly a somewhat variable [sat. product] term. The slope of the curve drawn gives $k_2/k_{-1} \sim 2000$.

For hydrogenations under constant $[H_2]$, $[Ru_{total}]$, and [acrylamide] (such that K[olefin >> 1]), a plot of $(rate)^{-1}$ against added [propionamide] should be linear according to equation 6, and Figure 3 shows the data at lower $[H_2]$, where the k_{-1}[propionamide] term can compete with the $k_2[H_2]$ term (equation 4). The intercept again gives $k_1/K = 1.6\ s^{-1}$ while the slope gives $k_2/k_{-1} = 3500$. The independence of the rate on added propionamide at high $[H_2]$ results from reaction 3 becoming relatively fast (the second term of equation 6 approximates to zero).

DISCUSSION

The mechanism for the acrylamide hydrogenation is novel and has interesting implications concerning the important process of alkane activation,[6,7] as exemplified by the k_{-1} step of reaction 2. Reactions between a metal-hydride and metal alkyl (the k_1 step) are known[8] for hydrogenations catalyzed by $Co(CN)_5^{3-}$, and have been postulated[9] for hydrodimerization of acrylonitrile catalyzed by $RuCl_2(PPh_3)_3$. A plausible scheme for reaction 2 which involves concomitant formation of a d^7 Ru(I) dimer, $[RuCl(PPh_3)_2]_2$ (or two such solvated monomers), is shown in reaction 7.

Formation of a metal-metal bonded Ru(I) complex seems reasonable since a Ru(I) derivative isolated previously from DMA solutions was diamagnetic and this was attributed to metal-metal bonding;[10] the chlorides might also be bridging. Catalyst regeneration, the k_2 step, involves homolytic splitting of H_2 by two Ru(I) centers (reaction 3); this is exactly analogous to hydrogen activation by $Co(CN)_5{}^{3-}$ and is formally an oxidative addition reaction to two d^7 metal centers.[11]

The kinetic dependences on [Ru], [H_2], [acrylamide], and [propionamide], all analyze well for a reasonably consistent k_1/K value. The olefin dependence (Fig. 2) reflects the relative amounts of ruthenium-hydride and -alkyl present. Qualitatively it can be seen that at very low olefin concentrations, little alkyl is formed and the reaction will be slow; at very high olefin concentration, no hydride will be left and again a low rate is predicted. At higher [H_2] when reaction 3 is fast, the mechanism predicts a maximum rate when the alkyl to hydride ratio is unity; this requires a free olefin concentration of $1/K$ or ca. 7×10^{-3} M. Thus the olefin dependence is predicted to be like the inset of Figure 2. The rates are difficult to measure quantitatively in the range of low olefin concentration due to the small total gas uptake, but qualitative agreement was found. Maleic acid was essentially not hydrogenated under the conditions found effective for acrylamide, and one factor is certainly the much higher olefin complexity constant K which was estimated to be 900 M^{-1} (cf. equation 5).

The kinetics give strong evidence for cleavage of a C-H bond in the propionamide product by the Ru(I) dimer (the k_{-1} step); this is again an oxidative addition process at d^7 metal centers.[11] The k_2/k_{-1} value indicates a strong kinetic preference for H-H over C-H bond cleavage. Oxidative addition of aliphatic C-H bonds at single d^8 metal centers, and to a lesser extent at d^{10} centers, to yield alkyl hydride species is well known,[6,7,12] but a two-site process, which has parallels in heterogeneous activation of alkanes,[7] is more attractive in terms of a search for new catalyst systems for alkane transformations. Dimeric species likely contribute to the catalytic activity of some platinum(II) complexes for H-D exchange in alkanes and alkyl sidechains.[6]

The ruthenium system described here involves C-H cleavage at an "amide-activated" sp^3 carbon. The cyano group plays a similar role in the d^8 metal systems that form hydrido cyanomethyl complexes with acetonitrile.[12] Studies are in progress using a range of olefinic substrates with our system to determine the degree of activation necessary for observation of C-H cleavage in the hydrogenated product. A likely competing reaction could be cleavage of the aromatic C-H bond of a phenyl group of triphenylphosphine, i.e. the ortho-metallation reaction.[13] We have reported previously on the formation of the orthometallated complex $[RuCl(PPh_3)(o-C_6H_4PPh_2)]_2$ via a stoichiometric reaction between the trisphosphine complex $HRuCl(PPh_3)_3$ and olefins such as maleic and fumaric acids,[14] and preliminary spectroscopic data suggest that the same orthometallated species is formed slowly from the bisphosphine hydride with, for example, fumaric acid (reaction 8):

$$HRuCl(PPh_3)_2 + \text{fumaric acid} \rightarrow RuCl(PPh_3)(o-C_6H_4PPh_2)$$
$$+ \text{succinic acid} \qquad (8)$$

The reaction presumably goes via the $RuCl(PPh_3)_2(alkyl)$ complex (cf. equation 1). The $HRuCl(PPh_3)_3$ complex is reported to catalyze exchange between benzene and D_2 under vigorous conditions;[13] this system must activate a benzene C-H bond, and bis(triphenylphosphine) species could be involved. Use of alkyl-substituted phosphines should eliminate side-reactions such as 8.

Addition of 1 mole of PPh_3 to the $HRuCl(PPH_3)_2$ catalytic systems readily generates the $HRuCl(PPh_3)_3$ system.[14] The trisphosphine-acrylamide system gives hydrogenation rates five times those of the bisphosphine system under corresponding conditions, and the trisphosphine system readily hydrogenates maleic acid. The kinetics and mechanism for the trisphosphine system are quite different from those of the bisphosphine system and the rate determining step involves reaction of an alkyl with H_2.[3,15] Since this step is not detected in the bisphosphine system, this implies that a trisphosphine alkyl reacts with H while the bisphosphine alkyl prefers interaction with the metal hydride (cf. equations 9 and 2).

$$RuCl(PPh_3)_3(alkyl) + H_2 \longrightarrow HRuCl(PPh_3)_3 + \text{sat. product} \qquad (9)$$

The extra phosphine likely increases electron density on the metal and favours reaction 9 via an initial oxidative addition step to give $H_2RuCl(PPh_3)_3(alkyl)$.

The fact that the maximum hydrogenation rates in the $HRuCl(PPh_3)_2$ - acrylamide systems are established after about 10 min within each run suggests that not all the hydrogenation follows the path of reactions 1-3, since the rate should increase as the acrylamide concentration falls, at least to the low level of 1/K (see, for example, rate laws 4 or 5 under zero-order conditions in H_2, when there is no inverse dependence on propionamide). The test for the acrylamide dependence (Fig. 2) is also not very satisfactory; there is no obvious reason why the data do not fit closer to a line such as the one drawn. There could be some contribution from a pathway similar to reaction 9 but involving bisphosphine species, and also from reaction 8. The ruthenium(I) catalyst could also generate a hydrogenation pathway via an unsaturate route, that is, one that involves acrylamide coordination at Ru(I) prior to a reaction with H_2. Testing of these possible contributions is now in progress, particularly via efforts to synthesize the postulated Ru(I) catalysts.

We thank the National Research Council of Canada for financial support and Johnson Matthey Ltd. for the loan of ruthenium.

REFERENCES

1. B.R. James, A.D. Rattray, and D.K.W. Wang, J.C.S. Chem. Comm., 792 (1976).
2. B.R. James and D.K.W. Wang, J.C.S. Chem. Comm., 550 (1977).
3. B.R. James, "Homogeneous Hydrogenation," Wiley, New York, 1973, p. 83.
4. B.R. James and G.L. Rempel, Discuss. Faraday Soc., 46, 48 (1968).
5. P.R. Hoffman and K.G. Caulton, J. Amer. Chem. Soc., 97, 4221 (1975).
6. D.E. Webster, Adv. Organometallic Chem., 15, 147 (1977).
7. A.E. Shilov and A.A. Shteinman, Coordin. Chem. Rev., in press.
8. Reference 3, p. 106.
9. J.D. McClure, R. Owyang, and L.H. Slaugh, J. Organometallic Chem., 12, P 8 (1968).
10. B.R. James, R.S. McMillan, and E. Ochiai, J. Inorg. Nuclear Chem. Letters, 8, 239 (1972); B.C. Hui and B.R. James, Canad. J. Chem., 52, 3760 (1974).
11. J. Halpern, Adv. in Chem. Series, 70, 1 (1968).
12. A.D. English and T. Herskovitz, J. Amer. Chem. Soc., 99, 1648 (1977).
13. G.W. Parshall, Accounts Chem. Res., 3, 139 (1970); 8, 113 (1975).
14. B.R. James, L.D. Markham, and D.K.W. Wang, J.C.S. Chem. Comm., 439 (1974).
15. L.D. Markham, Ph.D. Dissertation, University of British Columbia, Vancouver, 1973.

REFERENCES

OXIDATIVE ADDITION TO METAL ATOMS: MECHANISM, USE OF
RESULTANT REACTIVE RMX AND R_2M SPECIES IN HOMOGENEOUS
CATALYTIC PROCESSES, AND METAL ATOM--ALKANE INTERACTIONS

Kenneth J. Klabunde

University of North Dakota

Grand Forks, North Dakota 58202 U.S.A.

I. MECHANISM

The mechanisms of oxidative addition or reductive elimination
reactions[1] dealing with transition metal systems in solution are
quite complex and varied.[2] Detailed stereochemical and kinetic
studies of such systems have been prevalent in recent years.[3,4]
Stille and coworkers[3] believe backside attack mechanisms often are
important, particularly in benzyl halide systems:

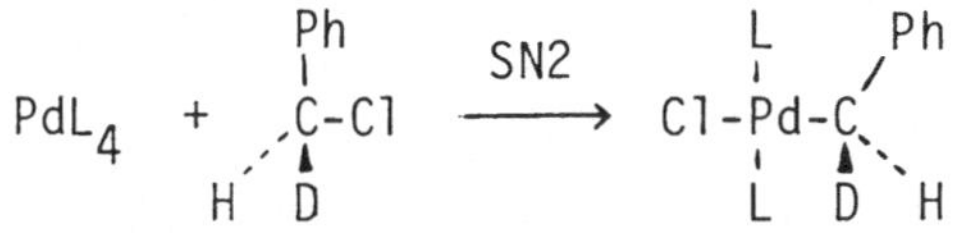

On the other hand, Osborn and coworkers[4] believe radical chain
processes can be important in many cases.

It has been found that a simple change from one alkyl halide
to another can cause a change in mechanism. Thus, steric effects,
reaction centers involved, and solvent effects are all important.

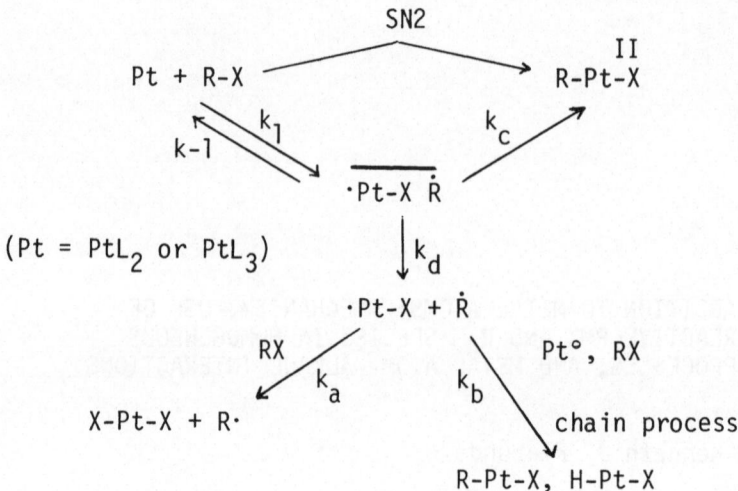

Our study of the mechanism of oxidative addition to palladium atoms was undertaken for several reasons: (1) it represents an example of a sterically unhindered metal center; (2) solvent effects and ligand effects would be eliminated; (3) the results may have implications regarding oxidative addition to surface metal atoms; (4) we have already obtained substantial chemical information about RPdX compounds that can be formed in this way,[5,6] and thereby have developed broad knowledge in this area. It must be pointed out

$$R\text{-}X + Pd \text{ atoms} \rightarrow R\text{-}Pd\text{-}X$$

$$R = CH_3, \; C_2H_5, \; (CH_3)_3C, \; (CH_3)_3CCH_2, \; CF_3, \; C_2F_5$$

$$X = Br, \; I$$

that there are many possible mechanisms, and it is very difficult to prove that just one is operable. This study shows at the very least however, that quite a number of possibilities can be eliminated, but a definite final choice cannot be made.

Possible Mechanisms for the
Pd atom & RX → RPdX   Reaction

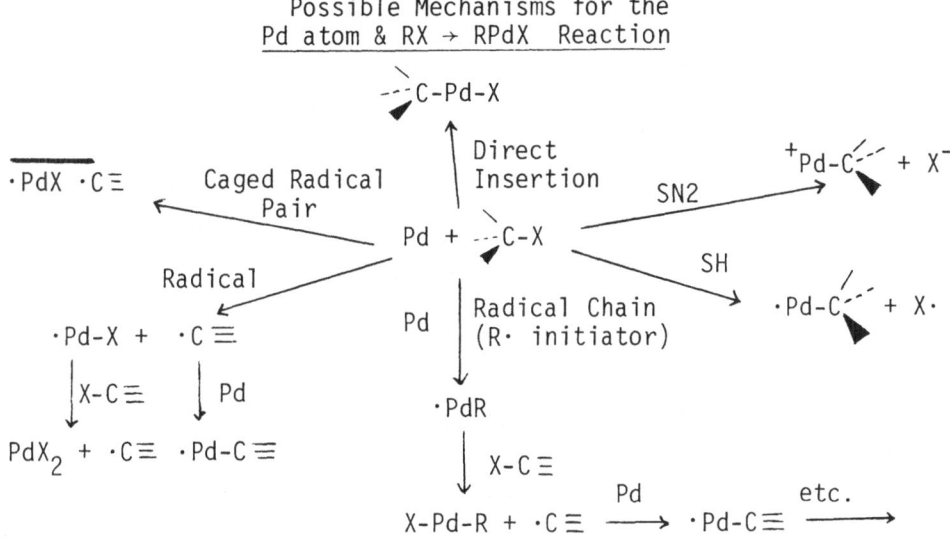

Isolable RPdX Species

Previous work in our laboratory showed the compounds $(CF_3PdI)_n$, $(C_2F_5PdI)_n$, and $(n\text{-}C_3F_7PdI)_n$ are stable, albeit reactive, compounds which can be prepared by codeposition of Pd vapor with the respective R_fI substrate.[5,7] A search for other products was made, in particular PdI_2 and gaseous coupling and disproportionation products, R_fF and R_fR_f. For the CF_3I-Pd system, no CF_4, C_2F_6, or C_2F_4 were formed when Pd vapor was condensed with CF_3I at $-196°C$, with subsequent slow warming to room temperature. Based on Pd vaporized, the products found were (CF_3PdI) (23% and 20%) and PdI_2 (29% and 33%) for two identical experiments. The high yields of PdI_2 were confusing in light of the apparent absence of CF_3 derived products other than CF_3PdI. We have not observed $(CF_3)_2Pd$ by nmr, and we believe CF_3 loss is due to polymer formation or decomposition and loss on the $Pd\text{-}PdI_2$ residue (cf. analogous behavior of CH_3I discussed later). For the analogous C_2F_5I-Pd system, again no gaseous products, C_2F_6, C_2F_4, or C_4F_{10}, were observed. The yields of C_2F_5PdI were quite high (66% and 49%), but still significant amounts of PdI_2 (17% and 6%) were formed. These experiments do point out that RPdX can be formed directly without formation of gaseous decomposition products. Therefore, in cases where gaseous products are formed, they probably come from decomposition of RPdX rather than from the process of formation of RPdX.

Non-Isolable RPdX Species

Many CH_3I-Pd depositions were studied. Yields of products, CH_4 and C_2H_6, were rather erratic, until it was discovered that not all of the gaseous products were released unless the reaction residue was pyrolyzed at approximately 100°C. A study was made in which the composition and yields of products were determined upon matrix warmup (-196°C → 25°; no CH_4 was formed during codeposition at -196°C) and then upon residue pyrolysis. Identical experiments showed good reproducibility. An average of three experiments is tabulated below:

Table 1:   Products from CH_3I-Pd Reactions

Product	% Yield	
	Released→25°C	Released→100°
CH_4	2.8	13
C_2H_6	0.21	2.3
C_2H_4	0.0	0.18
PdI_2	12	

We have not been successful in isolating the unstable intermediate, CH_3PdI, or even in trapping it as its Et_3P adduct. Therefore, the release of CH_4 upon pyrolysis probably is due to CH_3, CH_3I, or CH_4 being bound to or adsorbed tightly on the $Pd-PdI_2$ solid residue remaining after matrix warmup. Alternatively, CH_3PdI may decompose intermolecularly to $CH_3(Pd)n + PdI_2$ with just partial release of organic products, thus rationalizing the higher yield of PdI_2 than CH_4. A small amount of CH_3CH_2I also was observed as a product, which suggests that CH_2 may also be a decomposition product. Similar experiments utilizing CH_3Br gave similar product distributions, although the bulk of the product gases were released at 25°C or lower. (CH_4 9.4%, C_2H_6 1.8%, $CH_2 = CH_2$ 0.29%, $PdBr_2$ 7.0%) It is possible that $Pd-PdI_2$ residues "tie up" CH_3 fragments more effectively than $Pd-PdBr_2$ residues.

Reaction of ethyl iodide with Pd yielded more informative results than the CH_3I-Pd system. Three pertinent experiments were carried out: (1) codeposition of EtI with Pd followed by warming to room temperature and product removal, (2) deposition of Pd with EtI and HCl together and, (3) deposition of Pd with HCl followed by EtI addition. The latter two experiments were carried out in order to show whether HPdX (HPdCl or HPdI) could be a decomposition product enroute to the final products. The experiments do indicate that HPdX is important, particularly in formation of ethane. Similar results were obtained for EtBr employing the same experiments. In this case the presence of HCl caused increased ethane yields and _decreased_ ethylene yields (HCl and HBr behaved similarly).

Table II   Yields of Products from Various Reactions Involving Cocondensation of Palladium Vapors with Ethyl Iodide and Ethyl Bromide

Reactions	$C_2H_6(\%)$	$C_2H_4(\%)$	$PdX_2(\%)$
Pd + EtI	6.7	18	28
Pd + EtI + HCl	42	40	--
Pd + HCl followed by EtI	20	8.3	--
Pd + EtBr	13	11	10
Pd + EtBr + HCl	21	4.5	4.5
Pd + HCl followed by EtBr	3.6	1.3	--

In none of these experiments was butane observed as a product. This shows conclusively that ethyl radicals were not "free" in the matrix since if that were true, _some_ quantity of butane would be observed _even_ _if_ disproportionation were the favored process.

The deposition of _t_-butyl bromide with Pd vapor was studied since in this case formation of RPdX probably would not occur by an S_N2 type mechanism, and if free t-butyl radicals were formed, they would be expected to disproportionate to give about equal yields of 2-methylpropane and 2-methylpropene. Experiments with _t_-butyl bromide-HCl mixtures also were carried out as shown in Table III. These experiments again showed that HPdX could reduce R-X to RH,[16] but in this case the presence of HCl during the

Table III   Yields of Gaseous Products from the Cocondensation of t-Butyl with Palladium Vapor

Reaction	2-methyl propane	2-methyl propene
Pd + $(CH_3)_3CBr$	8.5	38
Pd + $(CH_3)_3CBr$ + HCl	16	40
Pd + HCl followed by $(CH_3)_3CBr$	7.0	0.0

deposition of $(CH_3)_3CBr$ did not affect the $(CH_3)_2C = CH_2$ yield. Overall, product yields with t-BuBr were quite high, which indicates the absence of an S_N2 process or other backside attack process. Even though a Pd atom is quite unhindered, at least a slight lowering in reaction efficiency would be expected if an S_2N were operating. To study this point further competition experiments between t-BuBr and n-BuBr were carried out. An equimolar mixture deposited with Pd vapor yielded a product mixture favoring $(CH_3)_2C=CH_2$ over 1 and 2-butenes by > 2:1. Thus, the Pd atom preferentially attacked the t-BuBr, again showing that an SN2 type process is not likely, and that some type of radical process is more likely. And as a final point for consideration, a Pd atom should not be nucleophilic anyway and a d^{10} configuration would be expected to be electrophilic.

Our most detailed studies involved neopentyl bromide (1-bromide-2, 2-dimethylpropane) and palladium vapor. An average of several determinations indicated products as shown below:

$$(CH_3)_3C-CH_2Br + Pd \xrightarrow[\text{to 25°C}]{\text{warm}}$$

$$CH_4 (16\%) + CH_3\overset{\overset{\displaystyle CH_3}{|}}{C}HCH_3 \ (3\%)$$

$$+ CH_3-\overset{\overset{\displaystyle CH_3}{|}}{\underset{\underset{\displaystyle CH_3}{|}}{C}}-CH_3 \ (4\%) + \overset{\displaystyle CH_3}{\underset{\displaystyle CH_3}{>}}C = CH_2 \ (51\%)$$

$$+ CH_2\!-\!\!\overset{\overset{\displaystyle CH_2}{\diagup}}{\underset{\underset{\displaystyle}{}}{C}}\!\overset{\displaystyle CH_3}{\underset{\displaystyle CH_3}{<}} \ (2\%) + PdBr_2 \ (30\%)$$

The main products of the reaction were CH_4 and $(CH_3)_2C = CH_2$. The actual CH_4 yield was probably somewhat higher, as we had difficulty in trapping this product quantitatively, and also have previously observed the tendency of CH_3 and/or CH_4 to be adsorbed on the $(Pd)_n$ residues. A study of product distribution changes when certain additives were incorporated was made. The additives used were NO (a good free radical scavenger), $C_6H_5CH_3$ (a good radical terminating agent since the CH_2-H bond is relatively weak), and PEt_3 (as a trap of radicals and/or RPdX). Yields are shown in Table IV.

Table IV Effect of Additives on the Product Yields for Neopentyl Bromide-Pd Vapor Reactions(%)

Additive	$C_2H_6 + C_2H_4$	$n\text{-}C_4H_{10}$	CH_4	$(CH_3)_2C{=}CH_2$	$(CH_3)_3CH$	$(CH_3)_4C$	$\begin{array}{c}CH_2\\ \diagup\diagdown\\ CH_2{-}C(CH_3)_2\end{array}$	$PdBr_2$
NO	0	0	15	48	2.3	4.3	0	33
$C_6H_5CH_3$	0	0	14	48	4.1	5.4	2.5	38
PEt_3	15	10	26	32	2.1	2.2	0.80	--

With NO as additive, no change whatever in product distribution was observed. This result precludes a radical chain process for the formation of the observed products. With toluene a slight decrease in $(CH_3)_2C=CH_2$ and a slight increase in $(CH_3)_3CH$ was observed. This change is in the expected direction if toluene were trapping t-butyl radicals by H donation.

However, the effect is quite small. And curiously, the neopentane yield was not increased, and in fact decreased slightly. The toluene results reinforce the idea that radicals, if formed, are not free to migrate in the matrix. Employing PEt_3 as an additive, the $(CH_3)_2C=CH_2$ yield was decreased moderately while the CH_4 yield increased greatly. This increased CH_4 yield may be due to displacement of adsorbed CH_3 or CH_4 from bulk $Pd-PdBr_2$. The C_2H_6 and C_2H_4 are most likely products of metal atom reactions with PEt_3, reactions which have been observed before.[8] (Note that $CH_3PdBr(PEt_3)_2$ was not formed, which indicates it is unlikely that neopentylpalladium bromide decomposed directly to yield methylpalladium bromide.)

The fact that the neopentyl moiety is fragemnted to give isobutylene and methane (from CH_3) indicates that a fast radical type process may be important. Careful studies on the thermochemistry of free radicals[9] and on the generation of "hot" free radicals[10] have shown that CH_3 eliminations are facile, but any C-C cleavage requires 26-40 kcal/mole. Thus, the starting radical must be significantly vibrationally excited to allow C-C cleavage to occur. However, at the same time we cannot eliminate a direct frontside insertion mechanism followed by RPdX decomposition or rearrangement, nor can we eliminate the possibility of a metallocycle intermediate such as $ClHPd\underset{CH_2}{\overset{CH_2}{<}}C(CH_3)_2$, and further work is in order.[10c]

We have studied several alkyl halides containing alkenyl or aryl functional groups, which have led to the synthesis of a number of unusual RMX and R_2M compounds, but which have been of little help in terms of mechanistic information regarding the RX/Pd reaction. For example benzyl chloride yields η-benzylpalladium chloride dimer.[11] In this case neither $PdCl_2$ or bibenzyl were formed, seemingly precluding an uncaged radical type mechanism.[12] Visual observations were such that it seemed likely that upon cocondensation of benzyl chloride and Pd vapor a π-complex formed first with subsequent C-Cl bond insertion during warmup. Thus, a deep red matrix immediately formed which changed in appearance considerably upon warming, a sequence followed very similiar to many M-arene matrices. (Whenever arene ligands (toluene, bromobenzene, bromopentafluorobenzene, xylene, etc.) are cocondensed with Pd (or other transition metals) colored complexes are believed to form. On warming C-X insertion can occur if C-X is available. If it is not, large metal clusters form (cf. later discussion)). We believe π-complexation, followed by C-X insertion on warming, occurs in aryl halide systems as well,[5,13] for example in the formation of C_6F_5PdBr.

The problem with the aryl and aryl-substituted alkyl halides in the present context is that these unsaturated groups very probably could cause a change in the C-X insertion mechanism compared to saturated R-X substrates. This idea is nicely illustrated by our work with b-bromo-1-hexene with Pd vapor. Here, it seemed likely that a prior π-complex formed since a clear condensate was produced which changed to yellow and then to brown on warming. The 6-bromo-1-hexene was converted catalytically to 6-bromo-2-hexene, but no $PdBr_2$ or other products attributable to R-Pd-Br formation were observed!

Since the heated crucible in which the metal atoms are generated is white hot and radiating much light during a metal atom reaction, we anticipated possible photoeffects in RX-Pd reactions. A shielded crucible (with Kaowool) was employed in a EtI-Pd reaction. Table V shows product yields for identical experiments employing unshielded and shielded crucibles. The results are essentially identical, and so at least in this case photolysis has no effect.[14]

Table V Ethane and Ethylene Yields for a Et-Pd Reaction Employing Shielded and Unshielded Crucibles

	Ethane(%)	Ethylene(%)
Shielded	5.6	19
Unshielded	6.7	18

Discussion and Conclusions

There are a number of points that have to be reconciled in order to predict a mechanistic pathway:

(1) Coupling and disproportionation reactions of R radicals is a minor process and usually totally absent. Therefore, radicals, if formed are not free to migrate in the matrix.

(2) Radical scavengers have no effect on products yields or distributions. Therefore a radical chain process is not important in product formations.

(3) Photolysis by the crucible has no effect, at least in the case of a saturated alkyl halide such as EtI.

(4) t-Alkyl halides react more efficiently than primary halides, as shown by direct competition experiments, indicating a non-backside attack mechanism, such as S_N2.

(5) Vibrationally excited radicals probably are formed.

(6) HPdX is an important, reactive intermediate product capable of reducing R-X to R-H.

(7) Gaseous products come from decomposition of RPdX rather than from the process of formation of RPdX. Thus, if RPdX is stable, that is the only organic product found.

(8) Unsaturated alkyl halides react first by π-complexation followed by C-X oxidative addition, probably on matrix warmup.

(9) Based on our previous work with R-X--Ag vapor reactions,[15] it is likely that complexation to form RX--M occurs first at -196°C, and on warmup oxidative addition occurs. This previous work[14] showed that in the reaction 2Ag + RX → AgX + RAg high RX/Ag ratios favored RAg formation, thus indicating prior RX-Ag complexation before slow reaction to form R· + AgX in warmup. The R· could then quickly trap Ag from still existing RX--Ag complexes.

A mechanism that fits all of these observations is shown below. In brief, upon cocondensation of an excess of RX with Pd atoms, a σ-complex forms[14,15] which is stable at that low temperature. Upon matrix warmup, a caged radical pair forms and at the exact moment of its formation, the R· portion possesses enough excess vibrational energy that decomposition processes may occur. The radicals do not separate, but recombine to form RPdX, which, if it is stable, can be isolated. If RPdX is not stable, it decomposes on warming, yielding various products, depending on the nature of RPdX. Thus, when possible, HPdX and alkene are formed, and the HPdX can react with RX to form RH and PdX_2. The formation of HPdX is so favorable that in the case of CH_3PdI, come CH_2 appears to be formed, which reacts with CH_3I to form CH_3CH_2I. In the case of neopentyl bromide, isobutylene and CH_3 are formed. It is unlikely that the CH_3 combines with PdBr in this case since there are significant differences in products and yields for the CH_3Br + Pd experiment when compared with the $(CH_3)_3CCH_2Br$ + Pd experiment (CH_4 9.4%, CH_3CH_3 1.8%, $CH_2 = CH_2$ 0.09%, $PdBr_2$ 7.0% for CH_3Br experiment and CH_4 15%, CH_3CH_3 0.0%, $CH_2 = CH_2$ 0.0%, $PdBr_2$ 33% for $(CH_3)_3CCH_2Br$ experiment).

Whenever a β-H elimination is possible, cleaner reaction product mixtures are found. In the case of CH_3X, complex and competing reaction patterns are observed. These include retention of ·CH_3 by the Pd-PdX_2 residue, particularly in the case of Pd-PdI_2. It also is noted that whenever RI is used, the PdI_2 yields are much higher than would be predicted from the stoichiometry: 2RI + Pd → PdI_2 + 2R (organic products). This further reinforces the idea that the Pd-PdI_2 residue retain organic species, with some of them possibly being converted to polymeric products which will be non-recoverable. Thus, the presence of <u>iodide</u> and $CH3$ group are complicating features. Note that for EtBr, t-BuBr, and $Me3CCH2Br$, the reaction stoichiometry is approximately correct, as shown in Table VI.

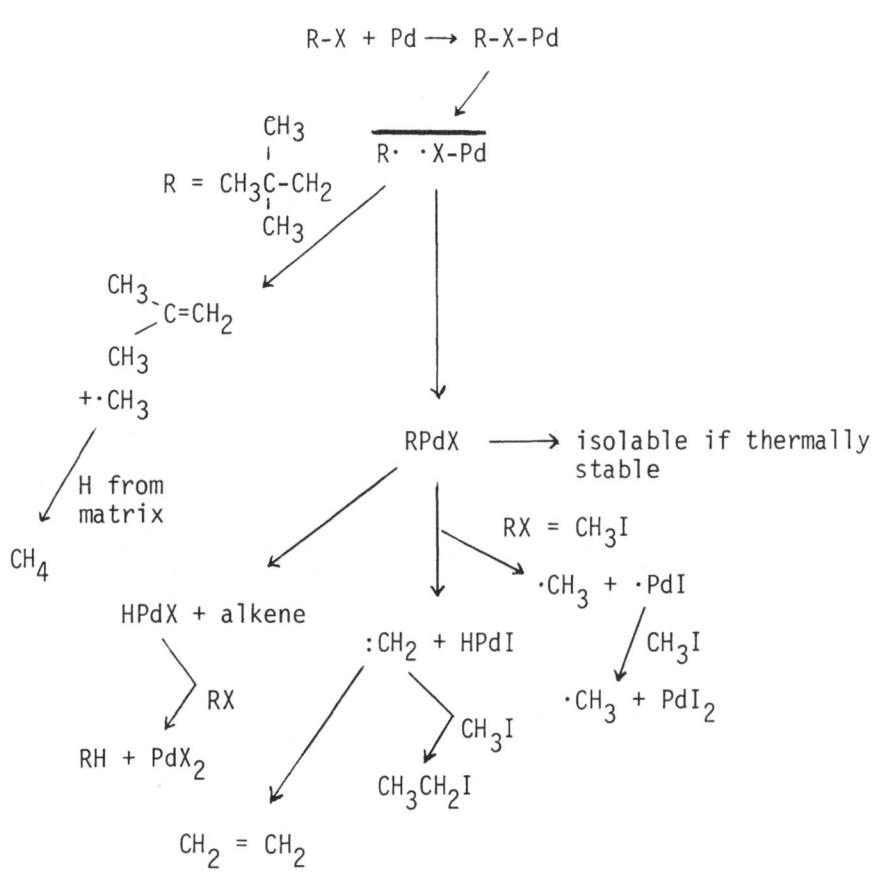

CH_3 combines with Pd-PdI_2 residues, as well as some coupling to form CH_3CH_3

Table VI: Stoichiometry Comparisions for
R-X + Pd Reactions (% Yields)

	PdX_2	Combined yield of organics
CH_3Br	7	12
CH_3CH_2Br	10	24
$(CH_3)_3CBr$	29	55[a]
$(CH_3)_3CCH_2Br$	30	60[b]
CH_3I	12	19[c]
CH_3CH_2I	28	25

[a]Average value.

[b]CH_4 not included since isobutylene and CH_4 come from same molecule.

[c]Heating needed to release most of organics.

II. Isolation of Novel Transition Metal RMX and R_2M Species

A. RMX Compounds

The main interest in our continuing studies of oxidative addition to metal atoms stems from our early discovery that stable, soluble (in organic solvents) "coordinatively unsaturated" transition metal RMX species could be generated.[17] The R group involved has a <u>remarkable</u> effect on the stability of the RMX species, particularly in the case of Ni or Pd, and thermal stabilities range from < -130° to 130°C! Table VII lists the non-solvated non-ligand stabilized RMX species we have prepared, and gives some brief comments about their properties. Some of these species have been proposed as reactive intermediates in Pd catalytic processes.[18]

From Table VII it is clear that perfluoro ligands yield the most stable "coordinatively unsaturated" RMX compounds. In particular, C_6F_5PdBr and CF_3PdI are of interest. These are very stable and soluble in acetone solution. Addition of donor ligands to these compounds results in immediate formation of the ligand stabilized form $L_2Pd(R_f)X$. Almost any ligand adds quantitatively

(L=NR$_3$, SR$_2$, PR$_3$, AsR$_3$, NH$_3$, dienes),[19] and <u>cis</u> complexes can be obtained with bidentate ligands. In the absence of stabilizing ligands, slow decomposition occurs upon standing in benzene solution.

One particularly interesting feature was noted for CF$_3$PdI, which weakly coordinated toluene.[19] Arene coordination has turned out to be quite important with these type of species, as we will see in later discussion.

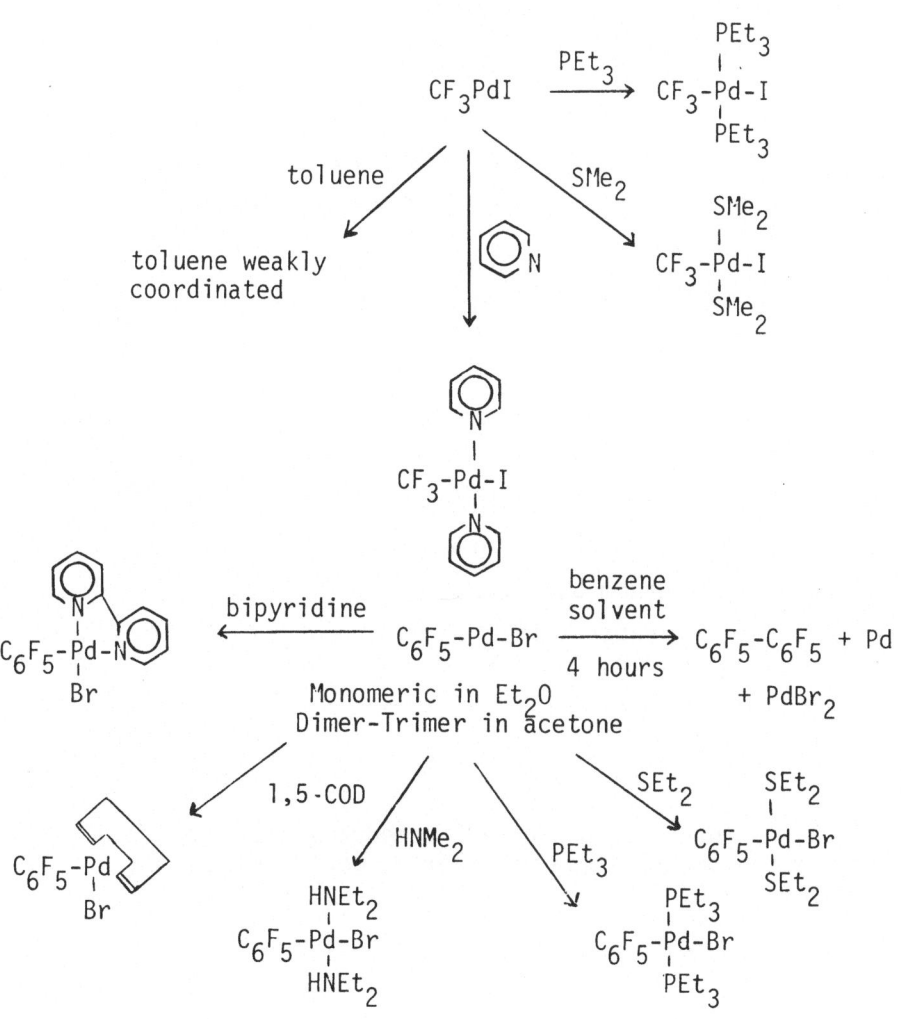

Table VII Unstable RMX (RCOMX) Species Produced

RMX	Comments	Reference
CF_3ZnI	CF_2 formation	20
$(CF_3)_2CFZnI$	readily hydrolyzed	20
$CF_3CF = CCaF(CF_3)$	eliminates CaF_2	21
C_6F_5CaF	readily hydrolyzed	21
$C_6H_5PdBr(Cl)$	stable < -100°C	5
$(CF_3)_2CFPdI$	could not be trapped	5
$CH_3PdBr(I)$	stable < -100°C	5
C_2H_5PdI	stable < -100°C	5
$C_6F_5NiCl(Br)$	trapped at -80°C	6
$CF_3NiBr(I)$	trace trapped at -80°C	6
CH_3NiI	CH_4 formation during deposition	6
C_2H_5NiI	CH_4 or H_2 formation during deposition	6

Stable, Isolated RMX Species Produced

RMX	Comments	Reference
$C_6F_5PdBr(Cl)(I)$	reacts with many ligands	5, 22
CF_3PdI	very labile, weakly coordinates toluene	5, 22
CF_3CF_2PdI	very stable in acetone	5
$CF_3CF_2CF_2PdI$	very stable in acetone	5
$C_6H_5CH_2PdCl$	decomposes 105°C	11
C_6F_5PtBr	reacts with Et_3P to give <u>cis</u> & <u>trans</u> adducts	6, 23

One of the more intriguing RPdX compounds that shows good stability is the benzylpalladium chloride derivative.[20] It is the only non-fluorinated RPdX that we found isolable and readily characterizable. There seems little doubt at this time that this interesting compound exhibits η^3-bonding and is dimeric.

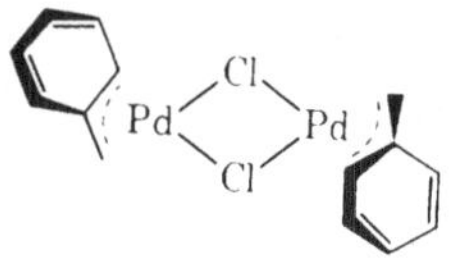

I

This conclusion about the bonding was based on the following reasoning: (1) The unusual stability of I vs. C_6H_5PdCl (105°C vs. < -100°C respectively).[11] (2) There is some literature precedent for η^3-benzyl bonding[24,25] such as in $(C_6H_5CH_2)Mo(C_5H_5)(CO)_2$. (3) Incremental addition of Et_3P to I in solution indicated that two equivalents of Et_3P (one per Pd atom) resulted in a rearrangement (loss of η^3-bonding), thus causing the CH_2 protons to be split into a doublet (by phosphorous)[11] and the ortho protons to be shifted downfield into the normal region for aryl protons in such σ-bonded compounds, i.e., $(Et_3P)_2Pd(Cl)CH_2C_6H_5(II)$. Addition of two equivalents of Et_3P resulted in the formation of II, and analogous results were obtained with pyridine-d5 in place of Et_3P.[11] These results are similar to those found for Et_3P additions to

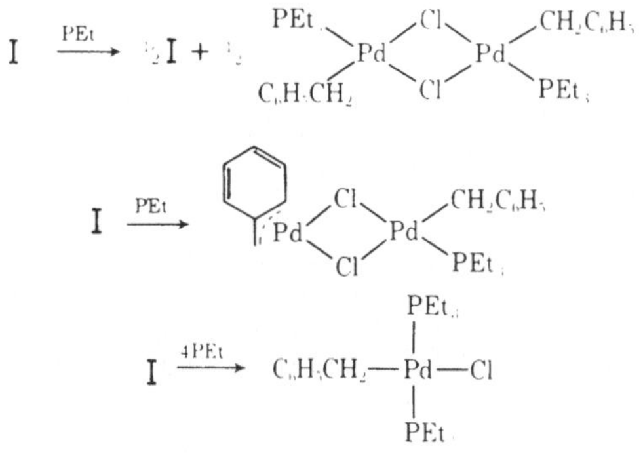

II

allyl-bonded compounds, except for one important feature: in allyl systems the halogen bridge is cleaved first and then rearrangement occurs. (4) The PMR spectrum of I is compatable with allylic type bonding. Thus, the spectrum of I is very similar to that reported by Stevens and Shier[26] for $[(Et_3P)_2PdCH_2C_6H_5]$.[+] The ortho protons are shifted upfield substantially: PMR (I in CD_2Cl_2, δ (ppm)) 7.92, 7.78, 7.65 (meta and para protons), 7.11, 7.00 (ortho protons),

and 3.58 (benzylic protons). For I, no preferred conformations
could be detected by PMR at a probe temperature variance of 40 to
-85°C. However, for a substituted version of I (3,4-dimethylbenzyl-
palladium chloride, III) temperature dependent PMR was observed.[11]

III

For the ortho protons, the doublet(proton 5) moves upfield and
the singlet (proton 3) moves downfield with decreasing temperature.
In PMR studies of palladium allyl complexes, generally, the protons
closest to the palladium atom are the most shielded and absorb
furthest upfield.[27] These results imply, not unexpectedly, that
the most stable low temperature configuration for III is that form
in which the methyl groups are furthest from the palladium.[11] At
lower temperature the two benzylic protons become non-equivalent.

The ortho protons continue to become more widely different in
their chemical shifts. These concurrent changes in chemical shift
imply that only one equilibrium process is being slowed, and hence
must explain both effects (benzylic proton and ortho proton changes).
One such process would be a $\pi \rightarrow \sigma \rightarrow \pi$ equilibrium, with a σ-bonded
benzyl group as a short lived higher energy intermediate. As a
C_1C_2 rotation would be rapid, H_1 and H_2 would be equivalent on an
nmr time scale as long as the $\pi \rightarrow \sigma \rightarrow \pi$ rearrangement was rapid.
As the temperature is lowered, it would become more difficult for
such a process to occur, and, in this case, a preferred conformation
would become favored. The π-system with the CH_2 groups farthest
away from Pd would be expected to be sterically favored, and would
finally become so favored at very low temperatures that H_A and H_B
would become nonequivalent. A C_1-C_2 rotation in the π-bonded
species would not necessarily occur, even at room temperature, when
H_1, H_2 equivalence is indicated, according to this proposal.

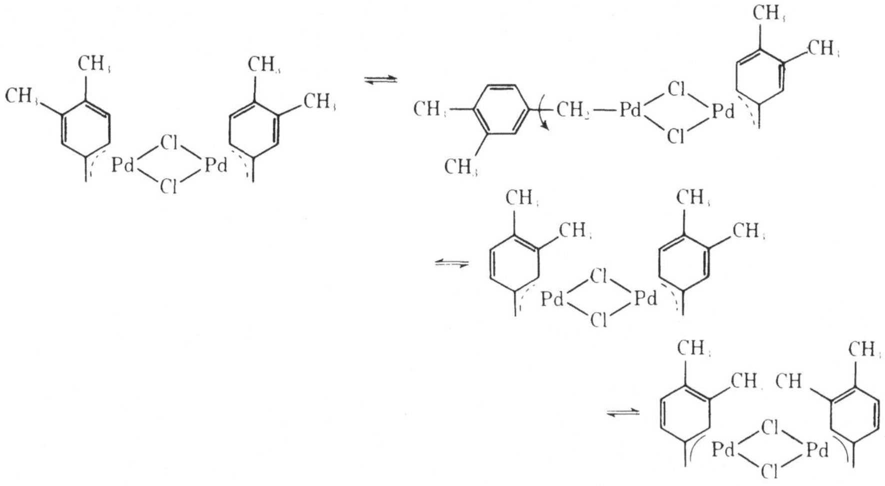

It seems unlikely that these low temperature pmr results could be explained by the occurrence of favored _cis_ or _trans_ isomers of these dimers, since in either case, the halves can be related by simple symmetry operations.

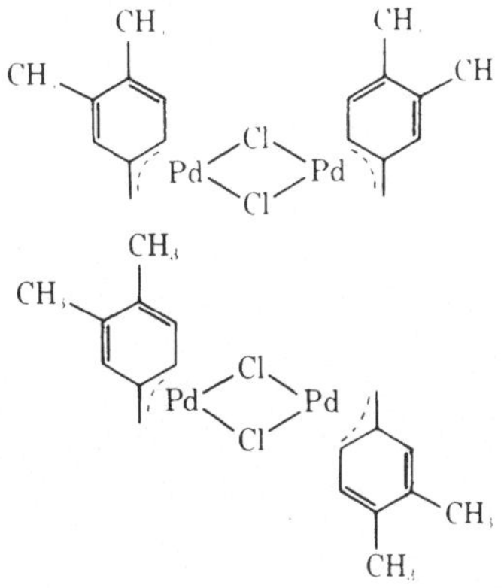

In general, these low temperature pmr results strongly support the η^3-benzyl type of bonding. And it is interesting to note that only nonsymmetrical benzyl systems we have studied show temperature dependent pmr. (5) Other spectroscopic investigations of I seem to support the η^3-bonding scheme. In the uv, large changes are observed between the uv spectrum of $C_6H_5CH_2Cl$ (λ_{max} = 2600 Å), II (2650 m, 3100 sh, 3440 s) and I (2380 s, 2900 m, 3850 w). Infrared spectra indicate that bands which would be attributed to aryl C-H in-plane bending or stretching modes for II (1155 and 1180 cm^{-1}) and benzyl chloride (1158 and 1182 cm^{-1}) were either not observed, or significantly shifted for I. The aryl C-C stretching frequencies for I are either missing or shifted for I (1448 and 1482 cm^{-1}) when this IR spectrum is compared to either benzyl chloride (1588, 1498 and 1452 cm^{-1}) or II. These changes in spectroscopic properties indicate that the aromatic character of the phenyl rings in I is greatly different from either benzyl chloride or II.

The chemistry of η^3-benzylpalladium chloride is quite interesting. It is reactive, combines readily with donor ligands, readily carbonylates at room temperature and atmospheric pressure, and can serve as a homogeneous catalyst for benzylation of methylacrylate,[11] (as proposed previously by Heck and Nolley).[28]

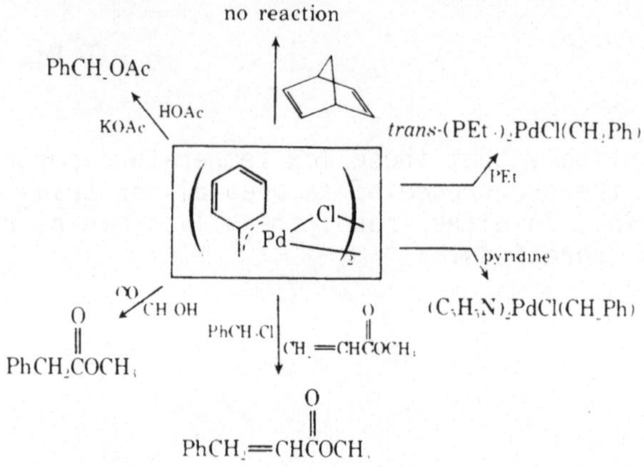

B. R$_2$M Compounds

Attempts to prepare C$_6$F$_5$CoBr and C$_6$F$_5$NiBr led to the discovery that these species have a strong tendency to disproportionate to yield (C$_6$F$_5$)$_2$M plus MBr$_2$. The (C$_6$F$_5$)$_2$M species are quite stable,

$$C_6F_5Br + M \text{ vapor} \longrightarrow [C_6F_5\text{-}M\text{-}Br]$$
$$\downarrow \quad M = Co, Ni$$
$$(C_6F_5)_2M + MBr_2$$

soluble in organic solvents, and coordinative unsaturation is readily attainable. Chemical studies of these compounds has led to the discovery of the first π-arene complexes of Co(II) and Ni(II). Thus, both (C$_6$F$_5$)$_2$Co and (C$_6$F$_5$)$_2$Ni coordinate toluene (or other arenes) in a η^6-fashion! The chemistry of these species is intriguing. For the Co system, $(C_6F_5)_2Co-\pi-C_6H_5CH_3$ (compound IV), the π-toluene ligand is very labile and readily exchanges with benzene or other arenes at <u>room temperature</u>. Powerful ligands, such as PEt_3 quantitatively displace the arene ligand. Water decomposes IV to yield one mole of $C_6H_5CH_3$, two moles of C_6F_5H, and

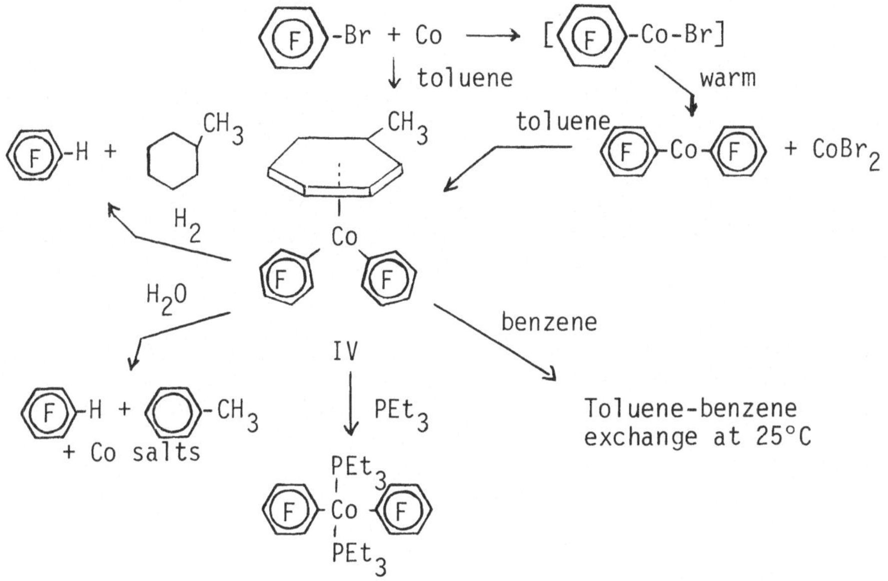

one mole of water soluble Co salts.[29] In toluene IV serves as a very short lived catalyst (homogeneous) for hydrogenation of toluene to methylcyclohexane. Catalyst turn over only reaches a maximum of two before destruction by hydrogenolysis of the Co-C σ-bond.

An x-ray structure determination of a crystal of IV grown in toluene was undertaken.[29] This compound crystallizes with four molecules of $[\eta^6-C_6H_5CH_3(C_6F_5)_2Co]$ in an orthorhombic unit cell belonging to space group <u>Pnma</u>. The lattice constants are <u>a</u> = 11.465 (9), <u>b</u> = 16.025 (12), and <u>c</u> = 9.503 (8) Å. The structure was solved using Patterson and Fourier techniques and refined by full-matrix least squares using 1335 diffractometer data (λ 0.7107 Å) observed in the range sin $\theta/\lambda \leq$ 0.626. Refinement of the asymmetric unit of structure, including coordinates for the hydrogen atoms obtained from a difference synthesis and anisotropic parameters for all other atoms, produced an <u>R</u> value of 6.1% and a weighted <u>R</u> of 5.8%.

The molecular structure consists of a cobalt atom σ bonded to two <u>F</u>-phenyl ligands and π-bonded to one toluene ligand. The Co-C_1 σ bond distance is 1.931 (5) Å and the C_1CoC_1' bond angle is 88.3

(3)$^\circ$. The Co atom is 1.627 (2) Å from the plane of the toluene
ligand and the average Co-C π-bond cistance of 2.141 (7) Å is similar to the results found in $(C_6H_5)CCo_3(CO)_6 \cdot \pi C_6H_3(CH_3)_3$ of 2.15
(3) Å.[30] The toluene ligand makes an angle of 86.3 with the plane
defined by cobalt and the two α-bonded carbons.

The molecule has m (Cs) symmetry which is fully utilized
crystallographically. The carbon framework of both F-phenyl ligands remains planar and the normal trend in bond angles for electron releasing and withdrawing substitutents is observed.[31] Carbon-
carbon bond lengths in the F-phenyl ligand average 1.373 (8) Å and
the observed C-F distance is 1.349 (7) Å. The average C-C length
in the toluene ligand is 1.391 (8) Å.

Formally IV is a 17 electron five coordinate system. It is
paramagnetic and exhibits a temperature dependent epr spectrum with
fine structure, and a magnetic moment of about 1.7 BM, thus indicating that IV is low spin in Co(II).

In the case of the Ni(II) compound (V), a diamagnetic closed
shell structure is formed. The π-toluene complex $(C_6F_5)_2Ni\pi-C_6H_5CH_3$
is formed very efficiently, and 3-4 gram quantities can be prepared in one afternoon using our very simple metal atom reactor
setup.[6]

An intermediate in the formation of V is C_6F_5NiBr which is stable
to approximately -80°, and can be trapped by addition of PEt_3. In
the absence of a stabilizing arene ligand, C_6F_5NiBr decomposed a-
bove -80°C to slowly yield decafluorobiphenyl, nickel, and nickel
dibromide. Bis-(pentafluorophenyl) nickel appears to be an intermediate enroute to final products as small amounts of $(C_6F_5)_2Ni$

can be trapped at temperatures > -80° by addition of PEt_3 or toluene.

The complex (V) is an air sensitive red-orange crystalline solid very soluble and stable in arene solvents. It is somewhat soluble in alkane solvents but very sensitive to decomposition in such media. It also is soluble in chloroform but with slow decomposition (several hours) yielding nickel and nickel halide precipitates. The arene ligand is exceedingly labile and can be replaced by many coordinating ligands. In fact, it is so labile that the toluene ligand can be rapidly exchanged with other arene ligands at room temperature. Analogous benzene, xylene, and mesitylene complexes can be prepared by exhaustive exchange with toluene or by carrying out the original metal vapor deposition procedure with bromopentafluorobenzene plus benzene, xylene, or mesitylene. Preliminary studies employing various other arenes such as fluorobenzene and anisole indicate that complexes like V are wide in scope with regard to the arene ligand that can be tolerated.

Attempts to form stable complexes like V where the C_6F_5 group was replaced with other similar σ-bonded ligands have not been successful. Thus, the following halides were simultaneously deposited with toluene and nickel vapor, and no stable organometallics were found:

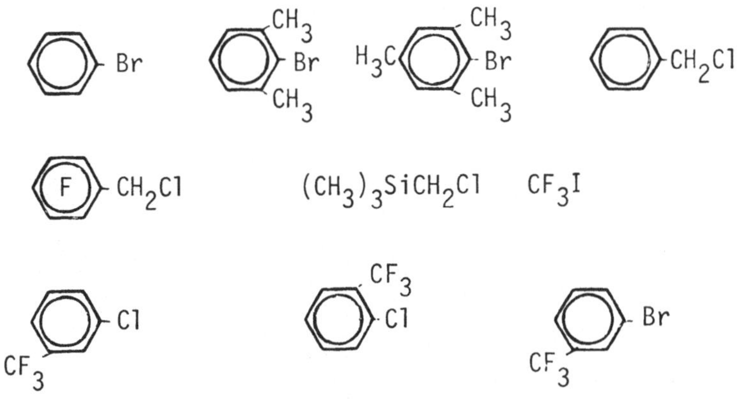

For most of these halides we know reactions take place and products are formed, but that the products do not have sufficient stability to be readily isolated. Therefore, the C_6F_5 group is quite unique in its stabilizing effects. From an X-ray crystal study we have determined that the Ni-C σ-bond is 1.891 (4) Å, a very short and therefore strong bond compared with other similar bond lengths in the literature.[32] Currently we are carrying out more detailed bonding analyses which will be the subject of later publications. At this time, in view of the electronegativity of C_6F_5 and the higher

stabilities of the complexes when more electron rich π-arenes are
employed, the stabilization of the complex is a delicate balance
of "push-pull" electronic configuration. This delicate stabilizing "balancing act' is needed in spite of the fact that V is a
closed shell diamagnetic 18 electron configuration. Thus, strongly
electronegative σ-bonded ligands with supposedly strong π-acceptor
characteristics are required, and another good candidate is the
pentachlorophenyl group, in light of the recent work of Wada and
coworkers.[33]

The extreme lability of the π-arene ligand allows complex V
to have a rich chemistry. As detected by NMR, V is about 10% dissociated at room temperature in deuterochloroform solution, which
illustrates the ease with which reactive open coordination sites
can be generated! Obviously there is an equilibrium set up between
dissociated and associated toluene in chloroform, with the chloroform serving as a weak ligand.

To ellucidate some of the chemistry of V we have carried out
three studies: (1) Arene competition studies were investigated by
adding equimolar amounts of various arenes to V in deuterochloroform. A coordination preference of mesitylene > toluene > benzene
> fluorobenzene > trifluoromethylbenzene was found, clearly showing
that more electron rich arenes serve as better ligands for this
unusual organometallic structure. (2) Kinetic analyses at various
temperatures of the π-toluene → π-mesitylene → π-mesitylene → π-
toluene exchange were carried out. Activation energies on the
order of 10 kcal were found (by Arrhenius methods) and the exchange
in either direction was found to be first order in complex and first
order in displacing arene. This indicates a transition state for
exchange involving both the incoming arene and complex and a symmetrical intermediate (almost "SN$_2$ like") is possible:

(3) V was studied as a homogeneous hydrogenation catalyst for arenes
at room temperature.[34] In toluene solution at 100 atmospheres hy-

drogen pressure V converts toluene to methylcyclohexane at 25°C. However, the catalyst turns over only 10 times (maximum) before it is destroyed through the hydrogenolysis of the Ni-C σ-bond to yield C_6F_5H and nickel metal residue. Competition experiments between benzene and toluene indicated a slightly greater hydrogenation rate for benzene. Although it is always difficult to prove homogeneity in such systems, we believe this process is homogeneous since we tried several ways of producing active nickel powder catalysts[35,36] none of which served as catalysts under these conditions, which leads us to assume the nickel residue or intermediate powder formed as V decomposes is not as active catalyst (under these conditions).

C. Metal Vapor - Alkane Reactions

We have previously reported that the codeposition of metal vapors with weakly coordinating solvents first yields unstable "solvated metal atoms".[35,36] On matrix warmup metal-metal bonds begin to form as the weak complexes break up. The final metal particles have large amounts of organic materials bound in them, and are very reactive sources of zero valent metal. Varying activities of the same metal are observed when the solvent is charged. Also, we have employed this process for many metals (Mg, Ni, Zn, Cd, Al, In, Sn, Pb, Te), and almost any type of solvent can be used. Thus, the metal vapor-solvent codeposition procedure constitutes an important new and versatile method for preparation of active metal powders, and can be carried out with very simple metal vapor reactor equipment[6] on large scale.[37]

Recently our studies have concentrated on the interactions of Ni with alkanes since we have found that very active Ni powder catalysts can be prepared this way.[36] We also found that the alkane is actually cleaved and broken up by the metal atoms and clusters at relatively low temperature ($\sim$ -50°C). Thus, the alkanes are "activated" under very mild conditions. As an example, n-pentane has been examined in detail. It has been found that codeposition of Ni/C_5H_{12} in the ratio 1:500 yields a final black nickel powder that contains about one strongly bound organic moiety to two Ni atoms. This dry powder is an extremely active catalyst for hydrogenation of aromatics or olefins, is stable to sintering up to 250°C, and has crystallite sized of 30-50 Å, and accordingly is non-ferromagnetic. The organic groupings (R) are apparently bound to the tiny Ni clusters as CH_3, C_2H_5, C_2H_3, and other C_2, C_3, C_4, and C_5 fragments (shown by analysis of products of additions of H_2 or D_2 at 25°C and at 300°C).

$$Ni + C_5H_2 \xrightarrow{-196°C} Ni(C_5H_{12})_n$$

$$\downarrow \text{slow warming}$$

$$\underset{R}{\overset{R}{\diagdown}}(Ni)_n\underset{R}{\overset{R}{\diagup}} \xleftarrow[\substack{\text{excess} \\ C_5H_{12}}]{\substack{\text{vacuum} \\ \text{removal of}}} \xleftarrow[\substack{C_5H_{12} \text{ about} \\ -50°C}]{\text{reaction with}} (Ni) \text{ clusters } (C_5H_{12})_n$$

(final Ni powder)

[R = organic fragments of C_5H_{12}]

Codeposition of Ni:C_5H_{12} ratio of 1:100 or 1:50 yield much larger crystallite ferromagnetic final Ni powders. The strongly bound organics are on the order of 1:5 for organics:Ni in the final dry powder. This material is also an active catalyst, but slightly less active than the smaller crystallite non-ferromagnetic powder. Also, this larger crystallite ferromagnetic powder is slightly more susceptible to thermal sintering ($\sim$ 220°C vs 250°C than the non-ferromagnetic material).

In summary, these results indicate: (1) Ni atoms and small transient (Ni) clusters react with alkanes at fairly low temperatures ($\sim$ -50°C). And in fact there is a competition between further clustering of Ni vs reaction of the clusters with alkane, which after reaction effectively stops or impedes further cluster growth. Thus, very high initial dilutions (Ni:C_5H_{12} ratio 1:500) favor reaction with C_5H_{12} yielding final Ni crystallites < 50 Å and containing large amounts of strongly bound organics; (2) the strongly bound organic materials are probably in the form of organic radical fragments, and serve to hold the Ni framework from sintering and further crystallization, and do so up to very high temperatures (> 200°C); (3) the (Ni)$_n$-(alkane) powders are very active catalysts for hydrogenations, more active than Raney Ni, and catalyst selectivity can be "tailored" by alkane (or other solvent) choice and amount.

Acknowledgements. Long standing generous support of the National Science Foundation is greatly appreciated. Students who have contributed in a major way are acknowledged for their excellent hard work and perseverance: Drs. J. S. Roberts and H. F. Efner and Messrs. B. B. Anderson, S. C. Davis and M. Bader. Also, collaborative work with L. J. Radonovich on single crystal x-ray studies is extremely valuable.

References

1. L. Vaska, Accts. Chem. Res., 1, (1968) 335.
   J.P. Collman, Acct. Chem. Res., 1, (1968) 136.

2. R.F. Heck, J. Amer. Chem. Soc., 85, (1964) 2796.

3. J.K. Stille and K.S.Y. Lau, J. Amer. Chem. Soc., 98, (1976)
   5832, 5841 and references therein.

4. A.U. Kramer and J.A. Osborn, J. Amer. Chem. Soc., (1974) 7832;
   A.U. Kramer, J.A. Labinger, J.S. Bradley, and J.A. Osborn, J.
   Amer. Chem. Soc., 96, (1974) 7145.

5. K.J. K.abunde and J.Y.F. Low, J. Amer. Chem. Soc., 96, (1974)
   7674.

6. K. J. Klabunde, Accts. Chem. Res., 8, (1975) 393; Angew. Chem.,
   87, (1975) 309; Angew. Chem. Int. Ed. Engl., 14, (1975) 287.

7. B.B. Anderson, K. Neuenschwander, and K.J. Klabunde, manuscript
   in preparation describing detailed preparation and chemistry of
   $(CF_3PdI)n$.

8. K.J. Klabunde, J.Y.F. Low, and H.F. Efner, J. Amer. Chem. Soc.,
   96, (1974) 1984; also W.J. Kennelly, unpublished results from
   this laboratory.

9. (a) A.S. Gordon and J.R. McNesby, J. Chem. Phys., 31, (1950)
       853.

   (b) H.M. Frey and R. Walsh, Chem. Rev., 69, (1969) 103.

10. (a) C.W. Larson and B.S. Rabinovitch, J. Chem. Phys., 50,
        (1970) 871; and E.A. Hardwidge, C.W. Larson, and B.S.
        Rabinovitch, J. Amer. Chem. Soc., 92 (1970) 3278, and
        references therein.

    (b) J. Kochi, Editor, "Free Radicals," Chapter by J.A. Kerr,
        (1973) pg. 29.

    (c) The author thanks Professor M.L.H. Green for initially
        suggesting a possible metallocycle intermediate. Work
        along these lines is now being carried out.

11. J.S. Roberts and K.J. Klabunde, J. Amer. Chem. Soc., 99,
    (1977) 2509. J. Organometal. Chem., 85, (1975) C-13.

12. J.S. Roberts, Ph.D. thesis, University of North Dakota, (1975).

13. J.Y.F. Low, Ph.D. thesis, University of North Dakota, (1973).

14. K. J. Klabunde, J. Fluorine Chem., 7, (1976) 95.

15. A similar proposal was brought forth by Skell and Girand in
    their RX-Mg vapor publication: P.S. Skell and J. E. Girand,
    J. Amer. Chem. Soc., 94, (1972) 5518; also private communica-
    tions with P.S. Skell and J. E. Girand.

16. Alkyl halides can be reduced in a similar way by $HPtBr(PEt_3)_2$;
    cf. A.V. Kramer, J.A. Labinger, J.S. Bradley, and J.A. Osborn,
    J. Amer. Chem. Soc., 96, (1974) 7145; W.R. Moser, Abstracts,
    163rd National Meeting of the American Chemical Society, Boston,
    MA, April, 1972, ORGN-14.

17. K.J. Klabunde and J.Y.F. Low, J. Organomet. Chem., 51, C33
    (1973); K.J. Klabunde and J.Y.F. Low, J. Amer. Chem. Soc.,
    96, 7674 (1974).

18. (a) R.F. Heck, J. Amer. Chem. Soc., 90, 5518, 5526, 5531, 5535,
    5538, 5546 (1968);

    (b) P.M. Henry, Tetrahedron Lett., 2285 (1968).

    (c) T. Hosokawa, C. Calvo, H.B. Lee, and P.M. Maitlis, J.
        Amer. Chem. Soc., 95, 4914, 4924 (1973).

19. B.B. Anderson, K. Neuenschwander, and K.J. Klabunde, manuscript
    in preparation.

20. K.J. Klabunde, M.S. Key and J.Y.F. Low, J. Amer. Chem. Soc.,
    94, 999 (1972).

21. K.J. Klabunde, J.Y.F. Low and M.S. Key, J. Fluor. Chem., 2,
    207 (1972).

22. B.B. Anderson and K.J. Klabunde, manuscript in preparation.

23. J.S. Roberts, unpublished results from this laboratory.

24. R.B. King and A. Fronzaglia, J. Amer. Chem. Soc., 88, 709
    (1966).

25. F.A. Cotton and M.D. LaPrade, ibid., 90, 5418 (1968); F.A.
    Cotton and T.J. Marks, ibid., 91, 1339 (1969); F.A. Cotton,
    Acc. Chem. Res., 1, 257 (1968).

26. R.R. Stevens and G.D. Shier, J. Organomet. Chem., 21, 495 (1970).

27. R.F. Hartley, "The Chemistry of Platinum and Palladium," John Wiley and Sons, New York (1973).

28. R.F. Heck and J.P. Nolley, J. Org. Chem., 37, 2320 (1972).

29. B.B. Anderson, C.L. Behrens, L.J. Radonovich and K.J. Klabunde, J. Amer. Chem. Soc., 98, 5390 (1976).

30. R.J. Dellaca and B.R. Penford, Inorg. Chem., 11, 1855 (1972).

31. A. Domenicano, A. Vaciago and C.A. Coulson, Acta Crystallogr., Sect. B., 31, 221 (1975).

32. cf. M.L. Churchill and M.V. Veidis, Chem. Comm., 1099 (1970).

33. M. Wada and K. Oguro, Inorg. Chem., 15 (10), 2346 (1976); M. Wada, Kusabe, and K. Oguro, ibid., 16 (2), (1977).

34. cf. F.J. Kirsekorn, M.C. Rakowski, and E.L. Muetterties, J. Amer. Chem. Soc., 97, 237 (1975); E.L. Muetterties and F. J. Hirsekorn, ibid., 96, 4063 (1974) and references therein.

35. K.J. Klabunde, H.F. Efner, T.O. Murdock, and R. Ropple, J. Amer. Chem. Soc., 98, 1021 (1976).

36. K.J. Klabunde, S.C. Davis, H. Hattori, and Y. Tanaka, J. Cat., submitted.

37. T.O. Murdock and K.J. Klabunde, J. Org. Chem., 41, 1076 (1976).

ORGANOMETALLIC DERIVATIVES OF THE TRANSITION ELEMENTS:

III. THE REACTION OF ALKYNES WITH TRANSITION METAL ATOMS

L. H. Simons and J. J. Lagowski

Department of Chemistry
The University of Texas
Austin, Texas 78712 U.S.A.

INTRODUCTION

Our previous interest in the nature of the products formed during the reaction of transition metal atoms with arenes, led naturally to an investigation of the reaction of such metal species with other unsaturated ligands, viz. alkynes. Previously published reports indicate that organometallic products are not detected in the reactions of alkynes with atomic nickel[2] and chromium[3], although there is some indication[4] that nickel condensed with alkynes does form a metal-containing substance; in all reported cases, oligomerization of the alkyne was observed. We report here the reactions of nickel, iron, and manganese atoms with a variety of alkynes; the majority of this report is devoted to the reactions of nickel atoms which not only leads to oligomers, but also produce nickel-containing substances with useful catalytic properties. Thus, we are concerned here with two general subjects: (1) the reactions of metal atoms with alkynes and (2) the properties of the metal-containing products obtained from this reaction.

EXPERIMENTAL

Metal atom reactions were conducted by the method and using the techniques described in detail previously.[1a] In effect, bulk metal was heated resistively in a vacuum (~10^{-6} torr) and allowed to co-condense with the ligand on the walls of a reactor which were cooled to 77° K. For the most part, the reactions appear to occur in the frozen ligand matrix after the liquid nitrogen bath is removed, but before the matrix melts. The reaction mixture was

worked-up using conventional vacuum line techniques. Where product
separation was required, conventional gas chromatography or high
pressure liquid chromatography was employed. Products were characterized by nmr(^{1}H and ^{13}C) spectroscopy and mass spectrometry.

Nickel Atom Reactions

Nickel atoms react with a variety of alkynes to produce cyclic
oligomers, the greatest proportion of which are trimers and/or
tetramers. In each case a dark-colored, nickel-containing solid is
also formed. These substances, which are uncharacterized at this
point, are soluble in most organic solvents and posses unusual
catalytic properties (<u>vide infra</u>).

<u>Reaction with phenylacetylene</u> ($C_6H_5C\equiv CH$). The reaction of
phenylacetylene with nickel atoms gives oligomers which were identified as a dimer (2% I, R = C_6H_5), a trimer (30% II, R = C_6H_5)
and a tetramer (25% III, R = C_6H_5);

(I)   (II)   (III)

40% of the polymerized products were higher polymers of undetermined composition. Among the products of this reaction is a black,
diamagnetic, nickel-containing substance of undetermined composition.
Typically 0.8 g. of metallic nickel will yield 2-3 g. of this material; very little elemental nickel remains in the reaction mixture.
We estimate that ~80% of the bulk nickel which is evaporated finds
its way into the matrix, the remainder being deposited on interior
nonreacting surfaces, such as ligand inlet tubing, electrode supports, etc.

<u>Reaction with propyne</u> ($CH_3C\equiv CH$). Since our reactor cannot be
pressurized conveniently, the matrix could not be warmed to room
temperature in the reaction of nickel atoms with propyne. Accordingly, the matrix, once formed in the usual way, was kept at -45°C with
a slush bath. A dark-colored solution eventually formed when the
system was brought to thermal equilibrium at -45°C. Removal of the

volatile products left a black, diamagnetic, nickel-containing solid. The volatile product proved to be unreacted propyne; no oligomers were detected in this reaction.

Reaction with 1-hexyne ($CH_3(CH_2)_3C{\equiv}CH$). The reaction of 1-hexyne with nickel atoms produced only a trimeric oligomerized product which proved to be a mixture of the symmetrically (45%) and unsymmetrically (55%) substituted tri-n-butylbenzenes. The usual, black nickel-containing, nonvolatile product was also obtained.

Reaction with 2-pentyne. Nickel atoms react with 2-pentyne to produce one oligomer, 1,2,4,7-tetramethyl-3,5,6,8-tetraethyl-cyclooctatetrene (IV), in good yield,

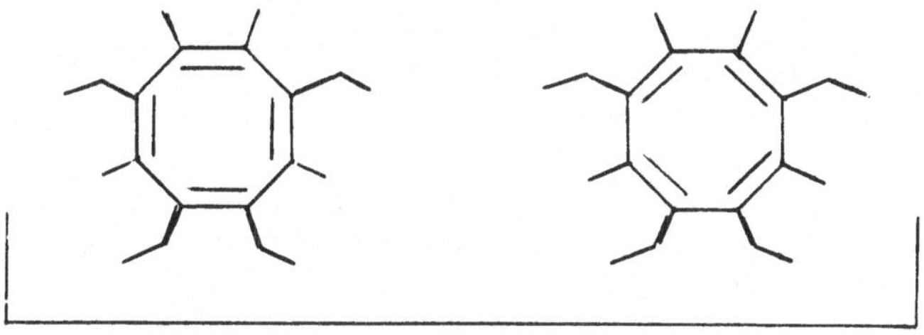

(IV)

together with a black, nickel-containing solid product. The ^{13}C nmr spectrum of the tetramer showed it was an equimoler of the two possible bond-shift isomers.

Iron Atom Reactions

Reaction with propyne. Iron atoms react with propyne giving a complex hydrocarbon product which appears to be polymeric. Neither oligomers nor metal-containing species are formed in this reaction.

Reaction with 2-pentyne. The reaction of iron with 2-pentyne produces only 1,2,4,7-tetramethyl-3,5,6,8-tetraethylcyclooctatet-raene (IV) as the oligomeric product and a dark red liquid (b.p. 55°C at 10^{-2} torr). The composition of the red liquid is $Fe(C_2H_5C{\equiv}CCH_3)_5$ using high resolution mass spectrometry. The ^{13}C nmr spectrum of the red liquid showed the compound to possess a nonhomogeneous paramagnetic behavior which made the resonances difficult to interpret; the solvent signals were, however, sharp.

Reaction with 2-Butyne. Iron atoms react with 2-butyne to produce hexamethylbenzene as the only oligomer and a yellow-orange solid. High resolution mass spectrometry showed the solid to be $Fe(CH_3C\equiv CCH_3)_5$; at present the constitution of the yellow-orange, iron-containing substance is unknown.

Reaction with methyphenylacetylene. This actylene is condensed to form two oligomers, 57% of the symmetrical trimer and 43% of 1,2,4,7-tetramethy-3,5,6,8-tetraphenylcyclooctatetraene, and a red-orange, iron-containing complex (softening point 140°) with a molecular formula of $Fe(C_6H_5C\equiv CCH_3)_5$ as determined by mass spectrometry.

Reactions of Manganese Atoms[5]

The co-condensation of 2-pentyne with manganese atoms did not lead to a detectable reaction under the normal conditions employed for other systems. However, both phenylacetylene and 1-hexyne gave oligomers when reacted with manganese atoms, viz. an unsymmetrical trimer (II) and a tetramer with an undetermined substitution pattern; higher polymers were also detected in the 1-hexyne reaction. In neither reaction was a metal-containing product detected.

TABLE I
The Products formed in the Catalytic Oligomerization
of Phenylacetylene.[a]

Catalytic System	Time for Completion[b]	% Oligomer			Highest Mass Polymer
		2*	3*	4*	
Ni[c]	1 hr.	2	30	25	> 700
Ni/propyne[d]	36 hr.	0	20	50	522
Ni/2-pentyne	0.33 hr.	2	60	20	522
Ni/1-hexene	36 hr.	0	60	20	> 700

*2=dimer; 3=trimer; 4=tetramer. [a]Reactions run with neat phenylacetylene at room temperature. [b]At the end of the reaction no free phenylacetylene was present. [c]This system is equivalent to a co-condensation reaction allowed to go to completion. [d]This symbolism indicates the source of the black, nickel-containing catalyst, e.g., the co-condensation of nickel atoms with propyne.

TABLE II
The Products formed in the Oligomerization of Acetylene.[a]

| | % Oligomer | |
Catalytic System	Trimer	Tetramer
Ni/propyne[b]	2%	98%
Ni/2-pentyne	100%	0%
Ni/1-hexene	100%	0%

[a]In cyclohexane solution at 1 atmosphere of acetylene and room temperature. [b]This symbolism indicates the source of the catalyst, e.g. the black, nickel-containing product in the co-condensation of nickel atoms with propyne.

TABLE III
The Products formed in the Oligomerization of Selected Alkynes.[a]

Catalytic System	Alkyne	Products
Ni/propyne[b]	$CH_3C{\equiv}CH$	1,2,4 Trimethylbenzene ($II,R=CH_3$)
		1,2,4,6 Tetramethylcyclo-octatetraene ($III,R=CH_3$)
Ni/propyne	$CH_3C{\equiv}CC_2H_5$	no reaction
Ni/2-pentyne	$CH_3C{\equiv}CC_2H_5$	no reaction
Ni/1-hexene	$CH_3C{\equiv}CC_2H_5$	no reaction

[a]Neat at room temperature and ambient pressure. [b]This symbolism indicates the source of the catalyst, e.g., the black, nickel-containing product formed in the co-condensation reaction of nickel atoms with propyne.

Reactions of Nickel-containing Products

Although we have not fully characterized the black nickel-containing products obtained from the reaction of alkynes with nickel atoms, we have discovered several reactions which may give some insight into their constitution, as well as being of practical value.

Oligomerization Catalysts. Nickel atoms, as well as the black, nickel-containing products serve as catalysts for the oligomerization of alkynes. The results for the oligomerization of phenylacetylene (neat) and acetylene are shown in Tables I and II; the acetylene reactions occur in hydrocarbon solvents (hexane and cyclohexane) as well as in ethers (dioxane and diethylether), ethylacetate, and ether/acetone mixtures. The results obtained with other alkynes are given in Table III.

Oligomerization of alkynes by the nickel-containing products also occur in different solvents, sometimes to produce unusual oligomers. Thus, propyne is oligomerized by the nickel/propyne product in dioxane at 60° to produce a mixture of trimers (45%) and tetramers (55%). The former consists of 1,2,4 trimethylbenzene (6%), 1,3,5-trimethylbenzene (5%), but the majority of the trimer (34%) is the linear diene-yne (V). The tetramer is a mixture of

(V)

1,2,4,6-tetramethycyclooctatetraene (30%), the 1,3,5,7 (10%), and the 1,2,4,7 (15%) isomers. The nickel-containing products will also bring about co-oligomerization of mixtures of alkynes. Thus, a propyne-acetylene (2:1) mixture at a total pressure of 20 p.s.i. is oligomerized by the Ni/2-pentyne product in diethylether/acetone mixture at room temperature to a mixture of benzene (47%) and toluene (53%). Increasing the proportion of propyne in the mixture to 3:1 and changing the solvent to a mixture of hexane-acetone under the same experimental conditions does not materially change the ratio of products or their identity.

Hydrogenation catalysis. The black, nickel-containing products also catalyze the hydrogenation of arenes. A solution of the Ni/1-hexyne product (50 mg.) in benzene at room temperature in contact with 2 atmospheres of hydrogen yields cyclohexane. The Ni/2-pentyne product produces the methylester of cyclohexylcarboxylic acid from methyl benzoate in ether under the same conditions, but acetophenone gives only phenylmethylcarbinol.

DISCUSSION

Although definitive evidence relating to the structures of the nickel-containing products obtained in the reaction of nickel with alkynes is not available at this time, some experimental results obtained with the product of the nickel/2-pentyne reaction suggest that this substance, **at least, contains Π-complexed derivatives of** cyclooctatraene. The nmr spectrum of this product shows the presence metal induced shifts of **tetramethyltetraethylcyclooctatet-raene (IV).** Its infra-red spectrum contains a band attributed to a complexed cis-olefin at 1590 cm^{-1} compared with the uncomplexed cis-olefin vibration occurring at 1710 cm^{-1}. Also, this product releases uncomplexed ligand upon air oxidation, dissolution in moist acetone, or on treatment with butadiene.

Although it is possible to formulate a mechanism (based on product analysis) involving either Π-complex metal intermediates or metallocycles, the isolation of the linear trimer V strongly sug-

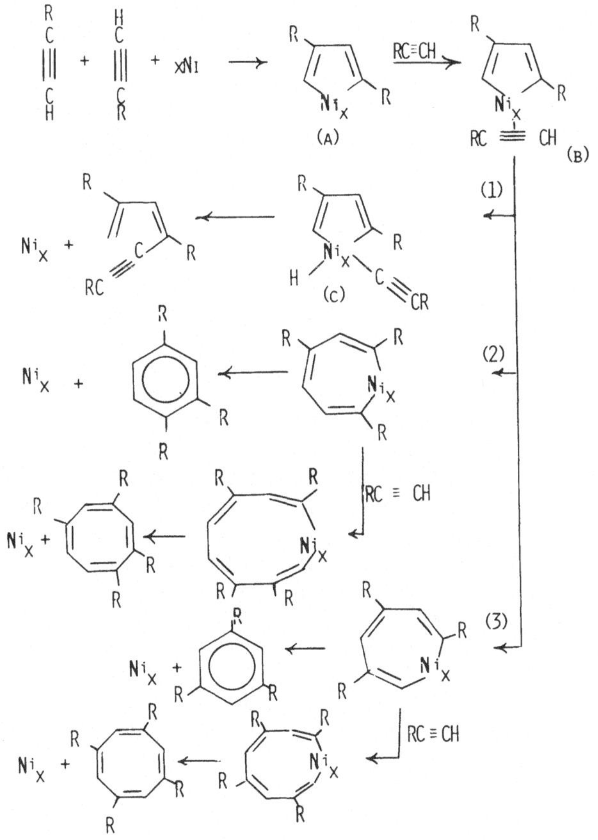

Figure 1: Oligomerization Scheme

gests the latter route. A generalized oligomerization scheme that
is consistent with the available data is shown in Fig. 1. We presume
that the key intermediate (b) is an unsaturated metallocycle carrying a Π-complexed alkyne; (b) can then yield oligomers by one of
three pathways. Route 1 leads to the linear trimer by way of a
metal-hydride intermediate (c). Routes 2 and 3 involve the ring
expansion of the metallocycle by incorporation of the complexed alkyne; 2 leads to the unsymmetrical cyclic trimer, whereas route 3
yields the symmetric cyclic trimer. In both routes 2 and 3, it is
possible to expand the metallocycle ring, possibly by a hydride
shift mechanism; those expanded metallocycle rings are then the
source of the substituted cyclooctatetraenes formed in these reactions. The collapse of the metallocycle to the cyclic hydrocarbon
is probably similar to that postulated for the saturated metallocycles.[6] Unspecified in our present mechanism is the stoichiometry
of initial nickel-alkyne reaction. We have no evidence on this
point; however, it seems reasonable that the metal-containing intermediates may be polynuclear to accommodate to the hydride shift
mechanism, although the ability of the solvent to form weak complexes
may be important in this regard as it is in the case of the saturated
metallocycles.[6]

We are continuing the investigation of the identity of the
intermediates in the oligomerization of alkynes using synthetic as
well as physical techniques.

Acknowledgements

We gratefully acknowledge the generous support of the
Robert A. Welch Foundation and the Natural Science Foundation.

References and Footnotes

[1](a) V. Graves and J. J. Lagowski, Inorg. Chem., 15, 577 (1976).
(b) V. Graves and J. J. Lagowski, J. Organometal. Chem., 120, 397
(1976). (c) L. H. Simons, P. E. Riley, R. E. Davis, and J. J.
Lagowski, J. Amer. Chem. Soc., 98, 1044 (1976).
[2]P. S. Skell, J. J. Havel, D. L. Williams-Smith and M. J. McGlinchey,
J. Chem. Soc. Chem. Comms., 1972, 1098.
[3]P. S. Skell, D. L. Williams-Smith, and M. J. McGlinchy, J. Amer.
Chem. Soc., 95, 3337 (1973).
[4]K. J. Klabunde, private communication.
[5]The reactions were carried out by J. Kachnick whom we thank for
communicating these results.
[6]R. H. Grubbs, A. Miyashita, M-I. M. Liu, P. L. Burk, J. Amer. Chem.
Soc., 99, 3863 (1977).

HYDROALUMINATION OF OLEFINS CATALYZED BY TRANSITION METAL COMPOUNDS

Fumie Sato

Department of Chemical Engineering, Tokyo Institute of
Technology
Meguro, Tokyo 152 (Japan)

It has long been recognized that organoaluminum compounds are
useful and versatile reagents and intermediates in organic synthesis
[1]. However, their application as sources of organic groups in
organic synthesis has largely been restricted to reactions of alken-
ylalanes which are readily obtainable by hydroalumination of alkynes
[2]. Organoaluminums are not obtainable by hydroalumination of
alkenes under mild conditions, and so they are most commonly prepared
from organolithium or Grignard reagents. Also, reactions using
organoaluminums often give less satisfactory results than the corresponding reactions of analogous derivatives of Li or Mg.

If the reaction of aluminum hydride and olefins could be made
to proceed under mild conditions, the corresponding organoaluminums
then would be readily available for application in organic synthesis
on the bench scale, and, therefore, these would be renewed interest
in the organoaluminums as intermediates for organic synthesis.

We will show here that some transition metal halides catalyze
the addition of lithium aluminum hydride or alane to olefinic double
bonds to afford the corresponding organoaluminums in excellent yield
under mild conditions, thus making them as readily available as the
organoboranes.

1. Hydroalumination of olefins

It is evident that the addition of $LiAlH_4$ or AlH_3 to olefins
to form organoaluminum compounds is closely related to the reaction
of boron hydride with olefins.

Hydroalumination

$$AlH_3 \quad + \quad 3 \ RCH{=}CH_2 \quad \longrightarrow \quad Al(CH_2CH_2R)_3$$

$$LiAlH_4 \quad + \quad 4 \ RCH{=}CH_2 \quad \longrightarrow \quad LiAl(CH_2CH_2R)_4$$

Hydroboration

$$BH_3 \quad + \quad 3 \ RCH{=}CH_2 \quad \longrightarrow \quad B(CH_2CH_2R)_3$$

In spite of their formal similarity, major difference exists. Thus, the hydroboration reaction is fast at room temperature [3], while the corresponding hydroalumination reaction is slow even at elevated temperatures [4]. Some substances (such as $AlCl_3$, $ZnCl_2$, $FeCl_3$, alkali metal halides, carbonates or phosphates), which had been claimed to catalyze the hydroalumination reaction [5], were found actually to have no effect at all [6]. Therefore, in cases where the synthetic route involves a choice between the organoaluminums or the organoboranes, the hydroboration procedure has proved preferable for laboratory work.

Recently it was found that some transition metal halides catalyze the addition of lithium aluminum hydride or alane to olefinic double bonds to give the corresponding organoaluminum compounds [7].

The reaction of $LiAlH_4$ with 1-hexene catalyzed $TiCl_4$ proceeded as follows :   1-Hexene (2.4 g, 29 mmol), $TiCl_4$ (86 mg, 0.45 mmol) and 30 ml of THF were placed in a 100 ml flask. $LiAlH_4$ (0.3 g, 7.9 mmol) was added all at once to the mixture at room temperature. The flask was permitted to remain for 30 minutes under nitrogen at room temperature. Then hydrolyzed to give n-hexane in nearly quantitative yield. In another experiment, bromine (5.3 g, 33 mmol) in 40 ml of benzene was added dropwise at 5 °C to the reaction mixture to give 1-bromohexane (4.3 g, 91% yield based on 1-hexene).

$$LiAlH_4 \ + \ 4 \ /\!\!\backslash\!\!/\!\!\backslash\!\!/ \ \longrightarrow \ LiAl(\!\backslash\!\!/\!\!\backslash\!\!/\,)_4 \left\{ \begin{array}{l} \xrightarrow{\ H_2O\ } \quad 4 \ R{-}H \\[2ex] \xrightarrow{\ X_2\ } \quad 4 \ R{-}X \end{array} \right.$$

R = n-hexyl
X = Cl, Br or I

This reaction shows the following characteristics.
1) In addition to $TiCl_4$, $ZrCl_4$, VCl_4, $(\eta^5\text{-}C_5H_5)_2TiCl_2$ [8] and $(\eta^5\text{-}C_5H_5)_2ZrCl_2$ also catalyze the $LiAlH_4$-olefin reaction. $TiCl_4$ is most effective.
2) The success of $TiCl_4$-catalyzed hydroalumination reaction is dependent on the combination of a solvent and an olefin.

Olefin	Effective solvent	Ineffective solvent
α-Olefin	THF, Monoglyme, Diglyme, Triglyme	Diethyl ether
Internal olefin	Diglyme, Triglyme	Diethyl ether, THF, Monoglyme

3) $LiAlH_4$ adds regioselectively to the terminal carbon atom of α-olefin. Even in the case of internal olefin, the main product is 1-alkylaluminum derivatives. Thus, for example, 1-bromohexane was obtained mainly from 2-hexene or 3-hexene by treating the reaction product with bromine.

2-hexene ⟶ 1-bromohexane(79%), 2- and 3-bromohexane(13% yield)
3-hexene ⟶ 1-bromohexane(66%), 2- and 3-bromohexane(13%)

In this way many olefins were hydroaluminated to 1-alkylaluminum compounds with the following decrease of the reaction rate ;
$RCH=CH_2$ > $R_2C=CH_2$ > $RHC=CHR$.

AlH_3, AlH_2Cl and $AlHCl_2$, which had been obtained from the reaction of $LiAlH_4$ with $AlCl_3$ in the ratio of 3:1, 1:1 and 1:3, respectively [9], also reacted with α-olefin in the presence of $TiCl_4$ to give the corresponding organoalanes in excellent yield [7c]. Relative rates for aluminum hydride reaction with α-olefin are : $LiAlH_4$ > AlH_3 > AlH_2Cl > $AlHCl_2$.
Organoalanes are well known to react with oxygen to give alcohols on subsequent hydrolysis in almost quantitative yield [1], so, this facile reaction offers a very convenient method for the preparation of terminal alcohols from olefins. A typical experimental procedure is illustrated by the reaction of 1-hexene with AlH_3 : After the mixture of $LiAlH_4$ (7.9 mmol) and $AlCl_3$ (2.8 mmol) in 30 ml of THF had been stirred at -10 °C for one hour. 1-hexene (26 mmol) and $TiCl_4$ (0.45 mmol) were added. The mixture was stirred for 10hr at 10~15 °C, oxygen was passed through it and hexan-1-ol was obtained in 85% yield after hydrolysis. Alcohols also were obtained in excellent yield by converting $LiAlR_4$ to AlR_3 by treating it with $1/3 AlCl_3$ [1] prior to the oxidation.

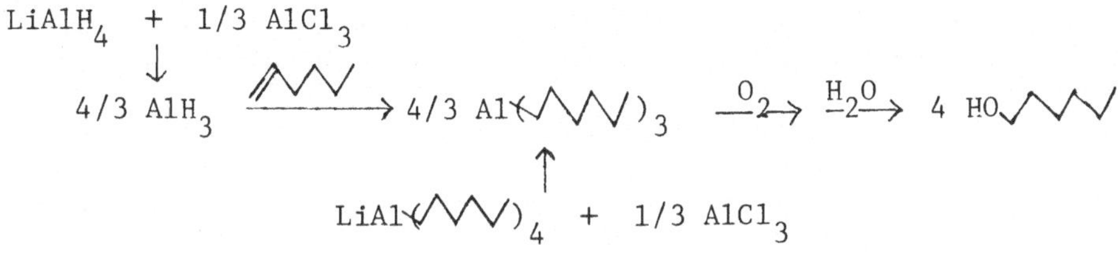

The precise mechanism of this hydroalumination reaction is not clear at present. But it can be assumed that a transition metal hydride, formed by the reaction of aluminum hydride with transition metal halides, plays an important role as evidenced by the following facts. (1) The reaction of an equimolar mixture of $ZrCl_4$ and $LiAlH_4$ in THF in the absence of olefin at room temperature evolved hydrogen during about 3 hours and precipitation of metallic compounds (presumably, of Zr and Al) was observed. (2) An equimolar mixture of $ZrCl_4$ and $LiAlH_4$ catalyzed the isomerization of 1-hexene to cis- and trans-2-hexene. In view of the recently developed hydrozirconation reaction [10], it is most conceivable that the reaction proceeds via hydrotransition metallation of the double bond followed by transition metal – aluminum exchange as shown in scheme 1 [7a]. The observation that the main product of the reaction of $LiAlH_4$ with internal olefins is 1-alkylaluminum is not entirely unprecedented because zirconium and titanium salts are known to catalyze the isomerization of secondary alkyl Grignard [11] and aluminum [12] reagents to give primary alkyl organometallics, moreover, stable 1-alkylzirconium compounds were prepared recently by the hydrozirconation of internal olefins [10].

Scheme 1

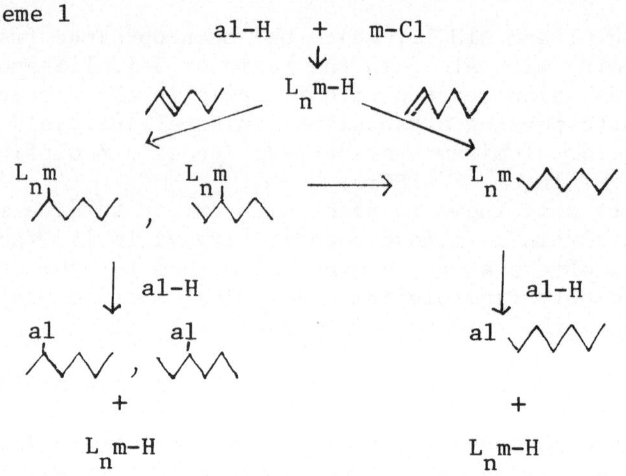

In summary, $TiCl_4$-catalyzed hydroalumination reaction offers a convenient laboratory method for the hydrogenation of olefins or for the preparation of 1-haloalkanes or terminal alcohols from olefins.
The discussions of these reactions presented in the following sections will further support strongly that this new hydroalumination reaction can contribute uniquely and substantially to the advancement of organic synthesis.

2. Hydroalumination of nonconjugated diolefins

The characteristic that the relative rate for hydroalumination reaction with olefin is highly sensitive to the structure of olefin permitted the selective addition of Al-H bond to the less hindered C=C bond of diolefin. LiAlH$_4$ reacted with 4-vinyl-1-cyclohexene, 1,4-hexadiene or 2-methyl-1,5-hexadiene readily to give the corresponding organoaluminate in which aluminum is connected to the less hindered double bond of diolefin as shown below [7b].

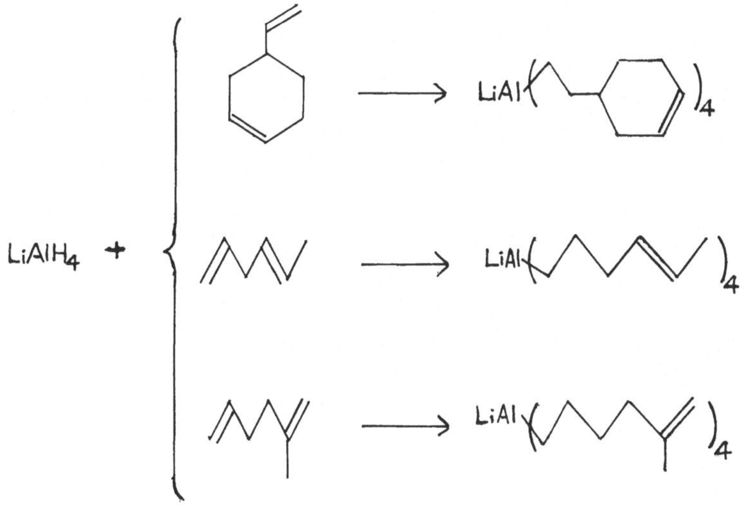

Thus, this facile reaction provides a convenient method for the selective reduction of nonconjugated diolefins or for the selective preparation of haloolefins. (Table 1)
The reaction of nonconjugated diene with AlH$_3$ proceeded in a similar manner as in the case of LiAlH$_4$ and gave unsaturated alcohols in excellent yield [7c].

    4-vinyl-1-cyclohexene → 2(4-cyclohexenyl)ethanol (95%)
    1,4-hexadiene   → 4-hexen-1-ol (83%)
    2-methyl-1,5-hexadiene → 5-methyl-5-hexen-1-ol (92%)

The characteristic point of this hydroalumination method is a high selectivity observed when there are two double bonds under different steric environments.

Table 1

$$\text{LiAlH}_4 \quad + \quad 4 \text{ moles of dienes} \quad \xrightarrow[\text{r.t., THF}]{\text{TiCl}_4} \quad \xrightarrow{\text{Hydrolysis or}} \text{halogenolysis}$$

Diolefin	Work-up	Product (Yield[a] %)
4-Vinyl-1-cyclohexene	H_2O	4-Ethyl-1-cyclohexane (97)
4-Vinyl-1-cyclohexene	NCS	2-(4-Cyclohexenyl)ethyl chloride (76)
4-Vinyl-1-cyclohexene	NBS	2-(4-Cyclohexenyl)ethyl bromide (79)
4-Vinyl-1-cyclohexene	I_2	2-(4-Cyclohexenyl)ethyl iodide (63)
1,4-Hexadiene[b]	H_2O	2-Hexene (94), n-Hexane (4), 1-Hexene (1)
1,4-Hexadiene	NBS	6-Bromo-2-hexene (70)
2-Methyl-1,5-hexadiene	H_2O	2-Methyl-1-hexene (90), 2-Methylhexane (7), 5-Methyl-1-hexene (3)
2-Methyl-1,5-hexadiene	NBS	6-Bromo-2-methyl-1-hexene (71)

[a] Yields were determined by GLC analysis and are based on the olefin.
[b] trans/cis = 97/3

Dialkylaluminum hydride [13], disiamylborane [3], 9-BBN [3] and $(\eta^5\text{-C}_5\text{H}_5)_2\text{Zr(Cl)H}$ [10] have been used for selective reduction of diolefins or for the selective preparation of haloolefins or unsaturated alcohols from diolefins. The LiAlH_4 or $\text{AlH}_3/\text{TiCl}_4$ system is an attractive alternative to these reagents. Moreover, the latter system has some advantages over the other reagents : (1) LiAlH_4 and TiCl_4 are commercially available and inexpensive. (2) One mole of aluminum is effective in bringing 4 moles (in the case of LiAlH_4) or 3 moles (in the case of AlH_3) of the diolefin into reaction, so the yield of product per mole of aluminum is high. (3) As this system contains no other organic ligands, it is easy to isolate the products after work-up. This is not always the case with the organo-metallic hydrides mentioned above.
Recently this new hydroalumination procedure was used successfully for the preparation of 1-octene-3-ol so called Matsutake alcohol, contained in Japanese mushroom, from the easily available 1,7-octadien-3-ol, i.e. the selective reduction of C_7 olefin without attacking C_1 olefin of 1,7-octadien-3-ol by $\text{LiAlH}_4/\text{Cp}_2\text{TiCl}_2$ system [14].

The reaction of α,ω-diene and $(i\text{-Bu})_2\text{AlH}$ was reported to give the different products, but no ω-alkenylaluminum compound, depending on the chain length of the starting dienes [15].

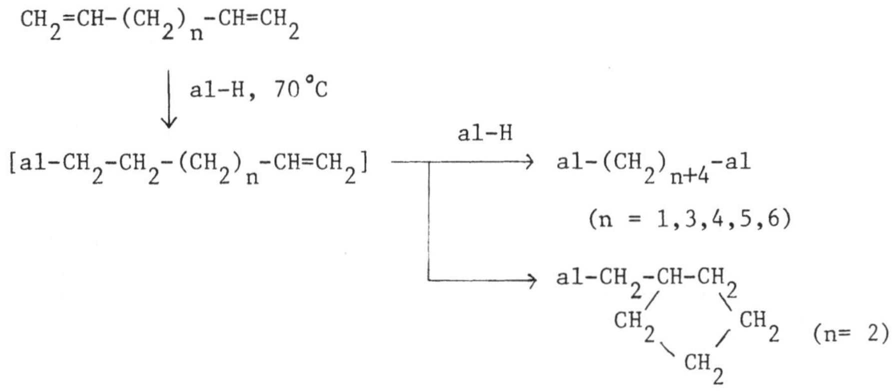

$$\text{CH}_2=\text{CH}-(\text{CH}_2)_n-\text{CH}=\text{CH}_2$$

$\downarrow$ al-H, 70 °C

$$[\text{al}-\text{CH}_2-\text{CH}_2-(\text{CH}_2)_n-\text{CH}=\text{CH}_2] \xrightarrow{\text{al}-\text{H}} \text{al}-(\text{CH}_2)_{n+4}-\text{al}$$

$$(n = 1,3,4,5,6)$$

$$\longrightarrow \text{al}-\text{CH}_2-\text{CH}-\text{CH}_2$$

$$\text{CH}_2 \quad \text{CH}_2 \quad (n=2)$$
$$\text{CH}_2$$

TiCl_4-catalyzed α,ω-diene-LiAlH_4 reaction have been found to proceed easily at room temperature to generate dialumino or monoalumino-derivative depending on the molar ratio of the reactants.

$$\text{al}-\text{H} + \text{CH}_2=\text{CH}-(\text{CH}_2)_n-\text{CH}=\text{CH}_2 \xrightarrow{\text{TiCl}_4} \text{al}-\text{CH}_2-\text{CH}_2-(\text{CH}_2)_n-\text{CH}=\text{CH}_2$$

$$2 \text{ al}-\text{H} + \text{CH}_2=\text{CH}-(\text{CH}_2)_n-\text{CH}=\text{CH}_2 \xrightarrow{\text{TiCl}_4} \text{al}-(\text{CH}_2)_{n+4}-\text{al}$$

This reaction thus provides a satisfactory synthetic route to α,ω-dihaloalkanes or ω-haloalkenes from α,ω-dienes [7c]. When this method is compared with the hydroboration method, the regioselectivity of $\text{LiAlH}_4/\text{TiCl}_4$ system is as high as disiamylborane [16,17].

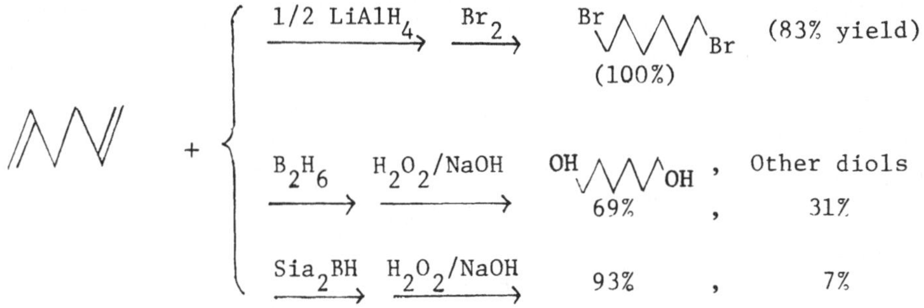

$\xrightarrow{1/2 \text{ LiAlH}_4} \xrightarrow{\text{Br}_2}$	Br〜〜Br	(83% yield)
	(100%)	
$\xrightarrow{\text{B}_2\text{H}_6} \xrightarrow{\text{H}_2\text{O}_2/\text{NaOH}}$	OH〜〜OH	Other diols
	69%	31%
$\xrightarrow{\text{Sia}_2\text{BH}} \xrightarrow{\text{H}_2\text{O}_2/\text{NaOH}}$	93%	7%

(50% yield based on
reacted diene)

(64% yield based on
Sia$_2$BH)

3. Hydroalumination of butadiene or isoprene

Aluminum hydrides were reported to react with butadiene or
isoprene sluggishly to give a mixture of products, so this reaction
did not appear to offer a promising synthetic route for the synthesis of organoaluminum compounds [18].
Hydroalumination of butadiene or isoprene have been found to proceed
easily by the catalysis of titanium or nickel compounds with the
following characteristics [19] (Table 2).
1) Butadiene and isoprene undergo monohydroalumination preferentially.
2) Cp$_2$TiCl$_2$ is most effective catalyst.
3) Halogenolysis or deuterolysis of the reaction product indicate
that the monohydroalumination product which give 1-butene or 2-
methyl-1-butene by hydrolysis is best formulated as 3-alumino-1-
butene or 2-methyl-3-alumino-1-butene, respectively.

(77 : 3 : 20)

4) The Cp$_2$TiCl$_2$-catalyzed reaction of isoprene and LiAlH$_4$ seems to
proceed to the trialkenylaluminum stage but not beyond, this must
be owing to the steric hindrance of catalyst and the methyl group
of isoprene.
5) The addition of TiCl$_4$ to the Cp$_2$TiCl$_2$-catalyzed LiAlH$_4$-isoprene
reaction mixture lead to bring 4 moles of isoprene into reaction.

Table 2. Hydrolysis products after reaction of LiAlH$_4$ with butadiene or isoprene

Diene	Catalyst	Reaction condition	Conv.[d] %	Product distribution %			
	TiCl$_4$	a	56	16	68	16	
	Cp$_2$TiCl$_2$	a	100	10	70	20	
	Ni(PBu$_3$)$_2$Cl$_2$	b	65	0	68	32	
	TiCl$_4$	a	57	16	58	18	9
	Cp$_2$TiCl$_2$	a	75	4	88	6	2
	Cp$_2$TiCl$_2$-TiCl$_4$	c	100	5	87	6	2
	NiBr$_2$	b	93	4	53	33	10
	Ni(PBu$_3$)$_2$Cl$_2$	b	100	2	56	35	6

a; Diene/LiAlH$_4$ (molar ratio) = 4, Temp; r.t., Time; 24hr
b; Diene/LiAlH$_4$ = 3, Temp; -10°C, Time; 12hr   c; Diene/LiAlH$_4$ = 4,
Temp; r.t., isoprene + 1/4 LiAlH$_4$ $\xrightarrow{\text{Cp}_2\text{TiCl}_2, \text{12hr}}$ $\xrightarrow{\text{TiCl}_4, \text{4hr}}$
d; Based on diene.

The formation of allylic aluminum compound suggests that the hydro-alumination of conjugated dienes proceeds through an allylic-transition metal intermediates (scheme 2), as is also proposed for many catalytic reactions of dienes. The finding that nickel compounds, which did not catalyze the hydroalumination of α-olefin, show high catalytic activity thus is not surprising, because the formation of a π-crotyl system from the reaction of the nickel-hydride with butadiene have been reported [20].

4. Preparation of trialkenyl boranes

In this section I will show that this new hydroalumination reaction also contributes to the chemistry of organoboranes. As diborane usually doubly metalate dienes, it is impossible to prepare trialkenylboranes by the hydroboration method [3]. The alkylation of boron with organoaluminum compounds has been shown to proceed readily and completely if sufficient boron valences are available [21]. Thus it is now possible to prepare trialkenylboranes through the sequence of hydroalumination of dienes followed by trans-

Scheme 2

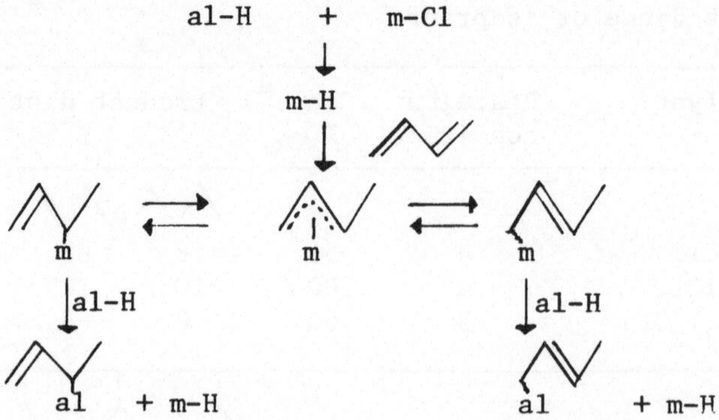

metallation in one-pot reaction [22].

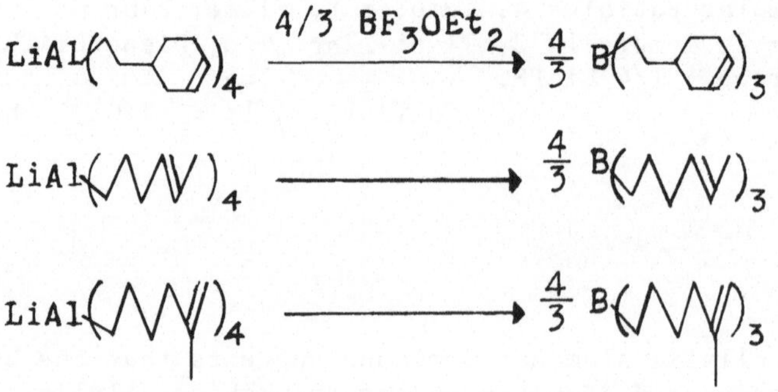

Preparation of trialkenylborane is accomplished in the following
illustration. After a 12 hour reaction of 1,4-hexadiene and 1/4
mole of LiAlH$_4$ in the presence of TiCl$_4$ in THF at room temperature,
1/3 mol of BF$_3$:OEt$_2$ was added all at once to the reaction mixture.
The mixture was refluxed for 5 hours and oxidized with alkaline
hydrogen peroxide to give 4-hexen-1-ol in 90% yield based on the
diene. Noteworthy here is the fact that trialkenylborane is obtained
from diene, LiAlH$_4$ and BF$_3$:OEt$_2$, because the reaction of LiAlH$_4$ with
BF$_3$:OEt$_2$ is known as one of the convenient hydroboration procedures
[23].

5. Conclusion

Hydroalumination of olefins catalyzed by transition metal
compounds proceed under mild condition, thus making the organoaluminum compounds as readily available as the organoboranes.
In many instances, the organoaluminum compounds and the organoboranes

exhibit individual characteristics not present in the other. In such cases the choice will necessarily be dictated by the characteristics of each group of reagents.

REFERENCES

1. (a) K. Ziegler, in H. Zeiss (Ed), Organometallic Chemistry, Reinhold Publishing Corporation, New York, 1960, p. 194-269.
   (b) A.N. Nesmeyanov, R.A. Sokolic, The Organic Compounds of Borane, Aluminum, Gallium, Indium and Thallium, North-Holland Publishing Company, Amsterdam, 1967, p.437-481.
   (c) H. Reinheckel, K. Haage, D.Jahnke, Organometal. Chem. Rev. A, 4 (1969) 47
2. E. Negishi, in D. Seyferth (Ed) New Applications of Organometallic Reagents in Organic Synthesis, J. Organometal. Chem. Library 1, Elsevier, Amsterdam, 1967, p. 112-122.
3. H.C. Brown, Organic Synthesis via Boranes, John Wiley & Sons, New York, 1975.
4. K. Ziegler, H.G. Gellert, H. Martin, K. Nagel, J. Schneider, Justus Liebigs Ann. Chem., 589 (1954) 91.
5. (a) G.R. Fulton, British Patent 757524, Sept. 19, 1956 ; Chem. Abstr., 51 (1957) P9673e.
   (b) H.W.B. Reed , W.R. Smith, U.S. Patent 2872470, Feb. 3, 1959; Chem. Abstr., 53 (1959) P7014g.
6. Reference 1a, 195.
7. (a) F. Sato, S. Sato, M. Sato, J. Organometal. Chem., 122 (1976) C25
   (b) F. Sato, S. Sato, M. Sato, J. Organometal. Chem., 131 (1977) C26
   (c) F. Sato, S. Sato, H. Kodama, M. Sato, J. Organometal. Chem., in press.
8. Recently Y. Otsuji and his collaborators have independently found that the low valent titanium compound obtained from the reaction of $LiAlH_4$ with Cp_2TiCl_2 catalyzes the addition of $LiAlH_4$ to α-olefin to give $LiAlH_3R$; K. Isagawa, K. Tatsumi, Y. Otsuji, Chem Lett., (1976) 1145
9. E.C. Ashby, J. Prather, J. Amer. Chem. Soc., 88 (1966) 729.
10. J. Schwartz, Angew. Chem. Int. Ed., 15 (1976) 333.
11. G.D. Cooper, H.L. Frinkbeiner, J. Org. Chem., 27 (1962) 1493.
12. F. Asinger, B. Fell, R. Janssen, Chem. Ber., 97 (1964) 2515.
13. (a) K. Ziegler, H.-G. Gellert, H. Martin, K. Nagel and J. Schneider, Justus Liebigs Ann. Chem., 589 (1954) 91.
    (b) R. Rienacker, G. Ohloff, Angew. Chem., 73 (1961) 240.
14. J. Tsuji, T. Mandai, Chem. Lett., (1977) 975.
15. (a) G. Hatta, A. Miyake, J. Org. Chem., 28 (1963) 3237.
    (b) K. Ziegler, Angew. Chem., 68 (1956) 721.
    (c) P.W. Chum, E. Wilson, Tetrahedron Lett., (1976) 1257.
16. G.Zweifel, K. Nagase, H.C. Brown, J. Amer. Chem. Soc., 84 (1962) 183

17. G. Zweifel, K. Nagase, H.C. Brown, J. Amer. Chem. Soc., 84
     (1962) 190.
18. Reference 1b, 418.
19. F. Sato, S. Haga, M. Sato, unpublished results.
20. C.A. Tolman, J. Amer. Chem. Soc., 92 (1970) 6777, 6785.
21. Reference 1a, 244
22. F. Sato, S. Haga, M. Sato, unpublished results.
23. (a) S. Wolfe, M. Nussim, Y. Mazur, F. Sondheimer, J. Org. Chem.,
     24 (1959) 1034.
     (b) F. Sondheimer, S. Wolfe, Can. J. Chem., 37 (1959) 1870.

OXIDATIVE ADDITION OF ACYL AND ALKYL HALIDES TO RHODIUM(I)

AND IRIDIUM(I) COMPLEXES AND ALKYL GROUP ISOMERIZATION

M. A. Bennett, * R. Charles, T.R.B. Mitchell, and J. C. Jeffery*
Research School of Chemistry, Australian National University,
P.O. Box 4, CANBERRA, A.C.T., Australia 2600

Decarbonylation of acid halides is catalyzed by planar d^8 metal complexes such as $RhCl(PPh_3)_3$, $RhCl(CO)(PPh_3)_2$ and $IrCl(CO)(PPh_3)_2$ and is a potentially useful process in organic synthesis.[1]   Key steps are believed to be oxidative addition at the metal atom, alkyl or aryl group migration and reductive elimination, as illustrated in Scheme 1 for the stoichiometric reaction of acid halides with $RhCl(PPh_3)_3$.[2-6]   As already reported briefly,[7] straight chain acyl halides react with $IrCl(PPh_3)_3$ to give octahedrally coordinated iridium(III)-alkyls, I, (R = C_2H_5, C_3H_7, C_5H_{11}, C_7H_{15}, CH_2CH_2Ph; L = PPh_3) in 70—80% yield.   The ethyl complex can also be made starting either from the dinitrogen complex $IrCl(N_2)(PPh_3)_2$[8] or from the solution which is said to contain $IrCl(PPh_3)_2$ prepared from the cyclooctene complex $[IrCl(C_8H_{14})_2]_2$ and triphenylphosphine in a 1:4 mole ratio.[9]

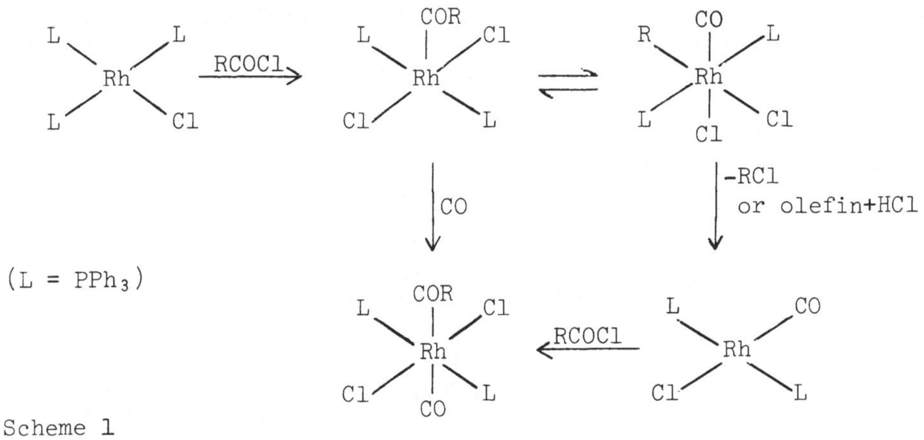

(L = PPh_3)

Scheme 1

$$
\begin{array}{ccc}
\text{L} & \text{R} & \text{CO} \\
 & \diagdown | \diagup & \\
 & \text{Ir} & \\
 & \diagup | \diagdown & \\
\text{Cl} & \text{Cl} & \text{L}
\end{array}
$$

I

The general sequence of reactions is the same for $IrCl(PPh_3)_3$ as for $RhCl(PPh_3)_3$, though the alkyliridium(III) complexes I are less prone to undergo reductive elimination and the five-coordinate acyliridium(III) precursors are less easily isolated than are their rhodium analogs.

α-Branched acyl halides react with $IrCl(PPh_3)_3$ in refluxing benzene to give the n-alkyl complex. Thus, $(CH_3)_2CHCOCl$, $CH_3CH_2CH(CH_3)COCl$, $(CH_3CH_2)_2CHCOCl$ and $PhCH(CH_3)COCl$ give, respectively, the n-propyl, n-butyl, n-pentyl and 2-phenethyl complexes (I; R = $CH_2CH_2CH_3$, $CH_2CH_2CH_2CH_3$, $CH_2CH_2CH_2CH_2CH_3$ and CH_2CH_2Ph) in 30—40% yields. Better yields are achieved starting from '$IrCl(PPh_3)_2$' $e.g.$ 2-ethylpropanoyl chloride gives the n-pentyl complex (I; R = $CH_2CH_2CH_2CH_2CH_3$) in 70% yield and 2-phenylpropanoyl chloride gives the 2-phenethyl complex (I; R = CH_2CH_2Ph) in 53% yield. This $sec \rightarrow n$-alkyl isomerization has not so far been observed directly in the corresponding $RhCl(PPh_3)_3$ reactions, probably owing to the ease with which the alkylrhodium(III) complexes undergo reductive elimination.

Although we have been unable to detect intermediate sec-alkyls in our system, they can be detected in the addition of acyl chlorides to the dimeric cyclooctene-iridium(I) complex $[IrCl(CO)-(C_8H_{14})_2]_2$.[10] Thus, in hot benzene, 2-methylpropanoyl chloride gives the dimeric isopropyl complex $[(CH_3)_2CHIrCl_2(CO)_2]_2$, but after 1.5 h an equilibrium mixture of isopropyl and n-propyl complexes ($ca.$ 2:3) is obtained. The same equilibrium mixture is obtained similarly starting from butanoyl chloride. Reaction of pentanoyl and 2-methylbutanoyl chlorides with $[IrCl(CO)(C_8H_{14})_2]_2$ give the expected n- and sec-butyl complexes, but the latter readily isomerizes to the former even at 34 °C.

The $sec \rightarrow n$-alkyl isomerizations undoubtedly occur via a series of reversible β-eliminations in an initially formed sec-alkyl (Scheme 2), the position of equilibrium being determined by the direction of addition of hydrogen to olefin in the presumed olefin-hydride intermediate. Models indicate that in type I complexes there is considerable steric hindrance between a sec-alkyl group R and the aromatic rings of the triphenylphosphine ligands, and we suggest that this is one important factor which determines the preference for the n-alkyl. It may also be that tertiary

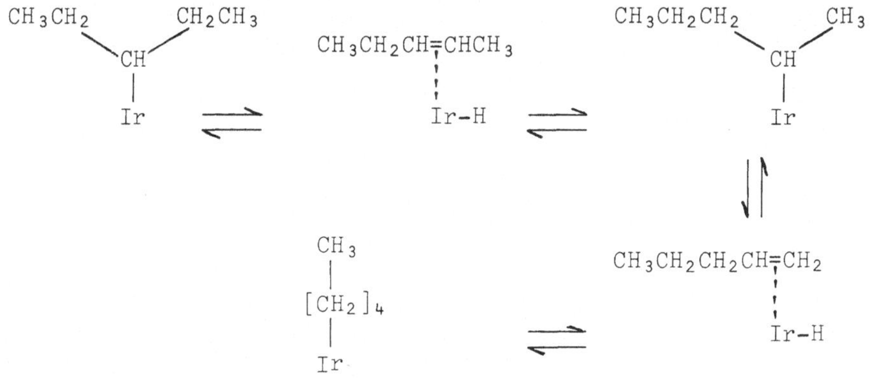

$$CH_3CH_2 \qquad CH_2CH_3$$
$$\diagdown CH \diagup$$
$$|$$
$$Ir$$

$$\rightleftharpoons$$

$$CH_3CH_2CH \overline{=} CHCH_3$$
$$\vdots$$
$$Ir-H$$

$$\rightleftharpoons$$

$$CH_3CH_2CH_2 \qquad CH_3$$
$$\diagdown CH \diagup$$
$$|$$
$$Ir$$

$$\updownarrow$$

$$CH_3$$
$$|$$
$$[CH_2]_4$$
$$|$$
$$Ir$$

$$CH_3CH_2CH_2CH \overline{=} CH_2$$
$$\vdots$$
$$Ir-H$$

$$\rightleftharpoons$$

Scheme 2

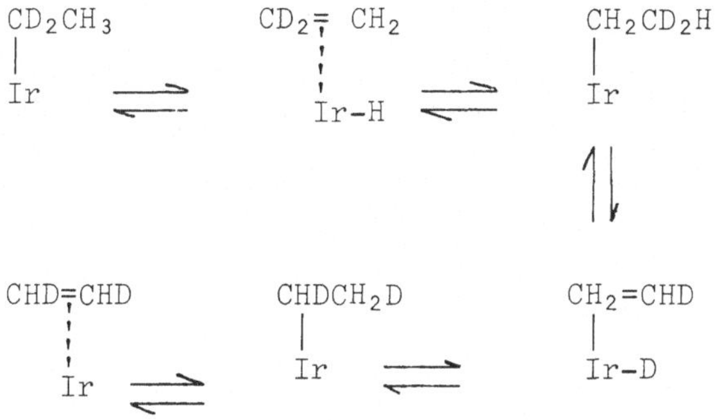

$$CD_2CH_3$$
$$|$$
$$Ir$$

$$\rightleftharpoons$$

$$CD_2 \overline{=} CH_2$$
$$\vdots$$
$$Ir-H$$

$$\rightleftharpoons$$

$$CH_2CD_2H$$
$$|$$
$$Ir$$

$$\updownarrow$$

$$CHD \overline{=} CHD$$
$$\vdots$$
$$Ir$$

$$\rightleftharpoons$$

$$CHDCH_2D$$
$$|$$
$$Ir$$

$$\rightleftharpoons$$

$$CH_2=CHD$$
$$|$$
$$Ir-D$$

Scheme 3

phosphines, which are good σ-donors, help to polarize the Ir-H bond in the sense $Ir^{\delta+}-H^{\delta-}$ and thus encourage anti-Markownikoff addition of the olefin.

Reversible β-eliminations are also responsible for the scrambling of deuterium atoms between the α- and β-carbon atoms of the ethyl group of CH_3CD_2COCl on addition to '$IrCl(PPh_3)_2$' or $[IrCl(CO)-(C_8H_{14})_2]_2$ (Scheme 3). Thus, addition of a 3:1 mixture of $CH_3CD_2-COCl/CH_3CHDCOCl$ (CH_3:CH ratio = 12:1) to $[IrCl(C_8H_{14})_2]_2/4PPh_3$ in refluxing benzene gives a deuterated ethyl complex $C_2H_3D_2IrCl_2-(CO)(PPh_3)_2$ in which the 'CH_3' to 'CH' ratio in the ^1H NMR is 5:1 after 1 h and 2.5:1 after 4.5 h. Addition of the same acid chloride to $[IrCl(CO)(C_8H_{14})_2]_2$ in refluxing benzene and isolation of the deuterated ethyl complex $[C_2H_3D_2IrCl_2(CO)_2]_2$ after 12 min gives a product with a CH_3 to CH ratio of 12:1 (*i.e.* no scrambling has occurred), but if the product is isolated after 4 h and 8 h under reflux the ratio is 5:1 and 1.5:1 respectively.

An obvious question is whether *sec*-alkyl intermediates can be observed in the oxidative addition of acyl halides to MClL₃ complexes of rhodium and iridium if the tertiary phosphines L are smaller and/or more basic than triphenylphosphine. The reaction of $(CH_3)_2CHCOCl$ with $IrCl(PMePh_2)_3$ follows the same course as that with $IrCl(PPh_3)_3$: the product is the *n*-propyl complex I (R = $CH_2CH_2CH_3$; L = PMePh₂) and no isopropyl complex is observed. In contrast, acyl chlorides react with $MCl(PMe_2Ph)_3$ {usually generated *in situ* from the ethylene or cyclooctene complexes $[MCl(olefin)_2]_2$ and PMe₂Ph in a 1:6 mol ratio} to give octahedral rhodium(III)- or iridium(III)-acyls II [M = Rh, Ir; R = CH_3, $CH_3CH_2CH_2$, $CH_3CH_2CH_2CH_2$, $(CH_3)_2CH$; L = PMe₂Ph]. These acyls do not rearrange to alkyls by CO migration, presumably because the firmer binding of PMe₂Ph relative to PPh₃ or PMePh₂ does not create a vacant site on the metal. Attempts to create such a site and promote alkyl migration by removal of Cl⁻ with NH₄PF₆ are successful in the case of iridium and R = CH_3, $CH_3CH_2CH_2$ or $CH_3CH_2CH_2CH_2$, the products being the cationic alkyl carbonyl complexes III. However, the 2-methylpropanoyl complex II [M = Ir, R = $CH(CH_3)_2$] is surprisingly unreactive towards NH₄PF₆ and no alkyl migration is observed. In the case of rhodium, chloride is abstracted by NH₄PF₆ but the alkyl group still does not migrate. The five-coordinate cation IV has been shown by single crystal X-ray study of its PF_6^- salt to be essentially square pyramidal with an apical acetyl group.

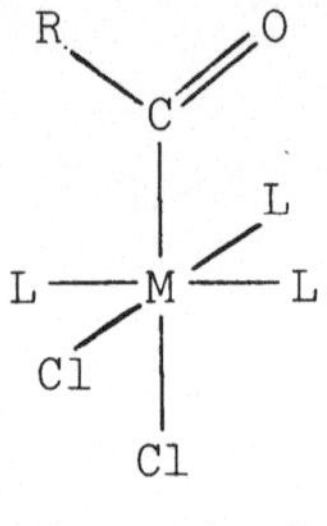

II

for M = Rh, L = PMe₂Ph, R = CH₃: ν(C=O) 1635s, 1710w sh (CH_2Cl_2)
1640s, 1680w sh (Nujol)

^{31}P NMR (CDCl₃) δ -5.6(dd, trans P's, J_{RhP} 107 Hz, J_{PP} 26 Hz)
δ +7.95(dt, P trans to Cl, J_{RhP} 158 Hz, J_{PP} 26 Hz)

for M = Ir, L = PMe₂Ph, R = CH₃: ν(C=O) 1610 (Nujol)

^{31}P NMR (CDCl₃) δ -36.5(d, trans P's, J_{PP} = 19 Hz)
-40.0(t, P trans to Cl, J_{PP} 19 Hz)

III $(L = PMe_2Ph)$

for R = CH_3, IR shows no $\nu(C=O)$ band; 2050 cm^{-1} [$\nu(CO)$],
310 cm^{-1} [$\nu(IrCl)$] showing Cl trans to CO.

IV (recrystallized
from methanol)

$^{31}P\{^1H\}$ NMR at -70 °C in CDCl$_3$
δ_p 1.29, dd, J_{Rh-P} 103 Hz, P trans
to P;17.13, dt, J_{Rh-P} 153 Hz,
$J_{P_cP_t}$ 23.4 Hz, P trans to Cl.

$\nu(C=O)$ 1720s (CH$_2$Cl$_2$) 1715s (Nujol).

A similar geometry has been reported for the neutral acyl
RhCl$_2$(COCH$_2$CH$_2$Ph)(PPh$_3$)$_2$,[6] although unexpectedly the corresponding
compound RhCl$_2$(COCH$_2$Ph)(PPh$_3$)$_2$ is stated[11] to be trigonal bipyra-
midal with an equatorial acyl group. In solution, **IV** is
fluxional: the ^1H NMR spectrum shows separate but broad resonances
for the different PMe$_2$Ph groups and the $^{31}P\{^1H\}$ NMR spectrum is
very broad. On cooling to -60 to - 70 °C the P-CH$_3$ resonances
appear as sharp but very complex multiplets, and the $^{31}P\{^1H\}$ NMR
spectrum corresponds to that expected on the basis of the solid
state structure.

In the presence of air, a solution of IV in acetone/dichloro-
methane decomposes to give a dinuclear acetylrhodium(III) salt
which we initially thought was the di-μ-chloro-bridged species V,
on the basis of the similarity of its spectroscopic properties to
those of the corresponding di-μ-bromo-species prepared[12] by
addition of methanolic $AgPF_6$ to $CH_3RhBr_2(CO)(PMe_2Ph)_2$. The same
complex V can also be obtained by treating $RhCl_2(COCH_3)(PMe_2Ph)_3$
with NH_4PF_6 in acetone/dichloromethane for 24 h, or by treating a
mixture of $[RhCl(C_2H_4)_2]_2$ (1 mol) and PMe_2Ph (4 mol) with acetyl
chloride to give what appears to be a mixture of isomers of mono-
meric, five-coordinate $CH_3CORhCl_2(PMe_2Ph)_2$, and allowing this to
react with NH_4PF_6. Single crystal X-ray analysis of the dinuclear complex obtained by decomposition of IV in acetone/di-
chloromethane shows the tri-μ-chloro-bridged structure VI with
each rhodium atom being octahedrally coordinated. The Rh-Cl
distances trans to the acetyl groups are *ca.* 0.15 Å longer than
those trans to the phosphine ligands, consistent with a high trans-
influence for the acetyl group. Remarkably, X-ray analysis
reveals the presence of 0.5 mol of 1,2-dichloroethane firmly held
in the lattice, even though this solvent was not used in the
preparation or recrystallization of VI. The source of the 1,2-
dichloroethane is at present unknown; it may be formed from di-
chloromethane during the decomposition of $[CH_3CORhCl(PMe_2Ph)_3]PF_6$
in the latter solvent, but this is only speculation. It seems
likely that Clark and Reimer's[12] di-μ-bromo compound (for which
only C and H analyses were provided) is really the tri-μ-bromo
analog of VI.

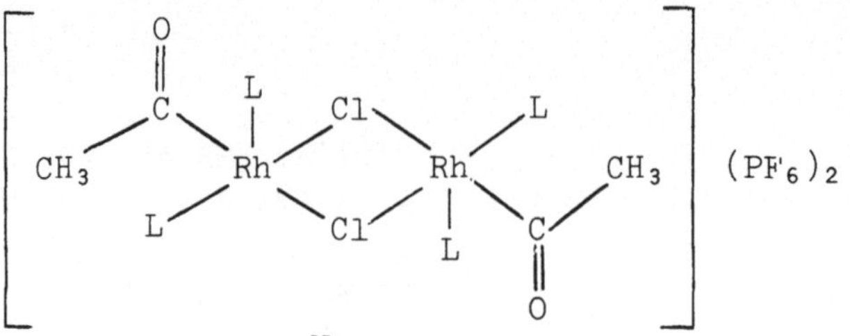

V

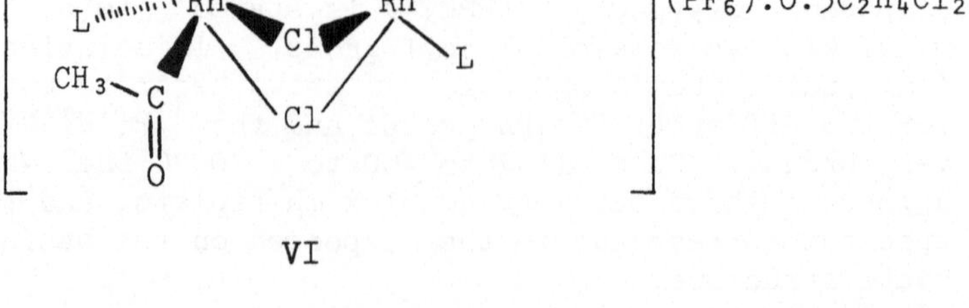

VI

Oxidative addition of acid chlorides to $MCl(CO)(PMe_2Ph)_2$ does not provide a convenient synthesis of octahedral alkyliridium(III) complexes because decarbonylation does not occur readily. However, in contrast with its PPh_3 and $PMePh_2$ analogs, $IrCl(CO)(PMe_2Ph)_2$ cleanly adds both primary and secondary alkyl iodides[13] in benzene at room temperature:

$$IrCl(CO)L_2 \quad + \quad RI \quad \xrightarrow[\substack{r.t. \\ 15\ min}]{C_6H_6} $$

VII

$L = PMe_2Ph; \quad R = C_2H_5, CH_2CH_2CH_3, CH(CH_3)_2, C_6H_{11}.$

The corresponding reaction with alkyl bromides is very much slower and there is no reaction with alkyl chlorides. On heating VII $[R = CH(CH_3)_2]$ at 45 °C for 20 h, a complex mixture is obtained which includes the isomeric n-propyl derivative, together with the corresponding dichloro- and di-iodo-complexes. However, if the oxidative addition of isopropyl iodide is carried out in methanol instead of benzene, the main product isolated is the n-propyl complex (VII; $R = CH_2CH_2CH_3$). So far, methanol is the only solvent of those studied which gives rise to this isomerization. It may be that the reactive iridium(I) species in this case is the cation $[Ir(MeOH)(CO)(PMe_2Ph)_2]^+$, but more work is needed to establish this. It seems that although sec-alkyls of iridium(III) containing dimethylphenylphosphine can be isolated, the n-alkyls are thermodynamically favored.

References

(1) J. Tsuji and K. Ohno, *Synthesis*, 1, 157 (1969).
(2) M.C. Baird, J.T. Mague, J.A. Osborn, and G. Wilkinson, *J.Chem.Soc.A*, 1347 (1967).
(3) K. Ohno and J. Tsuji, *J.Am.Chem.Soc.*, 90, 99 (1968).
(4) J.K. Stille, M. T. Regan, R.W. Fries, F. Huang, and T. McCarley, *Adv.Chem.Ser.*, No.132, 181 (1974).
(5) N.A. Dunham and M.C. Baird, *J.Chem.Soc., Dalton Trans*, 774 (1975).
(6) D.L. Egglestone, M.C. Baird, C.J.L. Lock, and G. Turner, *J.Chem.Soc., Dalton Trans*, 1576 (1977).
(7) M.A. Bennett and R. Charles, *J.Am.Chem.Soc.*, 94, 666 (1972).
(8) M. Kubota and D.M. Blake, *J.Am.Chem.Soc.*, 93, 1368 (1971).
(9) A. van der Ent and A. Onderdelinden, *Inorg.Chim.Acta*, 7, 203 (1973).

(10) B.L. Shaw and E. Singleton, *J.Chem.Soc.A*, 1683 (1967).

(11) K.S.Y. Lau, Y. Becker, F. Huang, N. Baenziger, and
     J.K. Stille, *J.Am.Chem.Soc.*, **99**, 5664 (1977).

(12) H.C. Clark and K.J. Reimer, *Inorg.Chem.*, **14**, 2133 (1975).

(13) Oxidative addition of methyl and substituted methyl halides
     to $IrCl(CO)(PMe_2Ph)_2$ has been described: A.J. Deeming and
     B.L. Shaw, *J.Chem.Soc.A*, 1128 (1969).

TRANSITION METAL FORMYL COMPLEXES AND TRANSITION METAL ANIONS;
SYNTHESIS BY TRIALKYLBOROHYDRIDE ATTACK ON METAL CARBONYL COMPOUNDS

J. A. Gladysz

Department of Chemistry, University of California

Los Angeles, California 90024[1]

Over the next few decades, an increasing emphasis is expected
upon the utilization of coal for the production of chemical feed-
stocks currently derived from petroleum. Coal can be readily converted into a 1:1 mixture of CO and H_2 (synthesis gas).[2] The CO
and H_2 can be subsequently transformed by heterogeneous catalysts
at high temperatures and pressures (Fischer-Tropsch process)[3,4] to
methane, gasoline, and alcohols. There is substantial current interest[5] in developing milder, homogeneous, and more selective methods
for accomplishing these conversions.

There is at present little direct experimental evidence which
bears upon the overall mechanism of Fischer-Tropsch type processes.
Consider the fate of the carbon atom in the heterogeneously catalyzed conversion $CO + 3H_2 \rightarrow CH_4 + H_2O$. The following series of reduction states has been postulated:[3,6]

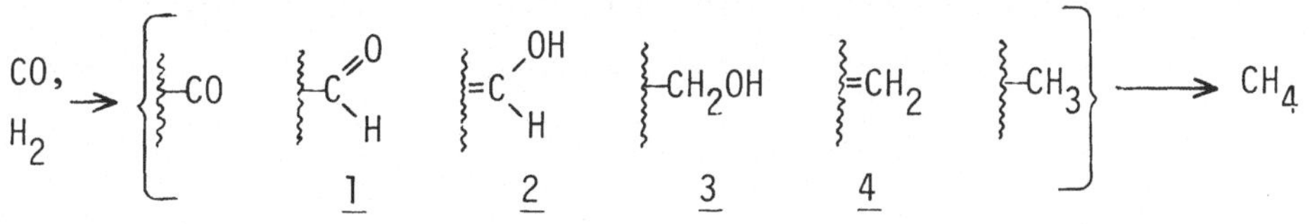

$$CO, H_2 \rightarrow \left\{ CO \quad \underset{1}{C\overset{O}{\diagdown}_H} \quad \underset{2}{=C\overset{OH}{\diagdown}_H} \quad \underset{3}{CH_2OH} \quad \underset{4}{=CH_2} \quad CH_3 \right\} \longrightarrow CH_4$$

The intermediates depicted above are not necessarily comprehensive nor sequential, although there is a consensus that the initially formed intermediate is a metal-bound formyl (1).[3,6] Some may not even be involved (consider 1 → 3 without 2). Nonetheless, it is evident that there are only a limited number of possible metal-bound intermediates between CO and CH_4.

To attempt to study the chemistry of intermediates 1-4 utilizing the actual catalysts employed for CH_4 production is at this time impractical. Reaction coordinate potential energy wells are too shallow and reaction conditions too severe to allow significant concentrations of any intermediate to accumulate. We therefore believe that mechanistic insight into the CO reduction process is best experimentally acquired through the study of each hypothetical step on model complexes related to 1-4. Consequently, we have embarked upon a program of preparation and characterization of reactive ligand types 1-4, all of which were unknown until recently, on homogeneous metal centers. In this paper, we will detail the synthesis of a variety of metal formyl complexes (ligand type 1), outline some of their reactions, and discuss other transformations in which metal formyl complexes may be intermediates.

The first metal formyl complex to be isolated was prepared by Collman and Winter in 1973 by reaction of $Na_2Fe(CO)_4$ with formic acetic anhydride (equation i).[7] Characteristic ^1H nmr (δ 14.95), ^{13}C nmr (260.1 ppm) and ir absorbances ($\nu_{C=O}$, 1607cm^{-1}) were reported. However all subsequent attempts to extend this methodology failed.[8]

Studies in our laboratory[9] and others[10-12] indicated that metal carbonyl acyls, $L_nM(CO)(COR)$, (5) react with carbon nucleophiles to give products resulting from nucleophilic attack at CO. The apparent superior kinetic electrophilicity of CO over the acyl group in 5 immediately suggested that anionic formyl complexes might be prepared by attack of a suitable hydride source upon 5 (equation ii).

$$Na_2Fe(CO)_4 + H-\overset{O}{\overset{\|}{C}}-O-\overset{O}{\overset{\|}{C}}-CH_3 \longrightarrow Na[(CO)_4Fe-\overset{O}{\overset{\|}{C}}-H] \quad (i)$$

$$\underline{5} \qquad L_n \overset{\overset{O}{\overset{\|}{M}}\text{--}C\text{--}R}{\underset{\overset{|}{\underset{O}{C}}}{}} \xrightarrow[\text{?}]{H^-} \qquad L_n \overset{\overset{O}{\overset{\|}{\overline{M}}}\text{--}C\text{--}R}{\underset{\overset{C}{H}\diagdown O}{}} \qquad (ii)$$

We sought a hydride source which was soluble in organic solvents and contained only one transferable hydride per mole. Reagents such as $LiAlH_4$ transfer variable stoichiometries of hydride. We found that electrophilic hydride donors such as $(i\text{-}C_4H_9)_2AlH$ and 9-BBN (9-borabicyclo[3.3.1]nonane) did not yield stable adducts when reacted with complexes of the type $\underline{5}$. However, trialkylborohydrides, which are commercially available from Aldrich as 1 M THF solutions, gave the desired formyl products.

To -50°C THF solutions of the manganese, iron and molybdenum carbonyl acyls depicted in Table I were added 1.2 equiv $Li(C_2H_5)_3BH$. New ^{1}H nmr signals were detected in the δ 12.40-13.10 range and as-cribed to products $\underline{6\text{-}13}$. These chemical shift values are extremely temperature dependent and have been found to be characteristic of anionic metal formyl complexes.[8,13,14] Yields were measured relative 1,2,4,5-tetrachlorobenzene internal standard.

The formyl complexes $\underline{6\text{-}13}$ are all unstable at room temperature, and although repeated isolation attempts failed, ir spectra were ob-tained for $\underline{6,7}$, and $\underline{10}$. All exhibit medium intensity formyl $\nu_{C=O}$ absorptions, and the manganese complexes are cis geometrical isomers (THF, cm^{-1}: $\underline{6}$: 2050(w), 1949(s), 1884(m), 1588(w,br); $\underline{7}$: 2055(m), 1962(s), 1929(s), 1863(w), 1604(m); $\underline{10}$: 1932(s), 1555(m)). Under optimized experimental conditions, only broad ^{13}C nmr resonances barely above baseline could be obtained for metal-bound carbons ($\underline{10}$: ca. 297, 293, 221 ppm, tentative, THF/C_6D_6, -40°C), although broad-ened arene, cyclopentadienyl, and alkyl carbon resonances were rou-tinely detected.

The rates of thermal decomposition of $\underline{6,7,9,10,11}$, and $\underline{12}$ were monitored by ^{1}H nmr and found to be first order and reproducible. The rate constants listed in Table II were obtained from plots with

Table I. Formyl Complexes Prepared

Starting Material		Product	^{1}H nmr, δ, -50°C	Yield,%
$(CO)_5Mn\overset{O}{\overset{\|}{C}}C_6H_5$	6	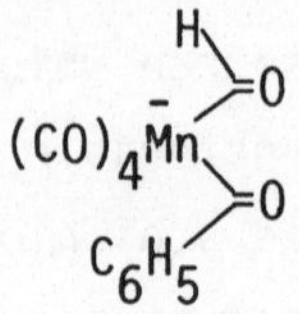 $(CO)_4\overset{-}{Mn}\diagup\overset{H}{\diagdown}C{=}O$, C_6H_5	12.80	98.4
$(CO)_5Mn\overset{O}{\overset{\|}{C}}CH_2OCH_3$	7	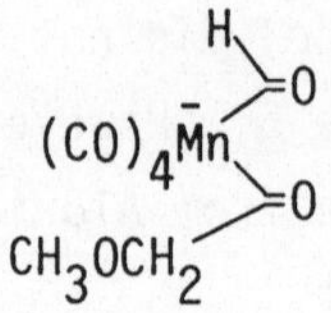 $(CO)_4\overset{-}{Mn}$, CH_3OCH_2	12.75	91.3
$(CO)_5Mn\overset{OO}{\overset{\|\|}{CC}}OCH_3$	8	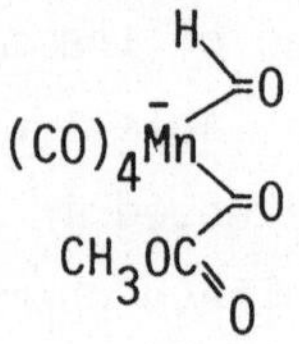 $(CO)_4\overset{-}{Mn}$, $CH_3OC{=}O$	12.65	34.0
$(CO)_5Mn\overset{O}{\overset{\|}{C}}CF_3$	9	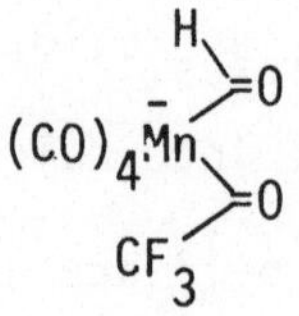 $(CO)_4\overset{-}{Mn}$, CF_3	12.43	25.6
$C_5H_5(CO)_2Fe\overset{O}{\overset{\|}{C}}C_6H_5$	10	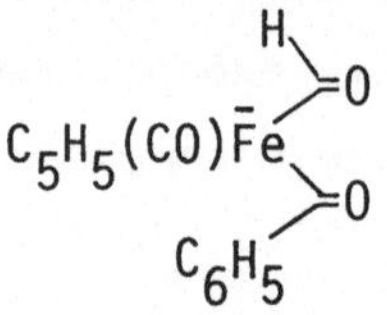 $C_5H_5(CO)\overset{-}{Fe}$, C_6H_5	12.91	76.9
$C_5H_5(CO)_2Fe\overset{O}{\overset{\|}{C}}C_6H_4OCH_3$	11	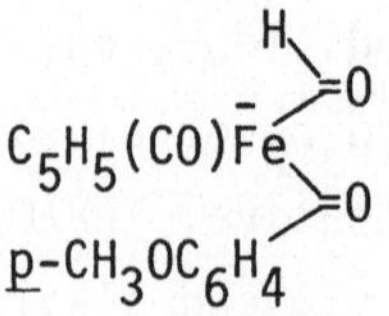 $C_5H_5(CO)\overset{-}{Fe}$, $\underline{p}\text{-}CH_3OC_6H_4$	13.10	96.0
$C_5H_5(CO)_2Fe\overset{O}{\overset{\|}{C}}CH_3$	12	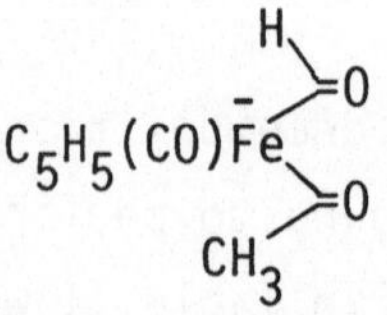 $C_5H_5(CO)\overset{-}{Fe}$, CH_3	12.83	96.8

$$C_5H_5(CO)_3Mo\overset{\overset{OO}{||\,||}}{C}COCH_3 \qquad \underline{13} \quad C_5H_5(CO)_2\overset{\overset{H}{|}}{Mo}\overset{O}{\underset{O}{\diagup}} \qquad 12.90 \qquad 16.7$$

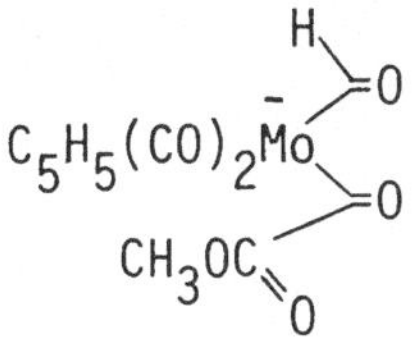

.981–.996 correlation coefficients to at least 2.5 half lives of experimental data. The addition of 2 equiv $(C_2H_5)_3B$ approximately doubled the rate of disappearance of $\underline{6}$, but did not effect k_{obs} for $\underline{10}$. In all cases, much poorer correlation coefficients were obtained with excess $(C_2H_5)_3B$ present.

The decomposition of $\underline{1}$ at room temperature was monitored by ir for 2 hr, and then $(C_6H_5)_3SnCl$ was added to the reaction mixture. It had been previously established[13,15] that the addition of $(C_6H_5)_3$-SnCl to metal anion (L_nM^-) solutions results in virtually quantitative formation of $L_nMSn(C_6H_5)_3$ adducts. Subsequently isolated were 0.42 equiv $(CO)_5Mn(COC_6H_5)$, 0.55 equiv $(CO)_5Mn[Sn(C_6H_5)_3]$, and 0.41 equiv benzyl alcohol. If the solution of $\underline{1}$ was allowed to decompose longer, all of the $(CO)_5Mn(COC_6H_5)$ became $LiMn(CO)_5$ (ir analysis), and some benzyl benzoate was formed.

These results can be accomodated by the decomposition mechanism depicted in Scheme I. We postulate hydride transfer from the formyl

Table II.   Decomposition Rates of Formyl Complexes

Complex	$k_{obs} \times 10^{-4}$/sec	Temperature, $°C^a$
$\underline{6}$	15.53 + .71	7
$\underline{7}$	1.25 + .17	7
$\underline{9}$	17.38 + .64	-23
$\underline{10}$	6.29 + .18	7
$\underline{11}$	5.67 + .15	7
$\underline{12}$	9.94 + .52	-16

$^a CH_3OH$ temperature calibration

to the acyl ligand (step a) as the rate determining step and that collapse to $(CO)_5Mn^-$ and benzaldehyde follows (step b).

We propose that the benzaldehyde produced from **6** is immediately reduced to alkoxide by unreacted **6** (step c). This requires that $k_{obs} = 2k_1$ and rationalizes the approximate 1:1:1 $(CO)_5Mn(COC_6H_5)$: $Mn(CO)_5^-$: $C_6H_5CH_2OH$ product distribution obtained. Accordingly, when 1 equiv benzaldehyde was added to **6** at 0°C, benzyl alcohol was obtained in 96% yield (gc) after aqueous workup. However, when benzoxide and $(CO)_5Mn(COC_6H_5)$ are allowed to stand, benzoylation occurs (step d). Pearson has previously reported on the analogous reaction between $(CO)_5Mn(COCH_3)$ and $NaOCH_3$.[16]

Scheme I

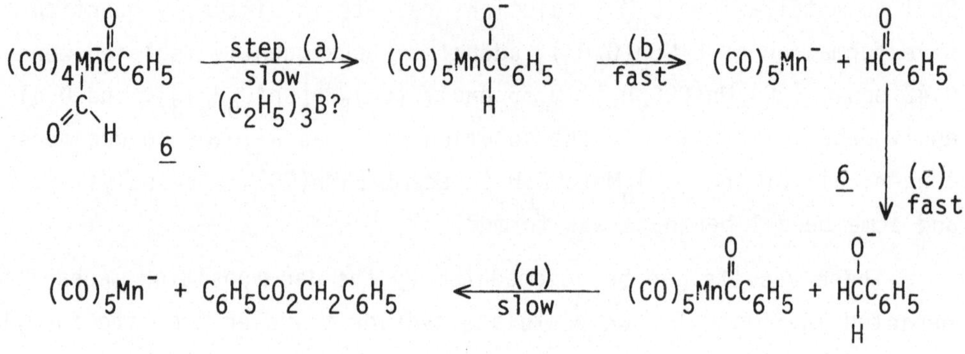

$$(CO)_4Mn\overset{\overset{\displaystyle O}{\|}}{C}C_6H_5 \quad \underset{\underset{(C_2H_5)_3B?}{slow}}{\xrightarrow{step\ (a)}} \quad (CO)_5Mn\overset{\overset{\displaystyle O^-}{|}}{\underset{\underset{H}{|}}{C}}C_6H_5 \quad \xrightarrow[fast]{(b)} \quad (CO)_5Mn^- + H\overset{\overset{\displaystyle O}{\|}}{C}C_6H_5$$

with $O\overset{C}{\diagdown}H$ below on the left structure, labeled **6**

$$\downarrow \text{**6** (c) fast}$$

$$(CO)_5Mn^- + C_6H_5CO_2CH_2C_6H_5 \quad \xleftarrow[slow]{(d)} \quad (CO)_5Mn\overset{\overset{\displaystyle O}{\|}}{C}C_6H_5 + H\overset{\overset{\displaystyle O^-}{|}}{\underset{\underset{H}{|}}{C}}C_6H_5$$

Since complexes **6-13** are too unstable to be isolated and purified we cannot definitively comment on the possibility of by-product $(C_2H_5)_3B$ mediating hydride transfer in step (a). The addition of 2 equiv $(C_2H_5)_3B$ approximately doubled the rate of disappearance of **6** but did not effect k_{obs} for **10**. Another possible initial decomposition step would involve CO loss from **6**: migration of -H or $-C_6H_5$ and reductive elimination of benzaldehyde could follow. However, preliminary $\Delta S^{\ddagger}$ for **6** and **7** are negative, which is inconsistent with a rate determining disassociative step.

The decomposition mechanism of the iron formyl **10**, while sharing some qualitative features with that for **6** above, appears more com-

plex and is the object of ongoing study. A sample of 10 was prepared
from 2.005 mmol $C_5H_5(CO)_2Fe(COC_6H_5)$ and allowed to decompose for 2
hr at room temperature, after which $(C_6H_5)_3SnCl$ was added. Subsequently isolated were 0.477 mmol $C_5H_5Fe(CO)_2(COC_6H_5)$, 0.630 mmol
$C_5H_5Fe(CO)_2Sn(C_6H_5)_3$, (formed from $C_5H_5Fe(CO)_2^-$) and 0.380 mmol
$[C_5H_5Fe(CO)_2]_2$, to account for 93% of the iron atoms. 0.392 mmol
benzyl alcohol (gc) was also formed. No other organic products
could be detected; specifically absent were benzaldehyde, benzoin,
benzil, benzyl benzoate, and stilbene.

Complexes 6-13 are structurally analogous to anionic bis acyl
complexes 14a prepared by Casey[10] and Lukehart.[11] Compounds 14a
can in general be protonated to afford formal enol tautomers of a
1,3-diketo-2-metallo system (15a).[12] However, protonation of 6 and
10 with one equiv CF_3SO_3H at -50°C gave immediate evolution of 1
equiv H_2 and the corresponding neutral metal carbonyl acyl (80-98%
isolated yield). Thus anionic metal formyls act as hydride donors
toward protons, and no evidence for a species such as 15b is observed. One resonance form of 15b contains a hydrido hydroxycarbene
(ligand type 2) moiety.

We decided to investigate reactions of $Li(C_2H_5)_3BH$ with other
types of metal carbonyl compounds. When $(CO)_5MnBr$ was reacted with
one equiv $Li(C_2H_5)_3BH$, it was half consumed and $LiMn(CO)_5$ was produced. This preliminary result suggested to us that metal carbonyl
monoanions might be prepared by hydride attack upon metal carbonyl
dimers.

The use of transition metal monoanions pervades organometallic
chemistry.[17,18] Numerous organometallic compounds have been prepared

through alkylation or acylation of these highly nucleophilic[17] species. Cluster complexes and other compounds containing metal-metal
bonds can be obtained by addition of an appropriate electrophile.[18]
However, conventional preparations of transition metal monoanions
require mercury-sodium amalgam[18] or other heterogeneous metal reduc-
tant.[15] The overall procedure is somewhat cumbersome, requires
special apparatus, and mercury containing by-products are sometimes
produced.[19]

As depicted in equations (iii)-(vii) we have found that the representative monoanions $[Co(CO)_4]^-$, $[C_5H_5Mo(CO)_3]^-$, $[Mn(CO)_5]^-$, and
$[C_5H_5Fe(CO)_2]^-$ are readily formed from $Li(C_2H_5)_3BH$ or $K(\underline{sec}-C_4H_9)_3BH$
and the corresponding metal carbonyl dimer.

$$(CO)_4Co-Co(CO)_4 + 2Li(C_2H_5)_3BH \longrightarrow$$
$$\underline{2LiCo(CO)_4} + 2(C_2H_5)_3B + H_2 \qquad (iii)$$

$$(CO)_5Mn-Mn(CO)_5 + 2Li(C_2H_5)_3BH \longrightarrow$$
$$\underline{2LiMn(CO)_5} + 2(C_2H_5)_3B + H_2 \qquad (iv)$$

$$C_5H_5(CO)_3Mo-Mo(CO)_3C_5H_5 + 2Li(C_2H_5)_3BH \longrightarrow$$
$$\underline{2Li[C_5H_5Mo(CO)_3]} + 2(C_2H_5)_3B + H_2 \qquad (v)$$

$$C_5H_5(CO)_2Fe-Fe(CO)_2C_5H_5 + 2Li(C_2H_5)_3BH \xrightarrow{\text{HMPA}}$$
$$\underline{2Li[C_5H_5Fe(CO)_2]} + 2(C_2H_5)_3B + H_2 \qquad (vi)$$

$$C_5H_5(CO)_2Fe-Fe(CO)_2C_5H_5 + 2K(\underline{sec}-C_4H_9)BH \xrightarrow{\text{3 hr}}$$
$$\underline{2K[C_5H_5Fe(CO)_2]} + 2 (\underline{sec}-C_4H_9)_3B + H_2 \qquad (vii)$$

We feel these rapid, quantitative (by ir), homogeneous, room
temperature reactions constitute a substantial improvement over existing methodology. Only volatile by-products [H_2 and $(C_2H_5)_3B$
when $Li(C_2H_5)_3BH$ is used] are formed, enabling in theory solvent
evaporation to an analytically pure product residue.

In typical procedures, 2.5 equiv $Li(C_2H_5)_3BH$ (1 M in THF) were
added to 1.0 equiv $[Co(CO)_4]_2$, $[Mn(CO)_5]_2$, or $[C_5H_5Mo(CO)_3]_2$ (0.02
M in THF). The corresponding lithium monoanions were formed in
quantative yields after a few minutes stirring. During this period,
1 equiv H_2 evolved, as determined by manometric and mass spectral
methods. Identical results were observed with $Li(\underline{sec}-C_4H_9)_3BH$, and
potassium monoanions could be obtained from $K(\underline{sec}-C_4H_9)_3BH$. Unfortunately, $Li[C_5H_5Fe(CO)_2]$ could not be formed in pure THF; at least
50% HMPA was required as cosolvent (equation vi). However, $K[C_5H_5-$
$Fe(CO)_2]$ was produced over a 3 hr period in THF when $K(\underline{sec}-C_4H_9)_3BH$
was employed (equation vii).

To the monoanions thus prepared were added THF solutions of
various electrophiles. The derivatives obtained are listed in Table
III and illustrate the general synthetic utility of our procedure.
The yields reported are for isolated, purified products, are based
upon metal, and are not optimized.

Investigation of the mechanism of $Li[Mn(CO)_5]$ formation has resulted in several remarkable observations. The following evidence
supports the reaction pathway depicted in Scheme II.

When $[Mn(CO)_5]_2$ was treated with <u>one</u> equiv $Li(C_2H_5)_3BH$ at -20°C
in THF in a ^1H nmr tube, a new complex was formed in 99% yield, as
indicated by the appearance of a signal at 13.66 δ (<u>p</u>-xylene reference and internal standard). This resonance is highly suggestive
of an anionic formyl complex, and our structural assignment <u>16</u> is

Table III. Monoanion Derivatives Prepared

Hydride Reagent	Electrophile	Product	Isolated Yield (%)
$Li(C_2H_5)_3BH$	$[(C_6H_5)_3P]_2N^+Cl^-$	$[Co(CO)_4]^-[(C_6H_5)_3P]_2N^+$	79
"	$(C_6H_5)_3SnCl$	$Co(CO)_4[Sn(C_6H_5)_3]$	83
"	CH_3I	$(C_5H_5)Mo(CO)_3CH_3$	77[a]
"	CH_3OCCCl (O O)	$(C_5H_5)Mo(CO)_3(CCOCH_3)$ (O O)	77[a,b]
"	$(C_6H_5)_3SnCl$	$(C_5H_5)Mo(CO)_3[Sn(C_6H_5)_3]$	76
"	C_6H_5CCCl (O O)	$Mn(CO)_5(CCC_6H_5)_3]$ (O O)	92[a]
"	$(C_6H_5)_3SnCl$	$Mn(CO)_5[Sn(C_6H_5)_3]$	88
"	CH_3OCCCl (O O)	$Mn(CO)_5(CCOCH_3)$ (O O)	81[a,b]
$K(\underline{sec}\text{-}C_4H_9)_3BH$	$(C_6H_5)_3SnCl$	$(C_5H_5)Fe(CO)_2[Sn(C_6H_5)_3]$	93
"	$C_6H_5CH=CHCCl$ (O)	$(C_5H_5)Fe(CO)_2CCH=CHC_6H_5$ (O)	72[a]
"	C_6H_5CCl (O)	$(C_5H_5)Fe(CO)_2CC_6H_5$ (O)	67[a]

(a) electrophile added at -78° instead of room temperature

(b) new compound; satisfactory spectral and elemental analyses were obtained

Scheme II.   Proposed Mechanism of Anion Formation

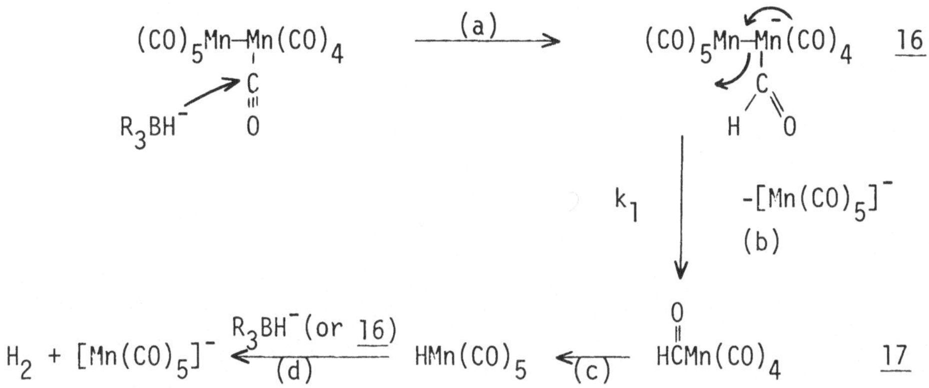

further supported by the observation of the formyl $\nu_{C=O}$ in the vibrational spectrum (1540 cm^{-1}, THF). CO stretching frequencies at 2076 (s), 1970(vs), 1941(m), and 1903(m) cm^{-1} are clearly present, but 16 is unstable at room temperature and assignment of a cis or trans geometry to the disubstituted manganese cannot at this time be made with confidence.[20]

Several reactions of 16 are depicted in Scheme III. Attempted O-methylation at -20°C with CH_3SO_3F[21] led only to CH_4 (80% by manometry and mass spectroscopy) and $[Mn(CO)_5]_2$ (98%). Treatment of 16 with one equiv $Fe(CO)_5$ yielded the iron formyl 18 (76-79% yield by nmr and ir) previously prepared by Collman (equation i).[7] The reduction of octyl iodide to octane by 18 has been previously noted,[7] and Casey[8] and Winter[14] have reported the formation of 18 from $Fe(CO)_5$ and trialkoxyborohydrides. These experiments provide further evidence that anionic formyl complexes can act as hydride donors, and the formation of $[Mn(CO)_5]_2$ in each case supports the formulation of a metal-metal bond in the starting material 16. The migration of a formyl ligand from manganese to iron in the conversion 16 → 18 is considered unlikely.

Scheme III. Reactions of Bimetallic Manganese-Formyl 16

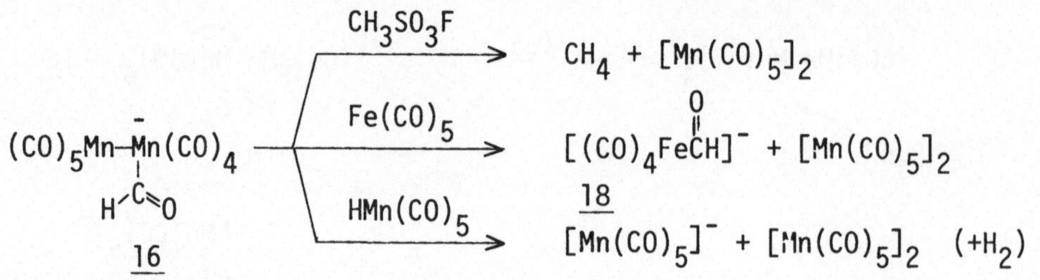

$$CH_3SO_3F \longrightarrow CH_4 + [Mn(CO)_5]_2$$

$$(CO)_5Mn-Mn(CO)_4 \quad \xrightarrow{Fe(CO)_5} \quad [(CO)_4Fe\overset{O}{\overset{\|}{C}}H]^- + [Mn(CO)_5]_2$$

$$\underset{H^{\diagdown C\approx O}}{\big|} \qquad\qquad\qquad \underline{18}$$

$$\underline{16} \qquad \xrightarrow{HMn(CO)_5} \quad [Mn(CO)_5]^- + [Mn(CO)_5]_2 \quad (+H_2)$$

Steps (b)-(d) At -20°C in THF, 16 is stable for hours, but at
room temperature it rapidly decomposes. One equiv $[Mn(CO)_5]^-$ and 0.5
equiv $[Mn(CO)_5]_2$ are formed, with no detectable intermediates by [1]H
nmr or ir. This experiment differs from preparative reactions in
that no trialkylborohydride is present at this stage. At 14°C, k_{obs}
for the decomposition of 16 is first order (2.03 $\pm$ .04 x 10^{-4}/sec;
Arrhenius E_a = 24.8 kcal/mol) which requires at least one intermediate
between 16 and the products. We propose that the unsaturated formyl
17 is formed in the initial α-elimination step (b). The acetyl homolog of 16 has been previously generated[22] and fragments analogously.
Rapid rearrangement of 17 to $HMn(CO)_5$ would be subsequently expected.[23]
In a separate experiment, $HMn(CO)_5$ was found to react instantly with
16 at -20°C to yield one equiv each of $[Mn(CO)_5]^-$ and $[Mn(CO)_5]_2$
(Scheme III). This accounts for the unobservability of intermediate
$HMn(CO)_5$ and is consistent with the first order disappearance of 16
($k_{obs} = 2k_1$). $HMn(CO)_5$ also reacted instantly with $Li(C_2H_5)_3BH$ at
-20°C to form $[Mn(CO)_5]^-$; hence in the preparative reactions, this
pathway may predominate. Formation of $[Mn(CO)_5]^-$ from $HMn(CO)_5$ (K_a
$\cong 10^{-7}$)[24] may occur as an acid-base process or via hydride attack on
CO. While the above body of results exclude many alternatives for
steps (b)-(d), mechanistic investigations are continuing.

We postulate that the other monoanions are formed by a mechanism
qualitatively similar to the one depicted for $[Mn(CO)_5]^-$. When
$[C_5H_5Fe(CO)_2]_2$ was treated with one equiv $K(\underline{sec}-C_4H_9)_3BH$ in THF at

20°C, an intermediate was formed ($\nu_{C=O}$ 1926(s), 1745(m), 1718(m), 1677(s) cm^{-1}) which slowly disproportionated to starting material and K[$C_5H_5Fe(CO)_2$]; no ^{1}H nmr resonance below δ 7.0 could be found. The less rapid formation of [$C_5H_5Fe(CO)_2$]$^-$ relative the other anions may be in part connected to the higher reduction potential of the parent dimer.[25] Anion formation from [$Co(CO)_4$]$_2$ and [$C_5H_5Mo(CO)_3$]$_2$ is too rapid to detect any intermediates, even at -50°C by ^{1}H nmr. In terms of limitations, [$Re(CO)_5$]$^-$ is the only common transition metal mono-anion which cannot be prepared in high yield by this methodology. However, an isolable homolog of 16 is formed which is the subject of current intense study in our group.

Since trialkylborohydrides can be readily prepared from trialkyl-boranes and alkali metal hydrides,[26] there exists the possibility that these reactions could be made catalytic in trialkylborane and stoichiometric in MH (M=Li,Na,K). We have also studied metalloid anion synthesis with Li(C_2H_5)$_3$BH; by adjusting the stoichiometry, either Li$_2$Se or Li$_2$Se$_2$ can be produced from Se$_x$ in THF.[27] However, the main thrust of our research effort is directed at the preparation and characterization of ligand types 1-4, and additional results will be reported in due course.[28]

I wish to warmly thank the following student collaborators who were responsible for the experimental work: Mr. Jay C. Selover, Wilson Tam, Gregory M. Williams, and Dennis L. Johnson. Finally, acknowledgment is made to the donors of the Petroleum Research Fund, administered by the American Chemical Society, for support of this research.

References

1. Contribution number 3917.

2. J.A. Kent, Ed., "Riegel's Handbook of Industrial Chemistry", 7th ed., Van Nostrand Reinhold Company, New York, 1974.

3. H.H. Storch, N. Golumbic, and R.B. Anderson, "The Fischer-Tropsch and Related Syntheses", Wiley, New York, 1951.

4. G.H. Olivé and S. Olivé, *Angew. Chem., Int. Ed. Eng.*, 15, 136
    (1976).

5. M.G. Thomas, B.F. Beier, and E.L. Muetterties, *J. Am. Chem. Soc.*,
    98, 1296 (1976); G.C. Demitras and E.L. Muetterties, *ibid.*, 99,
    2796 (1977).

6. W.A. Goddard, S.P. Walch, A.K. Rappé, T.H. Upton, and C.F.
    Melius, *J. Vac. Sci. Technol.*, 14, 416 (1977).

7. J.P. Collman and S.R. Winter, *J. Am. Chem. Soc.*, 95, 4089 (1973).

8. C.P. Casey and S.M. Neumann, *J. Am. Chem. Soc.*, 98, 5395 (1976).

9. J.A. Gladysz and J.C. Selover, Abstract INOR 41, 172nd ACS
    National Meeting, San Francisco, Calif., August 30, 1976.

10. C.P. Casey and C.A. Bunnell, *J. Am. Chem. Soc.*, 98, 436 (1976).

11. C.M. Lukehart, G.P. Torrence, and J.V. Zeile, *Inorg. Chem.*, 15,
    2393 (1976).

12. C.M. Lukehart and J.V. Zeile, *J. Am. Chem. Soc.*, 99, 4368 (1977).

13. J.A. Gladysz, G.M. Williams, W. Tam, and D.L. Johnson, *J. Organo-
    metal. Chem.*, in press, 139 (1977).

14. S.R. Winter, G.W. Cornett, and E.A. Thompson, *J. Organometal.
    Chem.*, 133, 339 (1977).

15. J.E. Ellis and E.A. Flom, *J. Organometal. Chem.*, 99, 263 (1975).

16. R.W. Johnson and R.G. Pearson, *Inorg. Chem.*, 10, 2091 (1971).

17. R.B. King, *Accounts Chem. Res.*, 3, (1970) 417.

18. J.E. Ellis, *J. Organometal. Chem.*, 86, 1 (1975).

19. R.B. King, *J. Inorg. Nucl. Chem.*, 25, 1296 (1963).

20. E.O. Fischer and E. Offhaus, *Chem. Ber.*, 102, 2449 (1969).

21. C.P. Casey, C.R. Cyr, and R.A. Boggs, *Synth. Inorg. Metal-Org.
    Chem.*, 3, 249 (1973).

22. C.P. Casey and R.L. Anderson, *J. Am. Chem. Soc.*, 93, 3554 (1971);
    C.P. Casey, C.R. Cyr, R.L. Anderson, and D.F. Marten, *J. Am.
    Chem. Soc.*, 97, 3053 (1975).

23. B.H. Byers and T.L. Brown, *J. Organometal. Chem.*, 127, 181 (1977).

24. D.F. Shriver, *Accounts Chem. Res.*, 3, 231 (1970).

25. R.E. Dessy, P.M. Weissman, and R.L. Pohl, *J. Am. Chem. Soc.*, 88, 5117 (1966).

26. C.A. Brown, *J. Am. Chem. Soc.*, 95, 4101 (1973); H.C. Brown, A. Khuri, and S.C. Kim, *Inorg. Chem.*, 16, 2229 (1977).

27. J.A. Gladysz, J.L. Hornby, and J.E. Garbe, *J. Org. Chem.*, in press, 43 (1978).

28. J.A. Gladysz and J.C. Selover, submitted to *Tetrahedron Letters*.

ELECTRONIC CONTROL OF STEREOCHEMISTRY IN CARBON-CARBON

BOND FORMATION AT METAL CENTERS

J. W. Faller

Yale University

New Haven, Connecticut 06520 U.S.A.

Product distribution in catalytic reactions can often be rationalized in terms of steric effects of the ligands. The relative thermodynamic stabilities of intermediates can be associated with steric factors and subsequently correlated with the stereochemistry of products. For example, the effect of phosphine bulk has traditionally been important in explanations of the isomer ratio in modified oxo processes.[1] The steric bulk of substituents in optically active phosphines is also often attributed a role in determining optical yield in catalytic asymmetric syntheses.[2] It is inevitably difficult to completely dissect the interactions into effects arising from steric interactions and those arising from electronic interactions. Nevertheless, a situation will be illustrated in which the direction of attack of an incoming nucleophile upon a coordinated olefinic moiety is controlled primarily by electronic factors. In this system the asymmetry arises from differences between a carbonyl and nitrosyl ligand; yet the difference is sufficient to produce stereospecific reactions.

We have shown in the past that the preferred orientation of substituted olefins[3], η^4-diene,[4] and η^3-allyl[5] ligands can be rationalized in the cases studied by considerations of steric interactions between cyclopentadienyl rings and substituents. These studies examined principally $\eta-C_5H_5M(CO)_2L$ complexes. Substitution of a carbonyl ligand by NO^+ can significantly alter conformational stabilities, but the most profound effect is observed in the rate of conformational interconversion. For example in the dicarbonyl-allyl complex 1, the exo and endo conformers interconvert by a rotation of the allyl group with a rate of approximately 10 sec^{-1} at 0°C. However, upon NO^+ replacement of a CO, the rotation rate is diminished to 10^{-6} sec^{-1}.

1 endo   **1 exo**

There are also some unusual effects on thermodynamic equilibrium of the conformers based upon the electronic stabilization
of various orientations. Generally a mono-substituted olefin such
as propene would tend to adopt an exo conformation such that the
substituent was oriented away from the cyclopentadienyl ring. For
example, as we have found in the iron dicarbonyl analog the ratio
of exo to endo is greater than 20:1. However in observing the

2 exo   **2 endo**

nitrosyl carbonyl molybdenum species one finds that there is a
strong shift in equilibrium such that the interaction with these
rings is offset by electronic stabilization from the nitrosyl.
That is, in this situation, the ratio of exo to endo conformers
is only 1.7:1. Thus, it is quite clear that the differing abilities of nitrosyl and carbonyl to stabilize a substituted olefin
in a particular position is an important factor in determining
equilibria. The introduction of a nitrosyl also favors an angular
deviation from the horizontal orientation, which lessens the interaction of certain endo isomers.

Since there was a profound effect on olefin orientation from
the electronic differential between CO and NO, we presumed that
there would be a similarly large effect on the stereochemistry in

the transition state and consequently a major directing influence
of the NO in addition reactions.   In an attack trans to NO, for
example, in the two $(R)-\eta^5-C_5H_5Mo(NO)(CO)(allyl)^+$ isomers the production of two diastereoisomers is possible.

O-N-Mo-C-O   O-N-Mo-C-O   **3**

CH$_2$Nu   NuH$_2$C

(R,R)   **(R,S)**

That is, if the attack were directed trans to NO in both isomers,
then a different product would arise from each.   As shown below,
attack trans to NO in the exo isomer would produce the stereomer
of RR configuration whereas attack trans to NO in the endo isomer
would produce the RS configuration.   [The relatively free rotation
about the metal-olefin bond does not invert chirality at C-2 of
the olefin.]

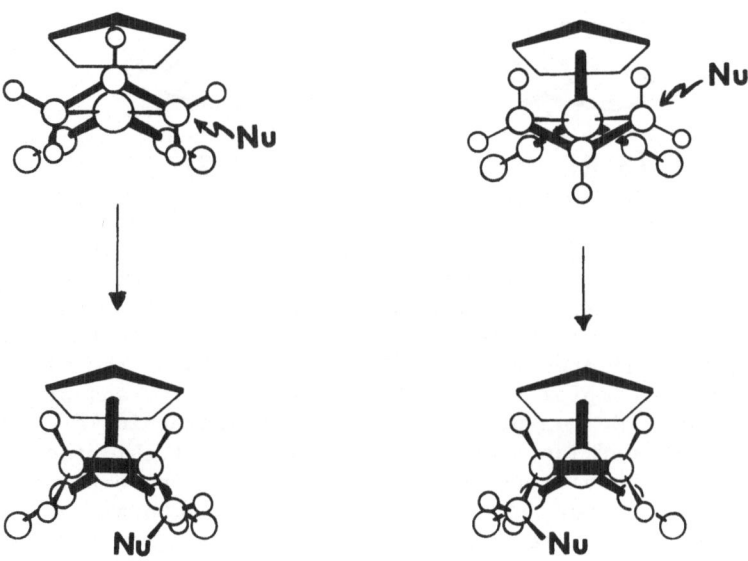

However, to our surprise it makes no difference whether one
starts with the exo or the endo allyl isomer because the same
diastereomer is produced in either case.[7] Thus, the endo and exo
isomerism makes no difference in the product distribution. In the
long run, this is fairly convenient because it simplifies synthetic pathways using these reagents. The nature of the additions
have been determined through extensive use of NMR spectroscopy by
synthetic correlations between the molecules and by x-ray
crystallography. We now believe that attack of the nucleophile
occurs cis to NO in the exo isomer and trans to NO in the endo
isomer. This assumption is consistent with all of the data which
we have at this time.

Once there are substituents attached to the allyl group, further complications could possibly arise. In addition to endo-exo
isomers one must consider the possibility of substituents being
located either cis or trans to the nitrosyl. Since rotation of
the allyl group in the precursor does not affect the product distribution, it is convenient to discuss isomers using the chirality
designation appropriate to the various centers of the molecule.
For example, we now have a chiral metal center at the molybdenum
from the Cp, CO, NO, and allyl moiety as well as a chiral center
at the terminal carbon atoms of the substituted allyl group. Thus,
an (S) metal center with an endo crotyl group would have SR and SS

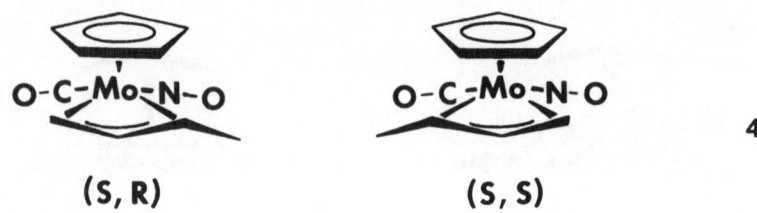

(S, R)   (S, S)   4

isomers. The advantage of this nomenclature is that it is invariant
on rotation of the allyl group. Thus, in the reaction of the
crotyl Mo(CO)$_2$Cp complexes with NO$^+$ there could be as many as
four different isomers formed. On addition of NO$^+$ to the crotyl
derivative we find predominantly one isomer. Thus, the nitrosyl
addition to the complex is over 90% stereoselective and produces
an endo isomer. We can prepare a different isomer in large yield
by using the silver ion promoted attack of crotyl chloride on the
molybdenum dicarbonyl nitrosyl.[8] In this case, one obtains the
exo isomer. At either case, one has a facile synthesis of either
an endo or an exo complex stereoselectively in high yield. A minor
complication in this process is that there is a small percent (~ 3%)
of the anti-methyl allyl compound produced as well. Regardless,

one is easily able to prepare a sample which is over 90% one isomer
before crystallization in these reactions.

The presence of the methyl group in the complex suggests the
probability of regioselectivity based upon the preference for
attack either at the substituted or unsubstituted end of the crotyl
moiety. This occurs in species which do not have asymmetry at the
metal atom.[9-11] We find however than the NO[+] appears to control the
selectivity in the complex. Thus in the crotyl case, we find that
the isomer with which we start adds the nucleophile such that one
produces a trans olefin and the reaction is stereospecific to produce one stereomer. The same isomer is produced whether the endo
form or the exo form is used. This, however, has been carried
out with only two of the four possible isomers. That is, on the
basis of the crystallographic work and inferences therefrom we
believe that the reaction of the endo-cis (RR-SS) and the exo-trans
(RR-SS) produce the isomer that we observe.

This reaction produces only chiral centers at substituents
attached to the metal. By using the 1,3 methyl allyl derivative
one produces a chiral center at a carbon atom in the process of
the attack. Again in this situation we find a stereoselective
attack to produce a single isomer and that isomer is shown in the
figure below. In collaboration with Dr. Adams we have determined
the structure of this molecule and find that the structure is consistent with our notions that attack occurs preferentially cis
to NO in the exo isomer.

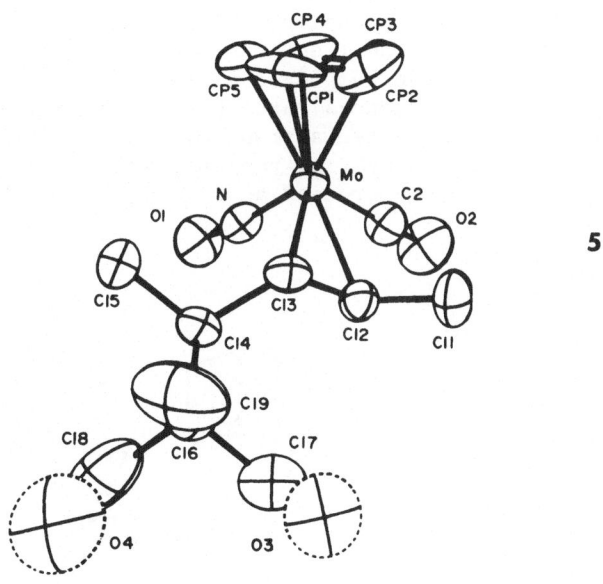

5

The stereochemistry observed in the crystal structure of this compound allows one to determine the direction of the attack and relate the stereochemistries of reactions of precursors.   The position of the new $C(CH_3)_2CHO$ substituent is consistent with attack cis to NO in the exo complex.   The orientation of the attack is indicated by the relative chirality of the carbon atom on which the attack has occured.   The most straightforward mechanism which would provide this stereochemistry is attack on the allyl on the side opposite to the metal (i.e., trans to the metal). Thus, our view of the reaction is that the nucleophile does not attack the metal directly, but on the fact of the allyl away from the metal and at the carbon atom which is cis to NO in the exo isomer.[13]

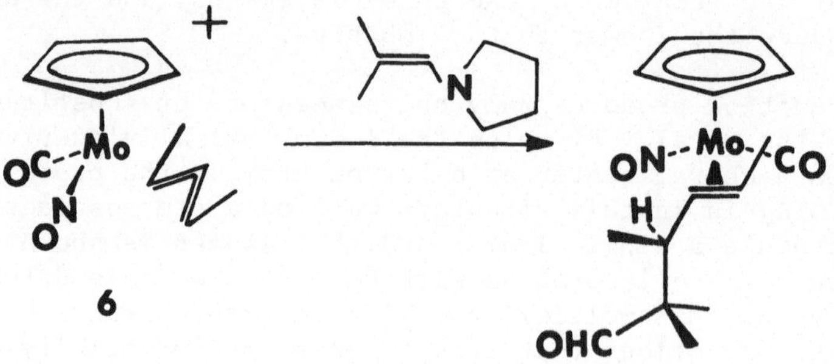

Further evidence is provided by the addition product of the 1,3-syn-anti-dimethylallyl complex.   Displacement of the olefin from this complex yields the identical product by NMR as displacement from 5.   However, the complexed olefin has a distinctly different pattern of resonances than those in 5 indicating that they are diastereomers.   Since a trans olefin is obtained, attack is required on the carbon bearing the anti-methyl substituent. Hence, this reaction would proceed as follows.

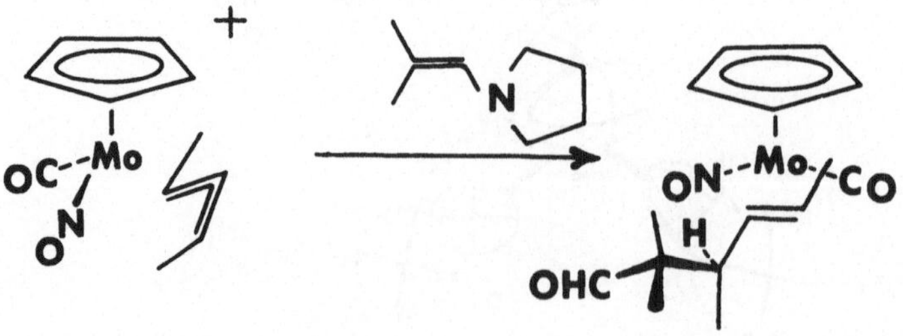

The origin of the preference for attack trans to NO in one isomer and cis to NO in the other is unclear. Fundamental differences in charge distribution in the ligand in the two conformations probably contribute extensively. Furthermore, the crystal structure of 5 shows a major deviation of the olefin from the horizontal orientation. In fact it is nearly parallel to the bond between the metal and the carbonyl carbon. Thus, considering the transition states in the attacks observed, the incipient olefins are nearly oriented in a stable position; whereas, attack at the opposite ends produces an incipient olefin in an unstable orientation. This latter source of the preference appears to be the most important and we are currently designing experiments to test this interpretation.

These results suggest that from the point of view of asymmetric induction and catalysis, stereochemistry can be controlled by what are apparently purely electronic effects. Thus in the future, the design of catalysts and asymmetric syntheses may profit from a greater emphasis on electronic factors.

Acknowledgement: I wish to gratefully acknowledge the assistance of Alan Rosan in the synthetic work and Richard Adams and Daniel Chodosh in the crystallographic work. This work was supported in part by the National Science Foundation Grant No. CHE77-14943.

Notes and References

1.   F. A. Cotton and G. Wilkinson, Advanced Inorganic Chemistry,
        3rd ed., Interscience Publishers, 1972, pp. 789-793.

2.   P. Pino, G. Consiglio, C. Botteghi, and C. Salomon, Homogeneous
        Catalysis-II Advances in Chemistry Series 132, ed. by
        D. Forster and J.F. Roth, American Chemical Society, 1974,
        pp. 295-323.

3.   J. W. Faller and B. V. Johnson, J. Organometal. Chem., 88,
        101 (1975).

4.   J. W. Faller and A. M. Rosan, J. Am. Chem. Soc., 99, 4858 (1977).

5.   J. W. Faller and B. V. Johnson, J. Organometal. Chem., 88, 101
        (1975).

6.   J. W. Faller and A. M. Rosan, J. Am. Chem. Soc., 98, 3388 (1976).

7.   The rate of endo-exo isomerism in the allylnitrosyl precursor
        is much slower than the rate of the addition reaction.

8.   N. A. Bailey, W. G. Kita, J. A. McCleverty, A. J. Murray,
     B. E. Mann, and N. W. J. Walker, J. Chem. Soc. Chem.
     Commun., 592 (1974).

9.   A. J. Pearson, Tet. Let., 3617 (1975).

10. T. Whitesides, R. W. Arhart and R. W. Slaven, Jr., J. Am. Chem.
     Soc., 95, 5792 (1973).

11. B. M. Trost and R. J. Fullerton, J. Am. Chem. Soc., 95, 293
     (1973).

12. J. W. Faller, A. M. Rosan, R. D. Adams, and D. F. Chodosh,
     unpublished.

13. This does not preclude the possibility of a prior equilibrium
     with intermediates in which attack has occured at other
     positions.

σ-π REARRANGEMENTS OF ORGANOTRANSITION METAL COMPOUNDS

IN HOMOGENEOUS CATALYSIS

Minoru Tsutsui* and Arlene Courtney

Texas A&M University

Department of Chemistry, College Station, Texas   77843

Since the isolation of triphenyltris(tetrahydrofuran)chromium(III)[1] as an intermediate in the synthesis of Hein's π-arene-chromium complex,[2] many examples of an important class of organometallic reactions, σ-π rearrangements, have been reported. These rearrangements are of both theoretical and practical interest since they appear to be very important in homogeneous catalytic processes, and therefore, a knowledge of the physical and chemical factors which govern this phenomenon is essential.

A $\sigma \rightleftarrows \pi$ rearrangement is a reaction in which an organic group σ-bonded (h^1) to a metal becomes π-bonded (h^n) to the metal. The reverse of this process, $\pi \rightleftarrows \sigma$ ($h^n \rightarrow h^1$), also occurs. It should be noted that the term rearrangement as utilized here has a somewhat different connotation than the normal view that a rearrangement involves the migration of a group from one atom to another in the same molecule with no net change in the number of constituent atoms in that molecule. The $\sigma \rightleftarrows \pi$ rearrangement being considered

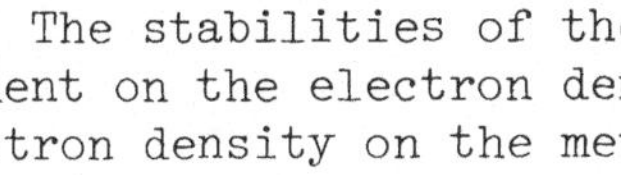

$$\overset{\text{W}}{\underset{\text{A--B}}{|}} \longrightarrow \overset{\text{W}}{\underset{\text{A--B}}{|}}$$

here concerns itself with a change in a ligand to metal bonding mode which may or may not involve a net change in the number of atoms in the molecule. These two bonding modes are illustrated in Figure 1.

The stabilities of the two bonding modes will be somewhat dependent on the electron density which is present on the metal. The electron density on the metal, in turn, is determined to a large

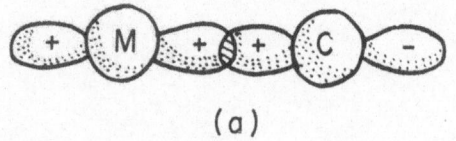

(a)

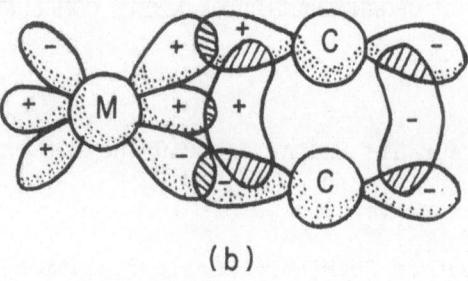

(b)

Figure 1. Metal-ligand bonding in (a) σ-complex; (b) π-complex.

extent by the other ligands which are present in the complex.[3-7]
Ligands which form strong dative π-bonds with the metal exert an
electron-withdrawing effect allowing the metal to better accept
electrons. Less electron density will be present for backdonation
thus stabilizing the σ-bonding mode. Similarly, ligands possessing
little dative π-bonding capacity will leave more electron density
on the metal stabilizing the π-bonding mode. These effects are
summarized in Figure 2. From the molecular orbital standpoint, π-
bonding ligands (L_1) that are trans to the ligand of interest (L_2)
stabilize filled and partially filled metal d-orbitals which might
otherwise interact with π-orbitals of the appropriate energy and
symmetry present on L_2. Thus, the net effect is stabilization of
the σ-bonding mode of L_2. Conversely, if L_1 is not capable of providing such stabilization (i.e. it is non π-bonding), π interaction
with L_2 is enhanced. Generally, this trans-effect dictates the
following stabilities

$$\sigma,\pi \quad \simeq \quad \sigma,\sigma \quad > \quad \pi,\pi.$$

The instability of "π,π" complexes is evidenced by the rapid isomerization of trans-$(h^2$-$C_2H_4)_2PtCl_2$ to the cis-isomer[8] (Figure 3).

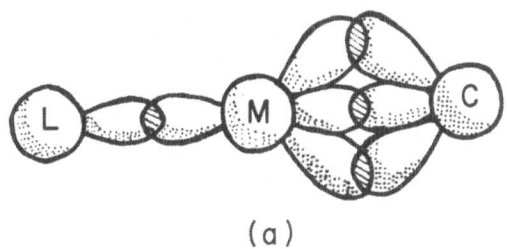

(a)

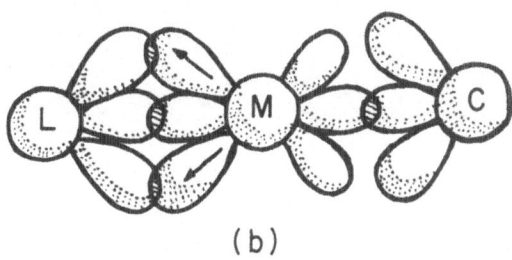

(b)

Figure 2. Effect of ligand L on metal orbitals when L is (a) non π-bonding (b) π-bonding.

Since the relative stability of the two species is dependent on the polarization of the metal center, it may also be correlated to the concept of "hard" and "soft" acids and bases.[9] If σ-organic groups are "harder" due to the small polarizability of the localized σ-orbitals than π-groups having the delocalized π-electron cloud, "hardening" the metal by placing electron-withdrawing groups on it will favor the σ-complex. Conversely, "softening" the metal with non-electronwithdrawing substituents will favor the π-complex.

FACTORS WHICH INDUCE REARRANGEMENT

Most σ-π rearrangement reactions are initiated in one of two ways -- reaction at the metal center or reaction on the ligand itself.

$$\textit{trans-}(h^2-C_2H_4)_2PtCl_2 \xrightarrow[-10\,^\circ C]{\text{ether}} \textit{cis-}(h^2-C_2H_4)_2PtCl_2$$

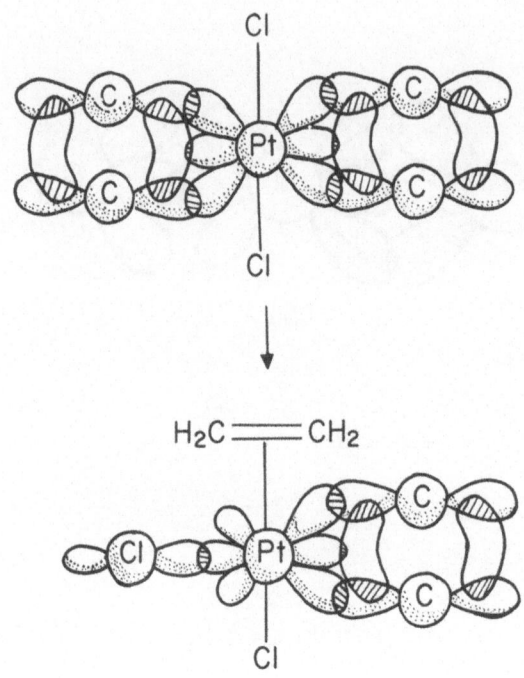

Figure 3

Reaction At the Metal Center

It is the change in electron density in the coordination sphere caused by addition of a ligand which acts as the initiator. Loss of a ligand which is capable of dative π-bonding generally results in a σ-π rearrangement while addition of such a ligand causes the reverse rearrangement. If loss of a ligand leaves a metal coordinatively unsaturated, and a rearrangement will satisfy the coordination requirement by donation of more than one electron pair, rearrangement will occur.

When σ-allyl complexes of h^5-cyclopentadienyltricarbonylmolybdenum[10] and tungsten[11] are irradiated with ultraviolet light, a σ-π rearrangement occurs yielding the π-allyl complex. Upon addition of carbon monoxide a σ-π rearrangement is observed.

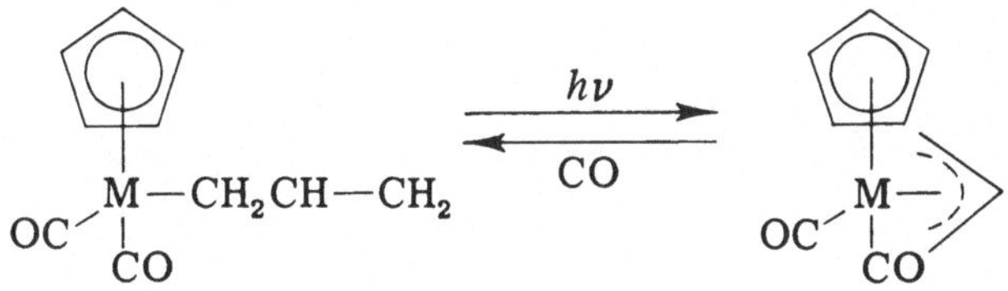

A similar rearrangement is observed with allyl complexes of cobalt cyanides.[12]

Some interesting results were obtained when the system $L_2Rh(h^5-C_3H_4R)Cl_2$ L = C_6H_5P, C_6H_5As or $C_6H_5Sb^-$ and R = H or CH_3 was reacted with sulfur dioxide, ethyelene and carbon monoxide.[13]

Reaction with sulfur dioxide yields a σ-allylic rhodium-sulfur dioxide complex (1).

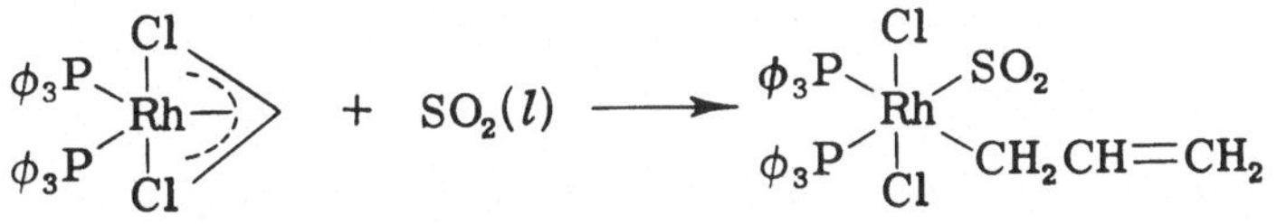

$$(1)$$

Reaction of the methallyl rhodium complex with ethylene leads to formation of a mixture of butenes with isobutene being formed from the methallyl group. The starting material is converted initially to an asymmetric π-methyallyl ethylene complex (2) which then reacts with another mole of ethylene to give a σ-methallylbis(ethylene) complex (3). Solvent reaction may form an hydride (4) which upon readdition of the ligand, L, yields isobutene, ethylene and a rhodium-ethylene complex (Figure 4). Reaction with carbon monoxide yields methallyl chloride and a rhodium-carbonyl complex.

σ-Cyclopentadienyl iron derivatives have been reported as

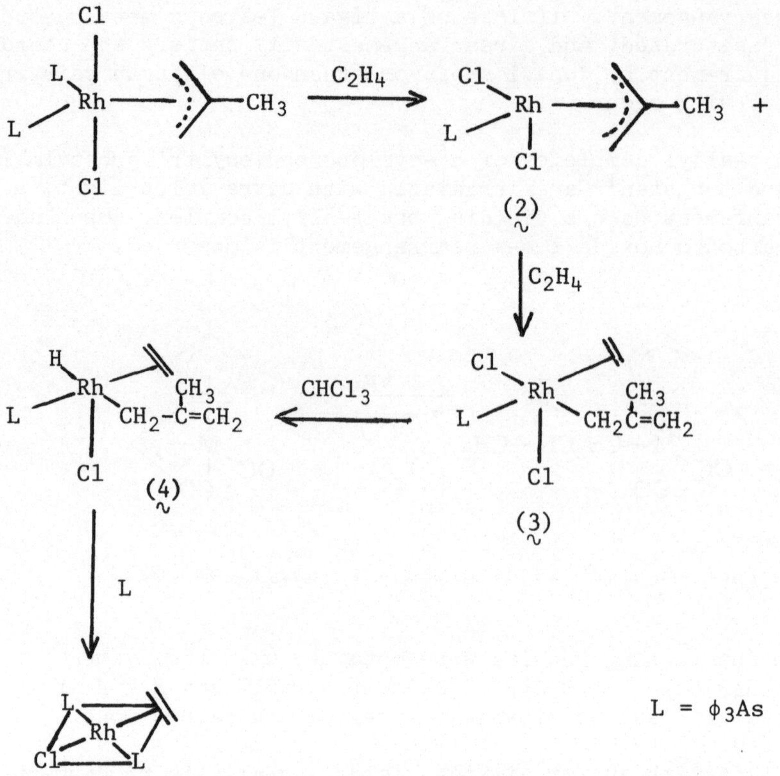

Figure 4.   Reaction of $(\phi_3As)Rh(h^3-C_4H_7)Cl_2$ with ethylene.

intermediates in the formation of ferricenium chloride and ferro-cene.[14] The reaction of sodium cyclopentadienide with iron(III)-chloride in a 2:1 molar ratio yields a σ-dicyclopentadienyl com-plex (5). This complex can be rearranged either thermally at -50°C or chemically by addition of organic solvents such as ether or

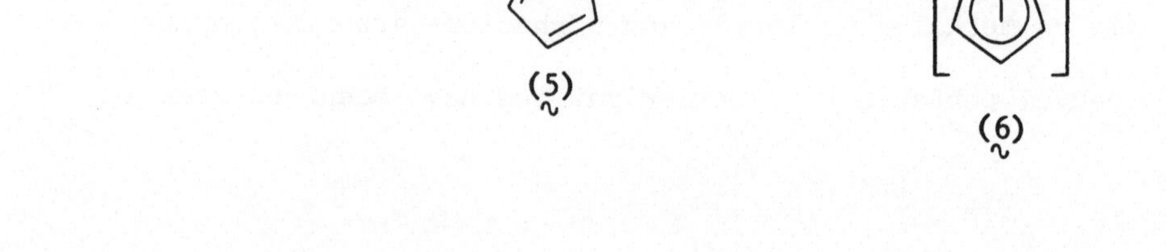

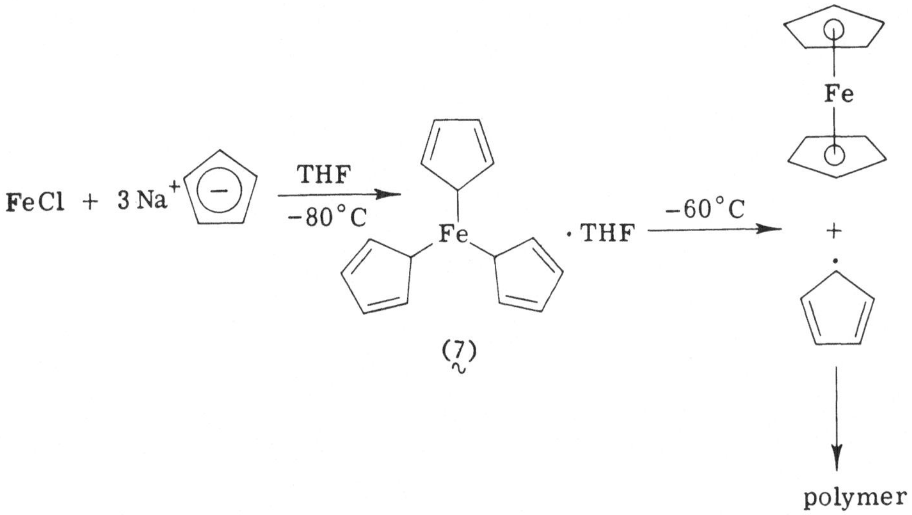

Figure 5.   Formation of Ferrocene.

pentane to ferricenium chloride (6). This rearrangement is initiated by loss of the labile THF ligand which leaves an open coordination site on the metal.

When the reactants are mixed in a 3:1 ratio, an intermediate (7) appearing to contain three σ-cyclopentadienyl groups is formed at -80°C. Upon allowing the temperature to rise to -60°C, rearrangement occurs which yields ferrocene and a polymeric cyclopentadiene. The proposed mechanism for this reaction is similar to that proposed in the formation of Hein's π-arene chromium complex[2] (Figure 5).

An irreversible π-σ rearrangement of cyclopentadienyl ligands has been observed in the reaction of titanocene dichloride and dimethyl sulfoxide.[15]

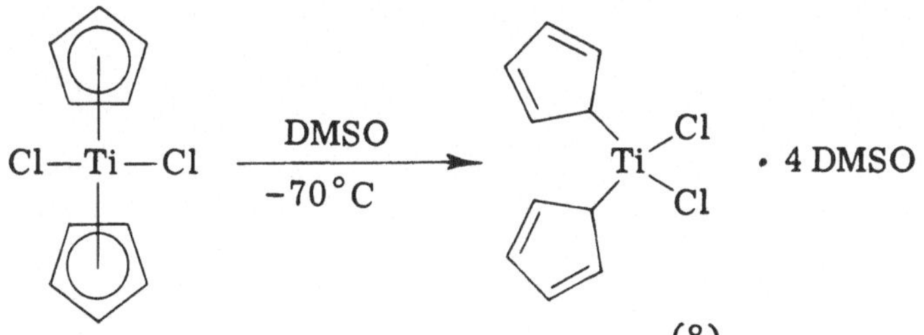

Evidence for such a rearrangement product as (8) is found in its
nmr spectrum and in isolation of a maleic anhydride addition product (9).

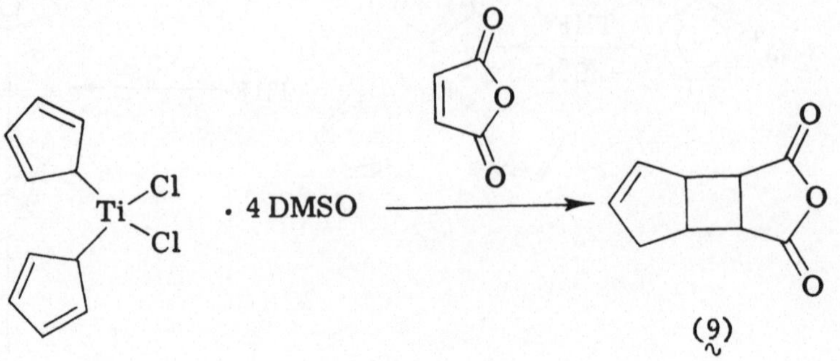

(9)

Reaction On the Ligand

Creating or removing unsaturation on the ligand can initiate
a σ-π rearrangement. Hydride abstraction and protonation reactions
are perhaps the most well elucidated examples.

Some transition metal σ-alkyl complexes with the alkyl group
being ethyl, n-propyl, or isopropyl react with the triphenylmethyl
cation forming an olefinic complex[16] (10).

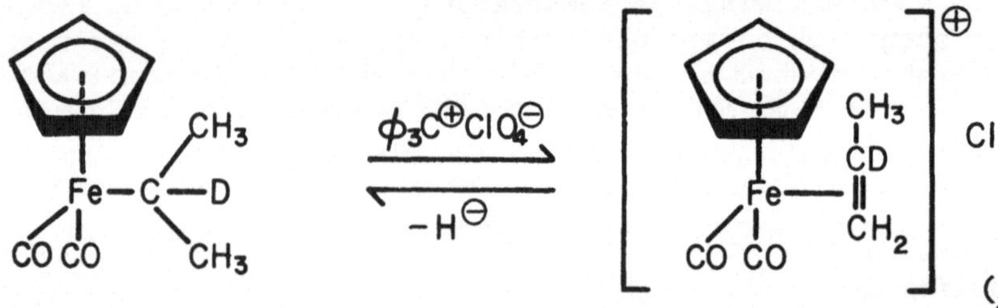

(10)

Other transition metal σ-alkyl complexes such as cyanoalkyl
complexes in which the alkyl moiety has an unsaturated group[17] undergo reversible protonation to form olefinic complexes (11).

Not all σ-π rearrangements require chemical initiation. Some
metal complexes exist in a dynamic equilibrium between the σ and π
forms.

$$\begin{array}{c} \text{Fe}-\overset{\displaystyle CN}{\underset{\displaystyle R}{C}}-H \end{array} \quad \underset{-H^{\oplus}}{\overset{+H^{\oplus}}{\rightleftharpoons}} \quad \left[\begin{array}{c} \text{Fe} \end{array} \right]^{\oplus} \qquad (11)$$

NMR spectroscopy has shown that the allyl ligand in both platinum and rhodium allyl complexes at room temperature in deutero-chloroform solution has all terminal hydrogens magnetically equivalent.[18] This phenomenon may result from an interchange of the four allyl protons via a short lived σ-allyl intermediate or transition state. As seen in Figure 6, for such a rearrangement to take place a rotation around the C(1)-C(2) bond occurs interchanging protons 1 and 2 concurrent with a rotation around the C(2)-C(3) bond interchanging protons 3 and 4.

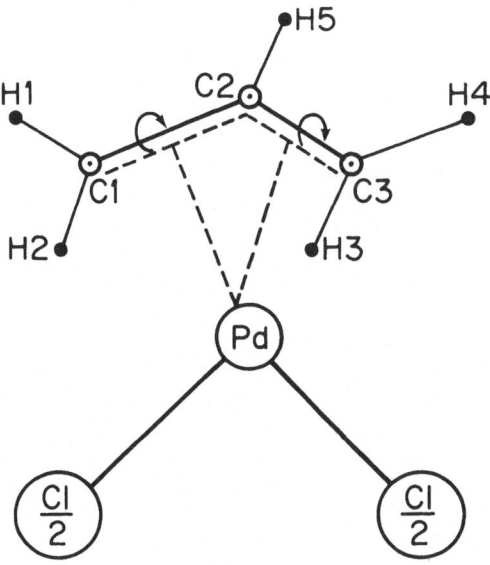

Figure 6. Rotation necessary for interchange of allyl protons in the idealized structure of one half of the dimer [(π-allyl)PdCl]$_2$.

Recently, the first organolanthanide complexes containing an allyl ligand $[(h^5-C_5H_5)_2LnC_3H_5$ where Ln = Sm, Er, Ho$]$ were reported.[19]

$$(h^5-C_5H_5)_3LnCl + C_3H_5MgBr \xrightarrow[-78°C]{THF-ether} (h^5-C_5H_5)_3LnC_3H_5 + MgBrCl$$

The bonding scheme in these complexes is very interesting. Spectral data indicate h^3-bonding in these lanthanide complexes in contrast to the h^1-bonding which is observed in the analogous actinide complex $(h^5-C_5H_5)_3UC_3H_5$.[20]

In transition metal complexes where the 18 electron rule is essentially obeyed, the bonding mode of the allyl ligand is mainly determined by the electronic requirements of the metal. However, this does not appear to be the case in the rare earth complexes. Instead, the bonding mode may be largely influenced by steric requirements. In the uranium case the coordination site necessary for π-bonding is highly constrained relative to that in the lanthanide complexes.

Temperature dependent nmr studies have shown that the uranium allyl complex is fluxional. This fluxionality has been proposed as a σ ⇄ π ⇄ σ interconversion in which the h^3-configuration lies approximately 8-9 kcal/mole higher in energy than the h^1-configuration.[20] Conversely, the lanthanide complexes do not appear to easily undergo π-σ rearrangements as evidenced by the fact that addition of tetrahydrofuran, which coordinates to the complex, does not cause rearrangement. Apparently, the steric requirements are not such that coordinative saturation is reached.

Thus, preliminary results indicate that in lanthanides and actinides σ-π rearrangements are dominated by a combination of electrostatic and steric requirements whereas in the other transition metals electrostatic considerations predominate.

Mechanistic Aspects

Although no one general mechanism has been established for the σ-π rearrangement, it appears that intermediates resulting from transition metal-β-interactions are often involved.

Even when a transition metal is coordinatively saturated, either filled or unfilled d-orbitals of suitable energy and orientation are available which may interact with orbitals at the β-position on the ligand. The existence of such interactions has been established by spectral studies.[21] In the systems where the β-position on the ligand is unsaturated, metal d-orbitals of suitable

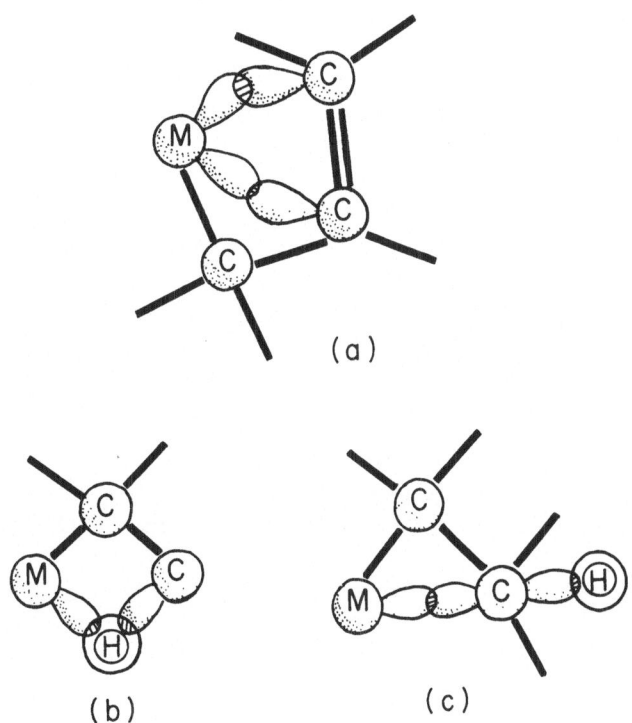

Figure 7. Possible Transition metal β-interactions (a) ligand with unsaturated β-carbon; (b) and (c) ligand with saturated β-carbon.

symmetry may be found which can weakly overlap with the π-orbitals of the β-substituent. If the β-position is saturated, metal d-orbitals may overlap with either one of the sp^3 carbon-substituent orbitals or an empty antibonding sp^3 lobe of the β-carbon. These interactions are summarized in Figure 7. The overall result of the β-interaction is the formation of an electron deficient β-atom or a cyclic system of orbital overlap having less electrons than orbitals. Thus, in going from the σ-species to the π-species, the electron deficient transition state will be stabilized by electron donation from the metal forming the π-bond and partial oxidation of the metal as shown in Figure 8. The reversible insertion of the

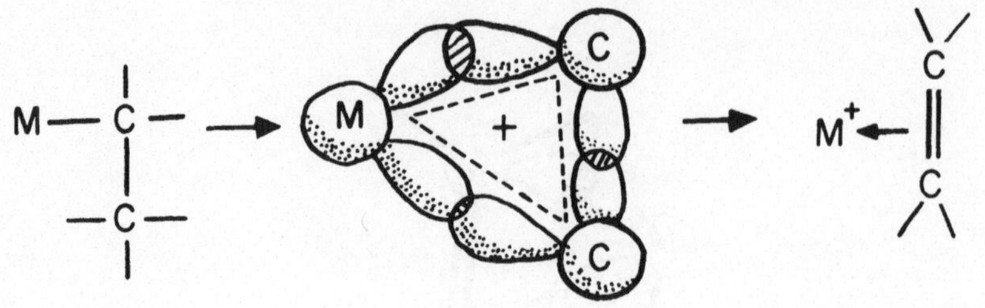

Figure 8. Possible route from σ-bonded ligand to π-bonded ligand.

$$trans\text{-}[PtHCl(PEt_3)_2] \xrightarrow[180°C]{\substack{CH_2=CH_2 \\ 90°C,\ 40\ atm}} trans\text{-}[PtClEt(PEt_3)_2]$$

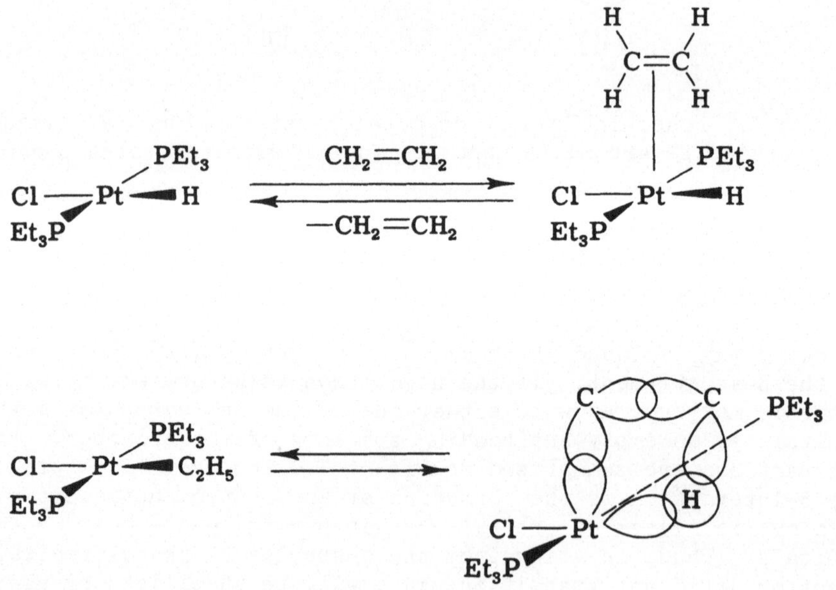

Figure 9. Insertion of ethylene into platinum-hydrogen bonds.

ethylene into the platinum-hydrogen bond[22,23] may proceed by this type of mechanism (Figure 9). Such a process can be viewed as a concerted, pericyclic type reaction.

σ-π Rearrangements and Catalysis

Woodward and Hoffmann have laid down some fundamental basis for the theoretical treatment of all concerted reactions -- "orbital symmetry is conserved in concerted reactions."[24] The mechanism of the concerted σ-π rearrangement $[\pi_s^2 + \sigma_s^2]$ would be strictly forbidden by the Woodward-Hoffmann rules in a metal-free system. In an effort to deduce the mechanism of "forbidden" metal catalyzed reactions, the transformation of reactants to products has been treated using third order perturbation methods,[25] triplet stability criterion[26] coupled with MO consideration,[27,28] and symmetry conservation arguments.[29] Metal d-orbitals seem to play a major role in overcoming the "forbiddenness" of these catalytic reactions. It is noteworthy that as a rule non-transition metal complexes are quite catalytically inactive.

The migration or insertion of coordinated unsaturated bonds into metal-ligand bonds,

$$M-L \quad + \quad X=Y \quad \longrightarrow \quad M-(X-Y)L$$

where X=Y may be CO, olefin, diene, RNC, RCHO, RCN, SO_2, O_2, etc., and L may be H^-, R^-, OR^-, NR_2^-, H_2O, X^-, etc., is an elementary process common to many catalytic reactions such as hydrogenation, hydroformylation, oxidation, and polymerization. The manner in which the metal can provide a template of atomic orbitals, through which electron pairs of transforming hydrocarbon ligands can interchange and flow into the required regions of space to overcome any "forbiddenness" of reaction, can be viewed by simple molecular orbital pictures. Figure 10 depicts those orbitals which must dominate ligand migration reactions of coordinated olefins.[30] Such a reaction sequence is observed for

$$(\pi-C_5H_5)MoH_2 \quad + \quad H_2C=CHX \quad \longrightarrow \quad (\pi-C_5H_5)_2MoH-CHXCH_3$$

where X = -CN or $-CO_2CH_3$, and addition takes place in a Markovnikov fashion.[31] The regioselectivity of migration can be accounted for by direct interaction of the metal HOMO and ligand LUMO and is consistent with the fact that LUMO's of olefins have the largest amplitude at the terminal carbon. It is also interesting that if L is an alkyl group containing an asymmetric carbon bound to the metal, the migration occurs with retention of configuration.

There are also cases where trans additions occur which are the

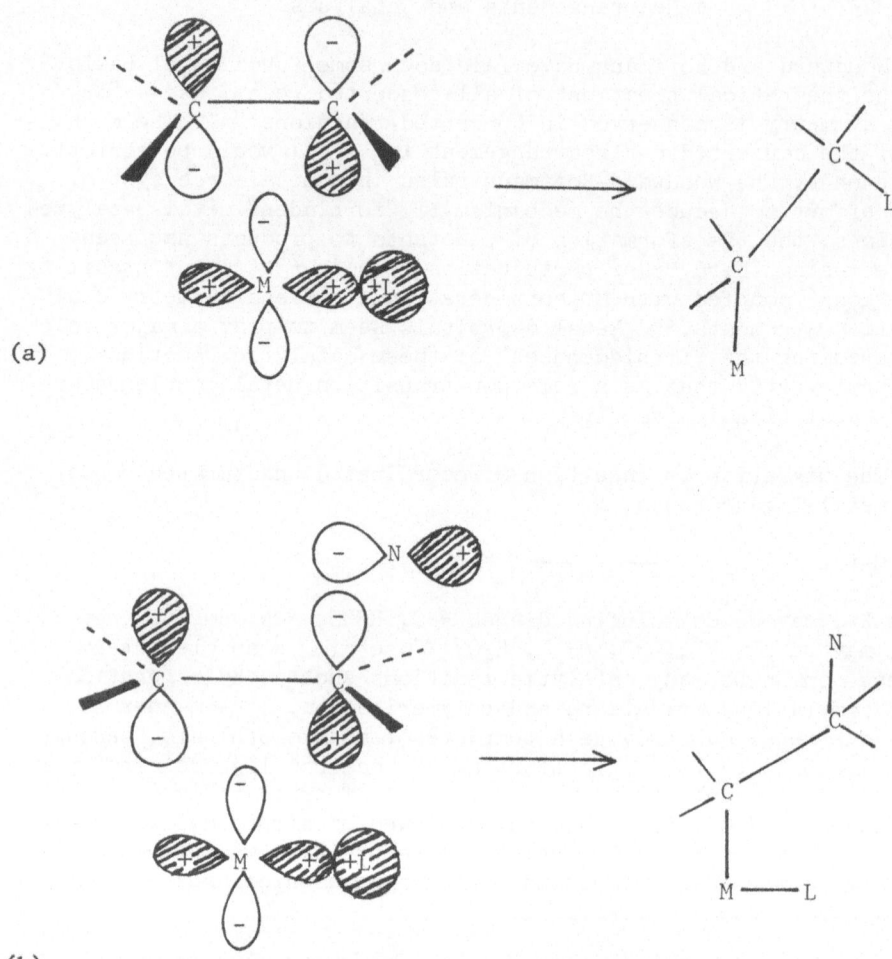

Figure 10.   (a) Ligand migration reaction-cis product; (b) Addition of an exterman nucleophile-trans addition product.

result of addition of an external nucleophile.[32,33,34] (Figure 10).

Figure 11 shows the orbital matching which occurs in systems where the metal has dominant s- or p-bonding. Some reactions catalyzed by these types of metals are shown in Figure 12. It can be concluded that a pericyclic reaction such as the σ ⇄ π rearrangement due to ligand migration will be allowed for metals utilizing p- and d-dominant bonding and forbidden for s-dominant bonding. Since in actuality all metal ligand bonds are hybrids of the s, p, and d orbitals, it can be inferred that the facility of reactions will be dependent on the amount of p- or d-orbital contribution to the hybrid, and the energy of activation is proportional to this contribution. Moreover, since the d- and s-orbital contributions act in opposite directions, it is reasonable to expect that the higher the stability of the d-orbitals relative to the s-orbitals the greater will be the influence of d-orbital bonding.

Members of the cobalt and nickel families have been found to undergo ligand migration reactions with palladium and rhodium showing the greatest ease of reaction.

Pd >> Pt, Ni
Rh > Co > Ir

Taking into account the atomic energy levels of these elements,[35] (Figure 13), it should be noted that both palladium (0) and rhodium(0) have inverted energy levels with the 4d lying below the 5s.

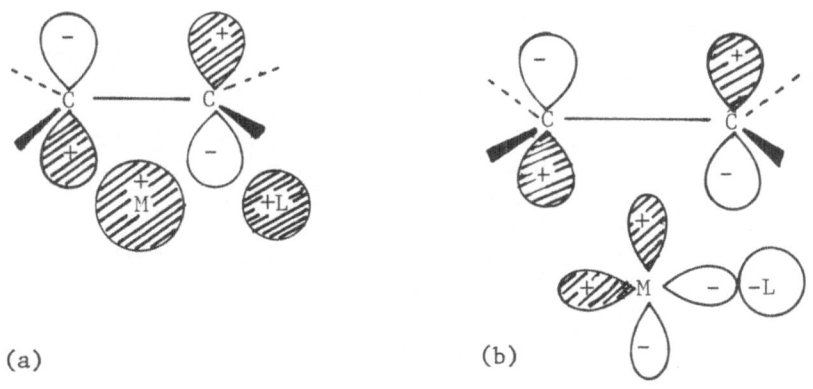

(a)   (b)

Figure 11. (a) Ligand migration for s-bonding metal-forbidden transformation; (b) Ligand migration for p-bonding metal-allowed transformation.

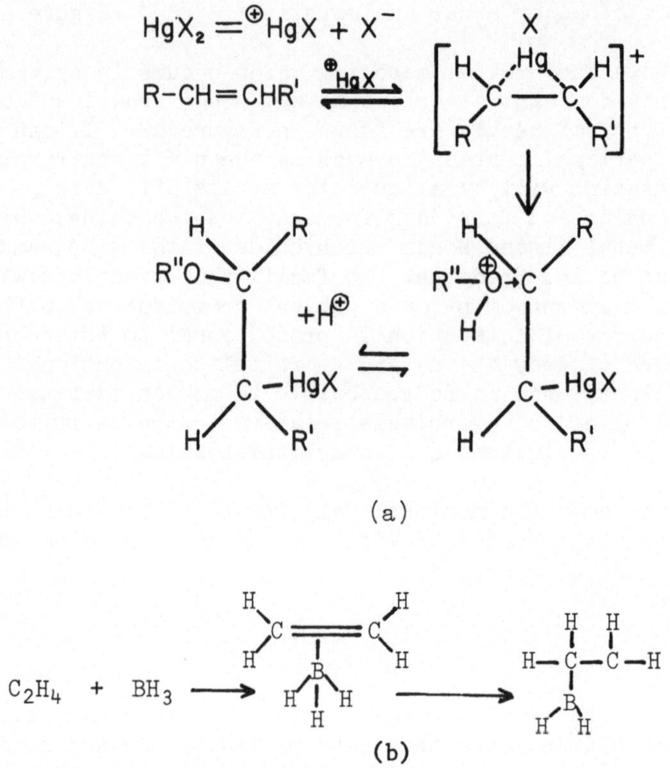

$$HgX_2 = {}^{\oplus}HgX + X^-$$

$$R-CH=CHR' \xrightarrow{\overset{\oplus}{HgX}}$$

Figure 12. (a) Trans addition of an external nucleophile catalyzed by the mercuric ion; (b) Cis hydroboration reaction.

Thus, these elements may exhibit orbital hybrids rich in d contribution. This in part may explain the unusual catalytic properties of these metals.

Recently, much work has been done both experimentally and theoretically in an attempt to better define the role of the σ-π rearrangement in catalytic processes. Two examples will be discussed in detail -- recent developments concerning the Wacker Process of olefin oxidation and the Ziegler-Natta type olefin polymerization.

The Wacker Process: Ions of platinum and palladium, of which the palladium chloride-copper(II) chloride[36,37] Wacker catalyst is an example, have been found to oxidize olefins to carbonyl compounds In contrast, mercury(II),[38] thallium(III),[39] and lead(IV)[40] yield glycols. Two explanations have been advanced. The difference may result from the ability of the transition metals to react through

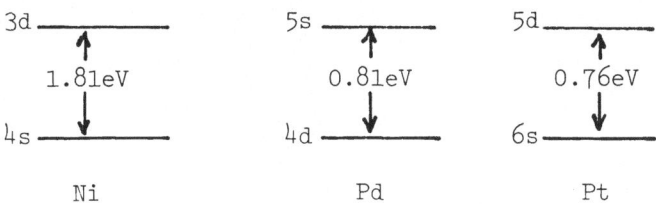

3d ———————   5s ———————   5d ———————
         ↑   ↑   ↑
    1.81eV   0.81eV   0.76eV
         ↓   ↓   ↓
4s ———————   4d ———————   6s ———————

        Ni   Pd   Pt

Figure 13. Energy levels of the nickel family metals.

a π-complex without proceeding through the σ-bonded intermediate
proposed for the main group ions[41] as shown in Figure 12a or that
the same σ-type intermediate decomposes by a different path. It
seems most reasonable that a σ-π type rearrangement does occur.

Experimental kinetic and isotope effect data[42] seem to indicate
that a π-olefin complex (12) is initially formed (Figure 14). This
is followed by the displacement of a chloride ion by water with
subsequent loss of a proton giving a hydroxo species which then rearranges to a σ-bonded β-hydroxy complex (13). From this point it
is unclear exactly how decomposition to the carbonyl product occurs.
One possible path involves a 1,2-hydride shift with direct elimination of the acetaldehyde. Another involves hydride abstraction
from the β-carbon by the metal forming a π-vinyl alcohol complex
(14). The metal hydride then may be transferred to the olefinic
group with elimination of acetaldehyde.

Preparation of h^5-cyclopentadienyldicarbonyl(h^2-ethenol)iron[43]
(15) by Ariyaratne and Green, the first stable π-vinyl alcohol
complex, generated interest in obtaining a static model for investigation of the mechanism of the Wacker Process.

$$Na\left[(h_5-Cp)Fe(CO)_2\right] + ClCH_2CHO \rightarrow (h_5-Cp)Fe(CO)_2(CH_2CHO)$$

$$+ \ NaCl \qquad\qquad \downarrow HBr$$

(15)

$$\left[(h_5-Cp)(CO)_2Fe \leftarrow \| \begin{array}{c} H \ \ OH \\ C \\ C \\ H \diagdown H \end{array} \right] Br \qquad \left[(h_5-Cp)(CO)_2Fe \begin{array}{c} H \\ O \\ CH \\ CH_2 \end{array} \right] Br$$

(15a)   (15b)

NMR 1 : 1.43(1H,s), 1.83(1H,t), 4.62(5H,s), 7.15(2H,d)

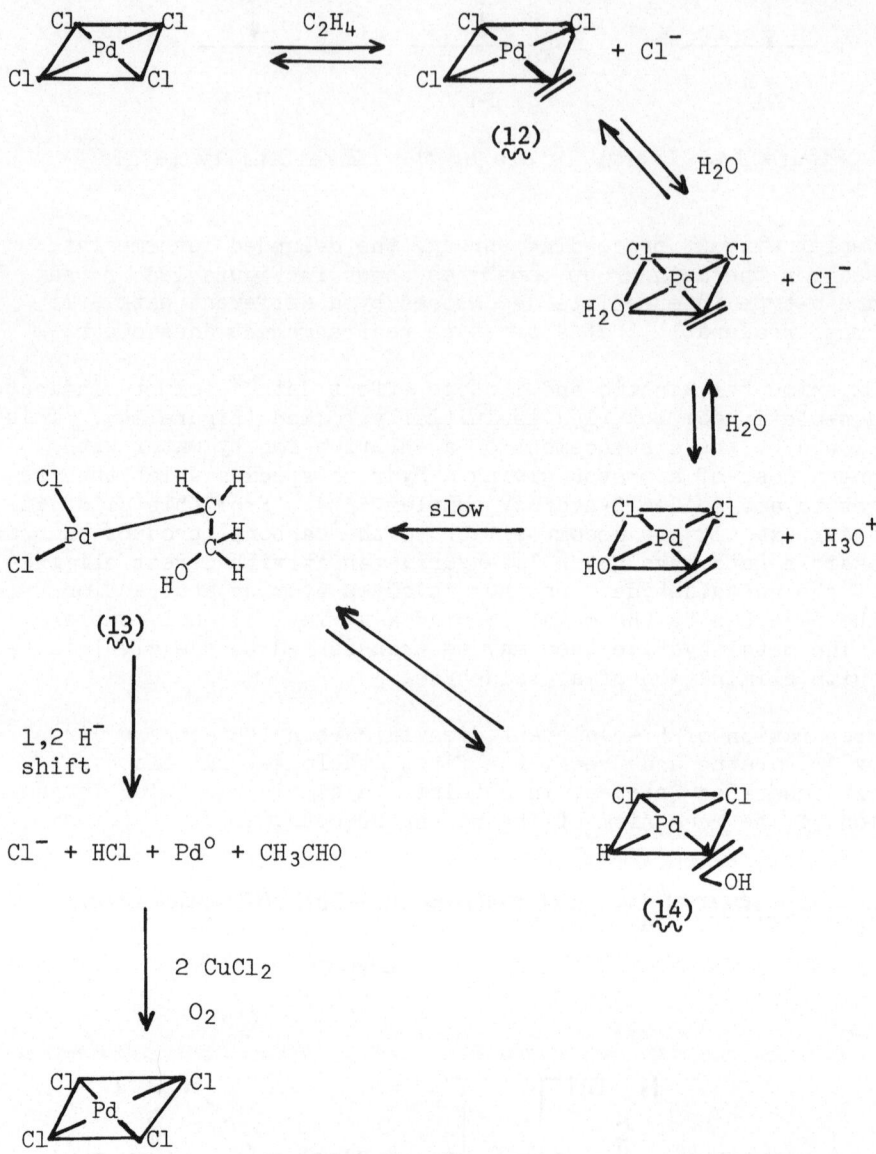

Figure 14.   Possible mechanism for the Wacker Process

The synthesis of 1,3-bis(h^2-ethenol)-2,4-dichloro-μ-dichloroplatinum(II) (16)[44] was reported by Wakatsuki, Nozakura, and Murahashi. However, Thyret[45] found the synthesis to be irreproducible.

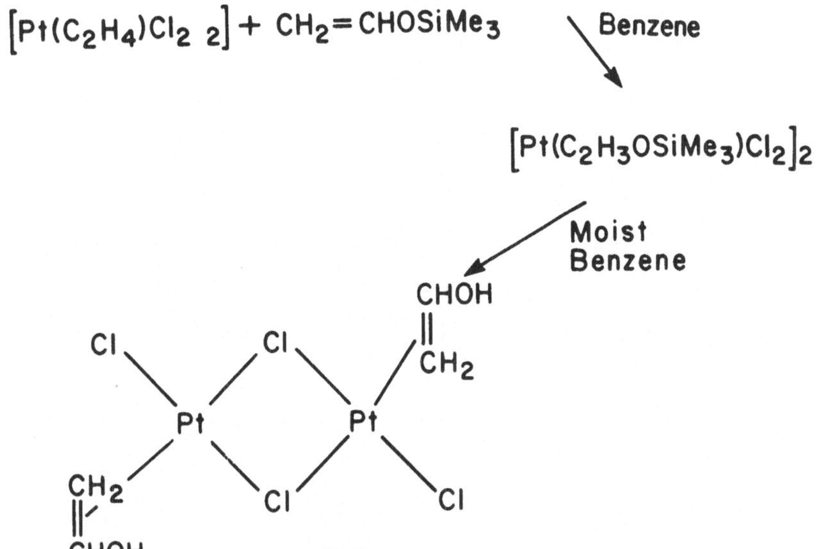
[Pt(C2H4)Cl2 2] + CH2=CHOSiMe3
Benzene
[Pt(C2H3OSiMe3)Cl2]2
Moist
Benzene
CHOH
CH2
Cl
Cl
Pt
Pt
CH2
CHOH
Cl
Cl

(16)

Chloro(acetylacetonato)(h^2-ethenol)platinum(II) (18) was synthesized via a trimethylsilyl ether complex[46] (17).

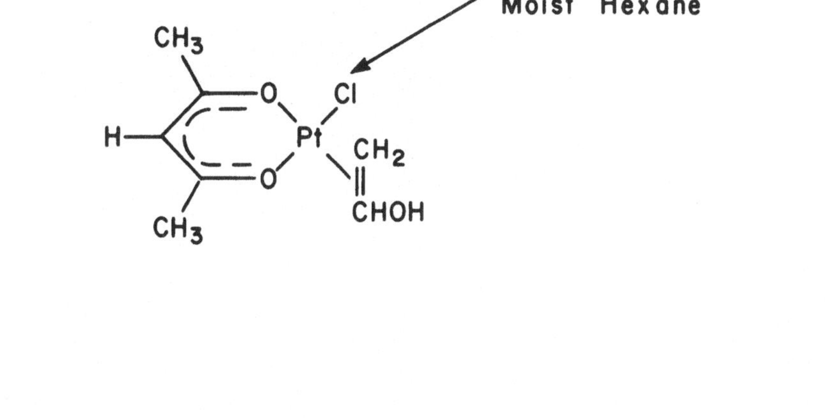
Pt(C2H4)(acac)Cl + CH2=CHOSiMe3
Pt(CH2=CHOSiMe3)(acac)Cl + C2H4
Moist Hexane
CH3
H
O
O
Pt
Cl
CH2
CHOH
CH3

This π-vinyl alcohol complex and the analogous π-propen-2-ol complex have since been more easily prepared through the formation of
the σ-bonded carbonyl complex (19) and subsequent protonation which
causes a σ-π rearrangement.[47] [48]

$$Pt(C_2H_4)(acac)Cl + CH_3\overset{\overset{O}{\|}}{C}R \xrightarrow{\text{KOH}} K\left[Pt(CH_2\overset{\overset{O}{\|}}{C}R)(acac)Cl\right] +$$

(19) $C_2H_4 + H_2O$

$$K\left[Pt(CH_2\overset{\overset{O}{\|}}{C}R)(acac)Cl\right] + HCl \longrightarrow Pt(CH_2=\overset{\overset{OH}{|}}{C}-R)(acac)Cl$$

$$+ \text{ KCl}$$

(19)   (18)

R = H or CH₃

The structure of this compound has been investigated by x-ray crystallography,[49] and the results are shown in Figure 13. All of the
structural features can be interpreted in terms of a bonding model
(21) which is intermediate between a conventional π-olefin complex
(20) and a σ-bonded aldehyde (22) (Figure 15).

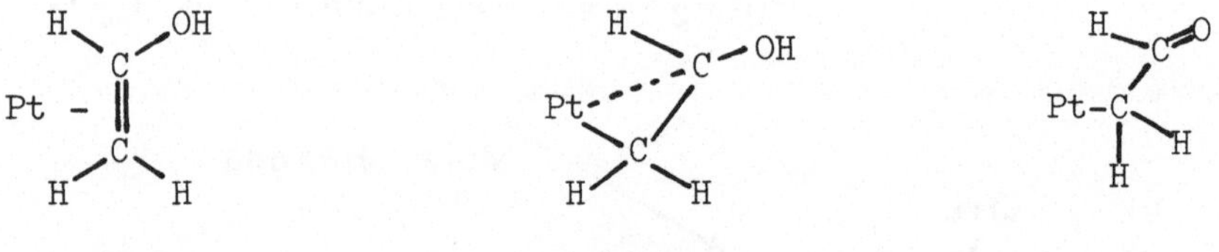

(20)   (21)   (22)

Figure 15

This scheme is supported by the following observations (Figure 16).
    (1) the principal coordination plane of the platinum atom
neither bisects the C-C bond nor includes the methylene carbon atom.
    (2) Both carbon atoms are within bonding distance of the
platinum, but the two Pt-C distances are significantly different.
    (3) The C-C bond length implies that the bond order is inter-
mediate between a single and double bond as is the C-O bond.
Additionally, the nmr spectrum[46,48] for this compound in polar
solvents shows an A_2X pattern for the vinyl protons rather than
the expected ABX pattern such as that found for the π-vinylsilyl
ether complex (17). Both the nmr and x-ray data yield a structure
which is in good agreement with evidence that the π-vinyl alcohol
complex is a moderately strong acid which dissociates giving the
β-oxoethyl complex the equilibrium being rapidly established
$(k > 50 \ sec^{-1})$.[48]

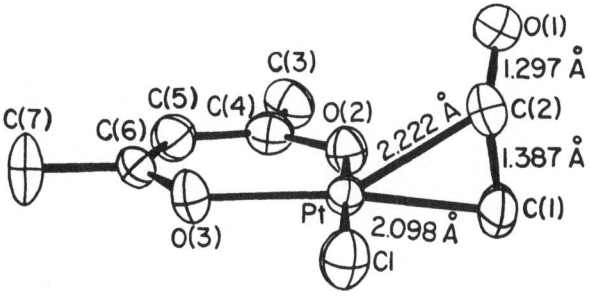

Figure 16. Structure of $(acac)Pt(h^2-C_2H_3OH)Cl$

These studies have shown that, indeed, π-vinyl alcohol complexes do exist and can undergo σ-π rearrangements. However, sufficient evidence has not been presented to verify whether a π-vinyl alcohol intermediate is present in the Wacker Process.

The Ziegler-Natta Process: The discovery of a catalyst system[50] consisting of a transition metal halide such as $TiCl_4$ and a non-transition-metal alkyl such as $AlEt_3$ for the low temperature polymerization of olefins by Ziegler and Natta has led to extensive experimental work in this area.

Although many mechanisms have been proposed, perhaps the most widely favored is that of Cossee.[51,52] Generally, as depicted in Figure 17, the process is seen as proceeding by initial alkylation of the titanium halide by the alkyl aluminum yielding an octahedral titanium complex having a vacant coordination site. An incoming ethylene molecule coordinates to the vacant site which is followed by alkyl migration to the ethylene and a π-σ rearrangement.

Recently, elucidation of the mechanism of this catalytic process has been attempted by using self-consistent field, all-valence electron calculations, in an effort to obtain a quantitative theoretical rationalization of the electronic structures and geometries of the intermediate complexes. Perkins, et. al.[28] provided the initial

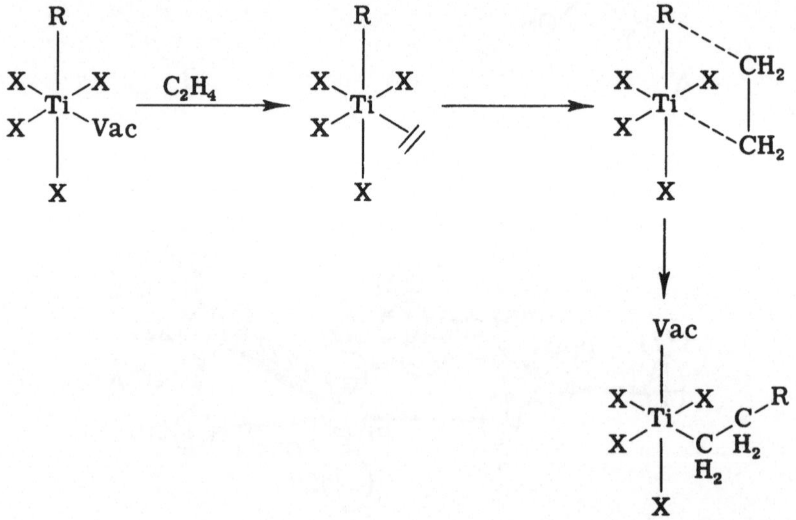

Figure 17. Cossee mechanism for Ziegler-Natta Process.

work in this area on the model of a "soluble" catalyst system sup-
posed to arise from a particular combination of $TiCl_4$ and $MeAlCl_2$.
In general, the results which were obtained were in agreement with
the Cossee mechanism. Some of the main features which appear from
the calculations were (1) at all the intermediate stages of the
process both the alkyl group and the ethylene ligand can remain
bonded to the titanium atom, and therefore, no large bond-breaking
energy is required and (2) a metal d-orbital is utilized as a transfer agent allowing the alkyl group to migrate and link to the
ethylene ligand.

More recently, calculations have been done to investigate the
characteristics of the addition of ethylene to the catalyst, the
nature of the driving force of the chain propagation, and the manner
in which the end products are liberated.[53,54] The system assumed
as model in this study was a particular combination of $Ti(OCH_3)_4$
and $AlEt_3$. This model is based on the actual catalytic system formed
by the reaction of $AlEt_3$ and $Ti(O\phi-pCH_3)_3O-nBu$ which oligomerizes
ethylene to 1-butene and 1-hexene, the structure of which is, however,
still not completely clarified.

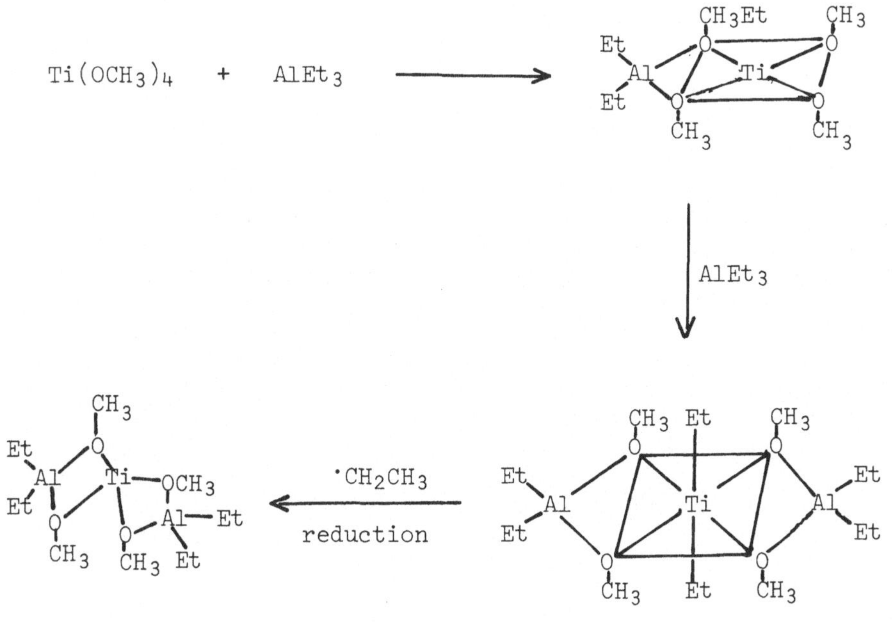

Figure 18

The formation of the initial catalyst complex has been postulated to follow the scheme in Figure 18 which has a 2:1 Al/Ti and
contains a reduced titanium(III) atom. This catalytic structure is
in keeping with the features of the EPR spectrum of the reaction
products obtained from $AlEt_3$ and $Ti(O\phi-pCH_3)_3(O-nBu)$.[53,55] Calculation has shown that the most stable configuration of the above
model would not be the octahedral complex having a vacant site as
the Cossee mechanism proposes, but rather, a trigonal-bipyramidal
complex which appears to be 1.43 eV lower in energy. This model
should have an unpaired electron which is strongly localized in
the highest stable orbital (d_{xz}) which should interact with the
titanium-alkyl bond.

Upon addition of ethylene according to a particular geometry,
the model assumes an octahedral configuration in which the d_{xz} orbital interacts with both the ethylene and alkyl groups. Thus, in
the model the titanium d_{xz} orbital acts as a transfer agent as it
favors movement of the labile alkyl ligand toward the electron rich
α-carbon of the ethylene. Therefore, an appreciable energy bar-
rier is not apparent on the model. The final oligomer liberation
has been attributed to a β-hydrogen abstraction from the alkyl chain
forming a carbon-carbon double bond while breaking the titanium-car-
bon bond. This appears to be a direct consequence of the fact that
as the chain length increases, the alkyl β-hydrogens approach close
enough to the C-2 ethylene carbon to form a pseudohydrogen bond.
Then, the negative character of the ethylene carbon abstracts the
hydrogen forming a carbon-carbon single bond and weakening the
titanium-alkyl bond. Thus, the oligomer is eliminated, and the
catalyst is regenerated. The important stages of the oligomerization are depicted in Figure 19.

According to the above calculations, the driving force of this
catalytic process appears to be the low energy pathway provided by
the titanium d_{xz} orbital. Also, both the propagation and liberation steps are strongly related, i.e. the propagation sequence has
an alternative at each point of liberating an α-olefin or lengthen-
ing the chain. This relationship fits the observed kinetic data
since rate constants for both chain propagation and chain liberation depend on the concentration of active sites, Ti*, and monomer.

$$r_p = k_p [Ti*] [C_2H_4]$$

$$r_1 = k_1 [Ti*] [C_2H_4].$$

The type of mechanism which is utilized by the Ziegler-Natta
process also seems to explain coupling and polymerization reactions
of a wide variety of alkenes by various transition metal compounds.

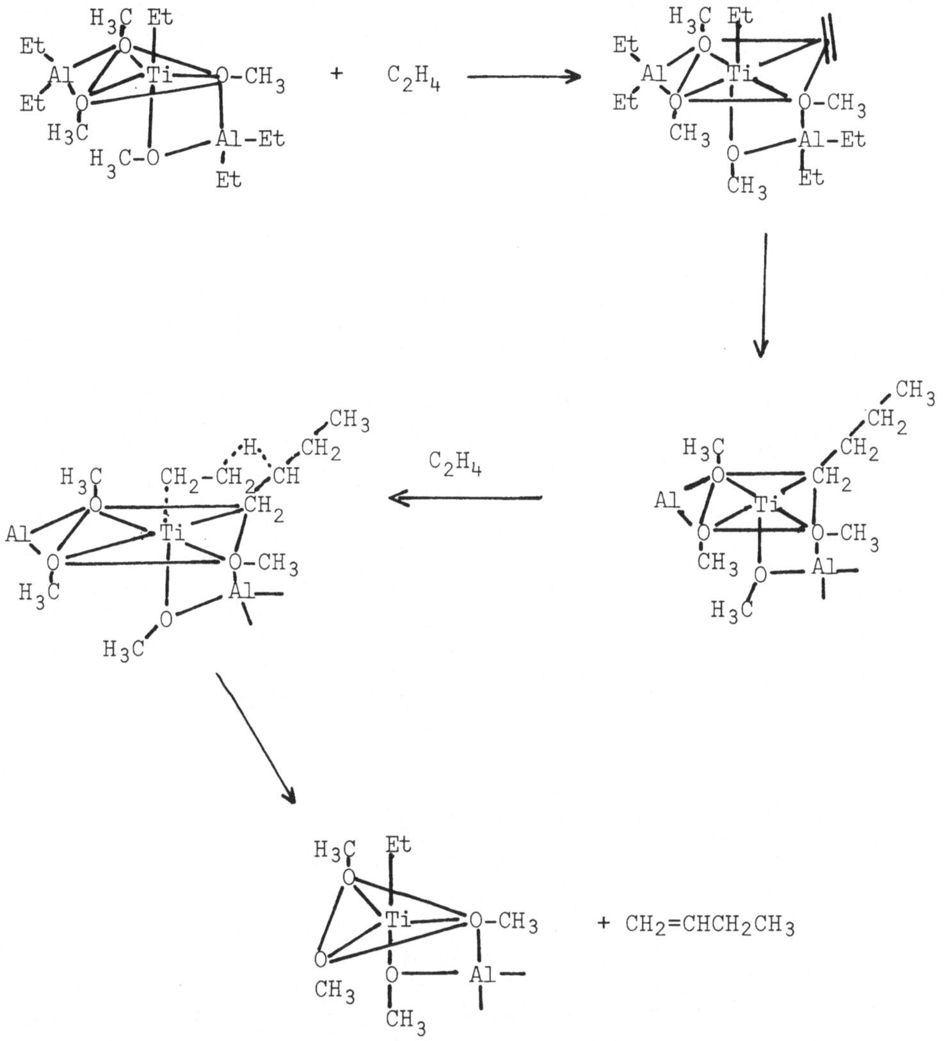

Figure 19

Stereospecific coupling of 1,2-disubstituted vinyl groups is attained by reaction of a vinyl Grignard with $CrCl_3$, $CoCl_2$, $PdCl_2$ and $NiBr_2$ catalysts.[57,58] The stereospecificity of the products appears to result from the selective configuration of the transition metal halide (Figure 20). A cis-coupling product (23) is obtained from the $CrCl_3$ or $CoCl_2$ catalyst which can take on an octahedral geometry (d^2sp^3 hybridization) whereas a trans-coupling product (24)

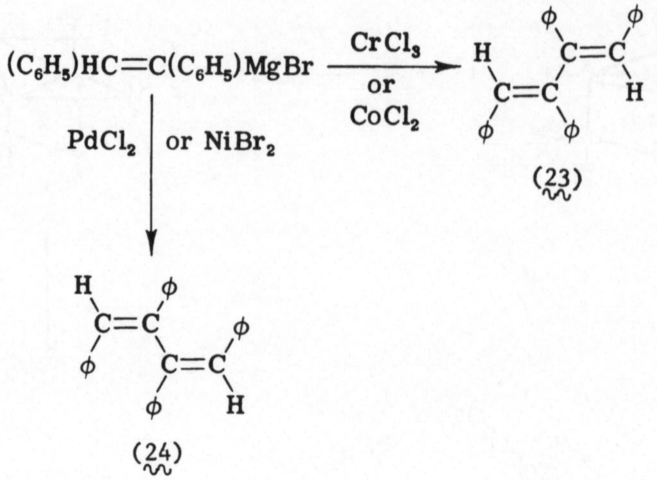

$$(C_6H_5)HC{=}C(C_6H_5)MgBr \xrightarrow[\text{CoCl}_2]{\begin{subarray}{c}\text{CrCl}_3\\ \text{or}\end{subarray}}$$

(23)

PdCl₂ | or NiBr₂

(24)

Figure 20

is obtained from $PdCl_2$ or $NiBr_2$ which can take on a square planar geometry (dsp^2 hybridization). The reaction probably proceeds from a σ-vinyl metal complex which rearranges to a coupled π-butadiene complex.

A "double coupling" reaction occurs when vinyl Grignard and titanium tetrachloride are combined.[59] One possible mechanism is shown in Figure 21.[60]

Activation of Small Molecules

One area which has recently become of interest is the activation of small readily available molecules such as N_2 and O_2 by metal complexes.

The activation of dinitrogen for NH_3 synthesis, oxidative fixation to nitric acid and fertilizers, organic amine and heterocyclic synthesis, and amino acid synthesis is a topic which has generated much interest. The sequence of filled molecular orbitals of the nitrogen is ($^1\Sigma_g{}^+$): $1\sigma_g{}^2$, $1\sigma_u{}^2$, $2\sigma_g{}^2$, $2\sigma_u{}^2$, $1\pi_u{}^4$, and $3\sigma_g{}^2$. The highest occupied molecular orbital, $3\sigma_g$, has an ionization energy of 15.58 eV (similar to the 15.75 eV or Argon). The lowest unoccupied molecular orbital, $1\pi_g$, lies 8.6 eV higher than the $3\sigma_g$. The low reactivity of molecular nitrogen has been attributed to the absence of orbitals in the energy range between the HOMO and LUMO.[61] Although many dinitrogen transition metal complexes have been iso-

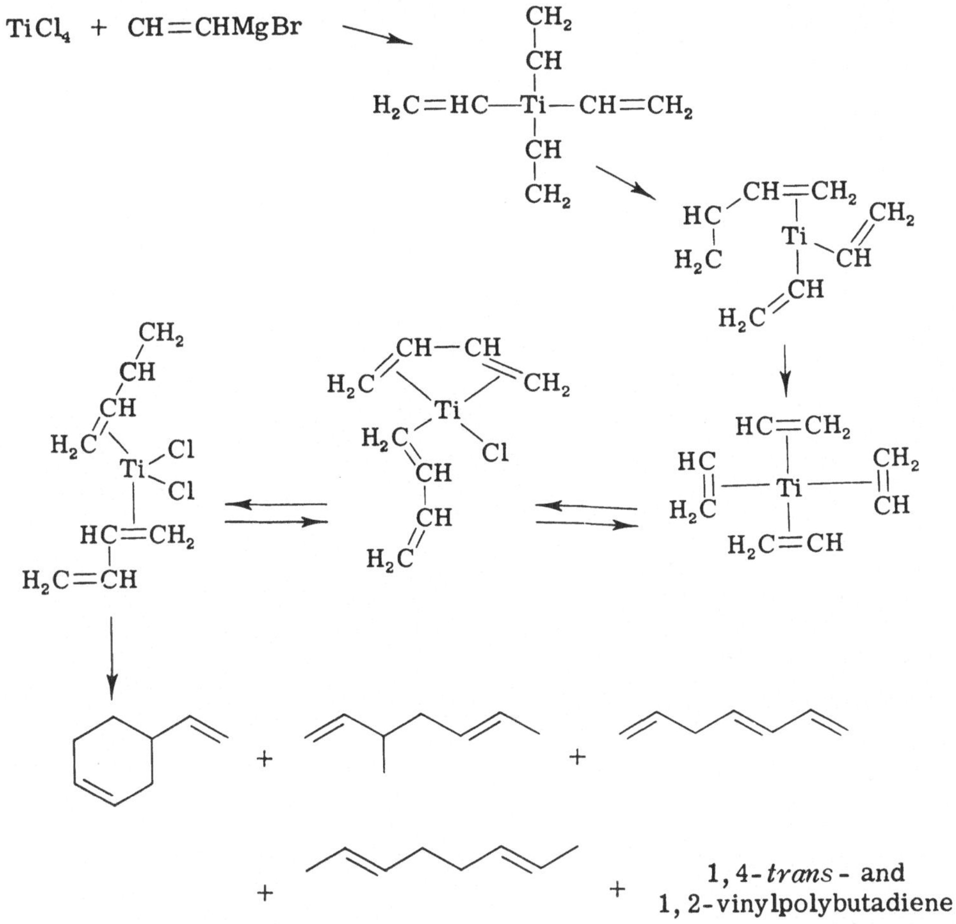

$TiCl_4$ + $CH \equiv CHMgBr$ ⟶

(reaction scheme with titanium organometallic intermediates)

$$CH_2$$
$$CH$$
$$H_2C = HC - Ti - CH = CH_2$$
$$CH$$
$$CH_2$$

products:

+ (methyl-branched diene) + (1,4-diene) + (diene) +

**1,4-*trans*- and
1,2-vinylpolybutadiene**

Figure 21. The "double coupling" reaction

lated, no efficient homogeneous catalyst for N_2 fixation has been developed.

The bonding of dinitrogen to transition metals can be separated into "end-on" and "side-on" categories as shown in Figure 22.

Most stable complexes of dinitrogen have "end-on" structures with only a few compounds showing reductive chemical reactivity. However, the compound
$\{C_6H_5\ Na \cdot O(C_2H_5)_2\ _2\ (C_6H_5)_2Ni\ _2N_2NaLi_6(OC_2H_5)_4 \cdot P(C_2H_5)_2\}_2$
having a "side-on" nitrogen has been isolated and characterized.[62-64]

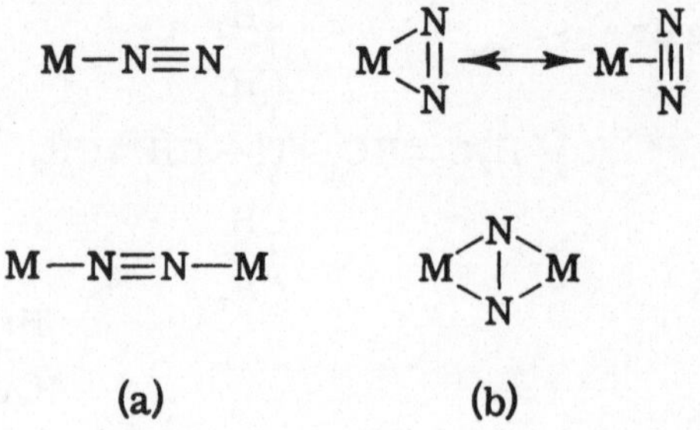

$$M—N\equiv N \qquad M{<}^{N}_{N}{\|} \longleftrightarrow M{-}^{N}_{N}{\|\|}$$

$$M—N\equiv N—M \qquad M{<}^{N}_{N}{>}M$$

(a)   (b)

Figure 22.   (a) "end-on" bonding mode; (b) "side-on" bonding mode.

Theoretical calculations show that the lower stability of the "side-on" mode with respect to the "end-on" mode can be attributed to the more antibonding character of the nitrogen-metal π-bond and the decrease in the total electronic population of the bond in the former.[65]

In the complex $Ru(NH_3)_5(N_2)^{+2}$ the dinitrogen ligand undergoes an end-over-end rotation

$$[(NH_3)_5Ru-^{15}N\equiv N^{14}]^{+2} \longrightarrow [(NH_3)_5Ru-^{14}N\equiv N^{15}]^{+2}.$$

This process has a half-life of 2 hours in aqueous solution at 25°C and occurs 45 times faster than the competing substitution reaction

$$[Ru(NH_3)_5(N_2)]^{+2} + H_2O \longrightarrow [Ru(NH_3)_5(H_2O)]^{+2} + N_2.$$

Therefore, it appears that this isomerization occurs via a "side-on" coordinated intermediate without ligand dissociation. The energetics of the isomerization and dissociation reactions indicate that the energy of activation of the dissociation of the nitrogen ligand in the "side-on" mode is only 7 kcal/mole as compared with 29 kcal/mole in the "end-on" mode.[66]

Recently, the unusual behavior of reaction systems which involve titanocene toward the normally quite inert molecule N_2 has been the subject of considerable interest. The fixation and reduction of molecular nitrogen has been observed in mixtures of titanocene dichloride and ethyl magnesium bromide and in other systems containing titanocene derivatives.[67-75]

It has been observed that when solutions of permethyltitanocene are exposed to molecular nitrogen three different dinitrogen complexes may be formed (25, 26, 27).[76] The sequence of possible reactions leading to these complexes is shown in Figure 23.

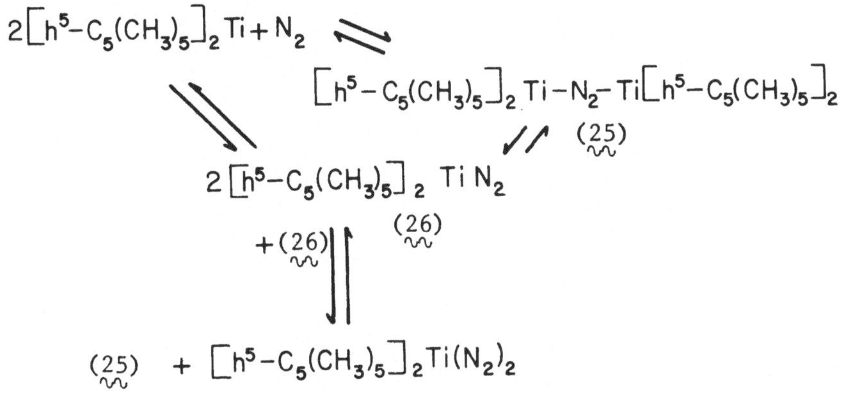

$$2\left[h^5-C_5(CH_3)_5\right]_2 Ti + N_2 \;\rightleftharpoons$$

$$\left[h^5-C_5(CH_3)_5\right]_2 Ti-N_2-Ti\left[h^5-C_5(CH_3)_5\right]_2$$

$$2\left[h^5-C_5(CH_3)_5\right]_2 TiN_2 \qquad (25)$$

$$+(26) \qquad (26)$$

$$(25) \;+\; \left[h^5-C_5(CH_3)_5\right]_2 Ti(N_2)_2$$

Figure 23.   Reactions of permethyl titanocene with N_2.

Spectral studies **are** indicative of a centrosymmetric Ti-N_2-Ti substructure (27, 28, 29) for compound (25).

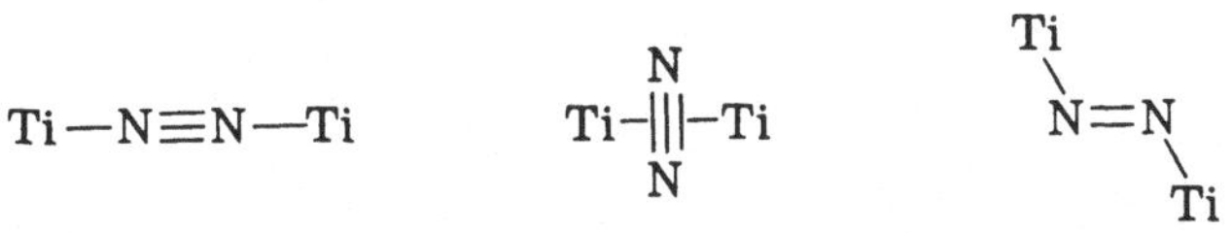

$$Ti-N\equiv N-Ti \qquad\qquad Ti-\overset{N}{\underset{N}{|||}}-Ti \qquad\qquad \overset{Ti}{\underset{Ti}{\diagdown}}N=N\diagup$$

$$(27) \qquad\qquad\qquad (28) \qquad\qquad\qquad (29)$$

$\sqrt{}$ {C nmr data indicate the presence of an "end-on" (h^1-N_2)-"side-on" (h^2-N_2) equilibrium for (26). There are three possible forms in which this equilibrium may occur.

In the first dinitrogne complex, h^5-$C_5(CH_3)_5$ $_2TiN_2$ which contains an "end-on" N_2 ligand, is in rapid equilibrium with a "side-on" isomer (Figure 24). An alternative explanation involves equilibrium between the h^1-species and a dimeric species.

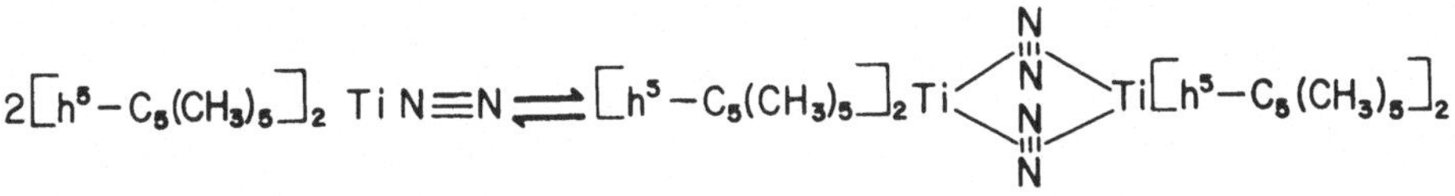

$$2\left[h^5-C_5(CH_3)_5\right]_2 Ti\, N\equiv N \rightleftharpoons \left[h^5-C_5(CH_3)_5\right]_2 Ti\underset{\underset{N}{\overset{N}{|||}}}{\overset{\overset{N}{|||}}{\diagup\diagdown}} Ti\left[h^5-C_5(CH_3)_5\right]_2$$

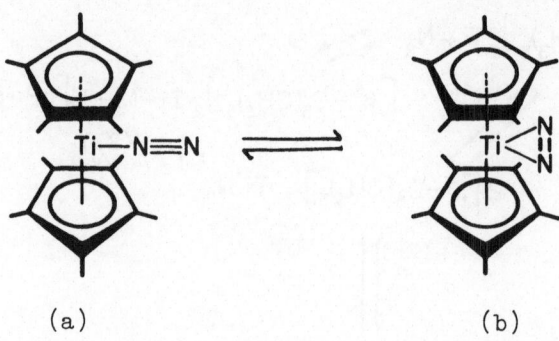

(a)   (b)

Figure 24.   (a) N_2 is "end-on"; (b) N_2 is "side-on".

Another alternative is an intramolecular fluxional equilibrium.

$$\left[h^5-C_5(CH_3)_5\right]_2Ti\!\!<\!\!^{N_2}_{N_2}\!\!>\!\!Ti\left[h^5-C_5(CH_3)_5\right]_2$$

$$\left[h^5-C_5(CH_3)_5\right]_2Ti-N\equiv N-Ti\left[h^5-C_5(CH_3)_5\right]_2$$
$$N\equiv N$$

OR

$$\left[h^5-C_5(CH_3)_5\right]_2Ti\overset{\underset{\displaystyle N}{\|\|\|}}{N}\!\!\diagdown\!\!Ti\left[h^5-C_5(CH_3)_5\right]_2$$
$$\diagdown N\equiv N$$

$$\left[\left[h^5-C_5(CH_3)_5\right]_2Ti\!\!<\!\!^{N_2}_{N_2}\!\!>\!\!Ti\left[h^5-C_5(CH_3)_5\right]_2\right]^{\dagger}$$

$$\left[h^5-C_5(CH_3)_5\right]_2Ti\overset{\underset{\displaystyle N}{\|\|\|}}{N}\!\!\diagdown\!\!Ti\left[h^5-C_5(CH_3)_5\right]_2$$
$$\diagdown N\equiv N$$

Presently, no clear distinction between these possibilities has been made.

     After a solution of the permethyltitanocene dinitrogen complex (26) is treated with HCl at -80°C, extraction of the reaction yields hydrazine.

$$(26) + 2HCl \longrightarrow \left[h^5-C_5(CH_3)_5\right]_2TiCl_2 + \tfrac{1}{2}N_2 + \tfrac{1}{2}N_2H_4$$

Reaction of the dimeric dinitrogen complex (25) with HCl gives a very small yield (<5%) of ammonium chloride.

Although the mechanisms of these reactions are not thoroughly understood, it has been suggested on the basis of nmr and ir results that the reactive configuration which ultimately leads to hydrazine is that of the h^2-dinitrogen complex of (26).

Therefore, it remains to be clarified in which way such N_2 complexes enter into reduction reactions -- i.e. in which bonding mode is the dinitrogen activated and in what way are the two bonding modes interrelated.

Presently, catalysts are being sought which can cause selective, direct oxygenation of organic substrates. Direct incorporation of one or both atoms of the oxygen molecule into organic substrates with great selectivity occurs in nature via mono and dioxygenases. Generally, these oxygenases are metalloenzymes, and activation of the O_2 involves coordination of dioxygen to the transition metal center. Knowledge of how metal species bind O_2 is essential in understanding such processes.

Currently, two geometrical structures for coordinated dioxygen are known -- the "side-on" which corresponds to a π-bonding mode and the "end-on" corresponding to a σ-bonding mode as seen in Figure 25. For example, the dioxygen complexes from iron(II) porphyrins[77] have the "end-on" configuration whereas the dioxygen complex of titanium octaethylporphyrin[78] has the "side-on" configuration. However, it is not known whether one bonding mode is more catalytically favorable or whether the two modes are interconvertable.

Although many parallels have been drawn between the σ-π rearrangement and catalytic activity, the exact nature of this relationship is still very unclear. The preceeding discussion has attempted to present some of the factors which are believed to influence σ-π rearrangements in organotransition metal complexes. However, much work is needed in this area. A detailed theoretical

Figure 25. (a) "side-on" O_2 coordination; (b) "end-on" coordination.

study is necessary which, coupled with experimental data, may better
elucidate the driving force of this reaction as well as the specific
role of the various factors which appear to affect the rearrangement.

ACKNOWLEDGEMENT

The authors are indebted to the Robert A. Welch Foundation by
whose support the preparation of this manuscript was made possible
and Dr. K. Tatsumi and Dr. P. Roling for their help.

REFERENCES

1. W. Herwig and H. H. Zeiss, J. Amer. Chem. Soc., 79 (1957) 6561.
2. H. H. Zeiss and M. Tsutsui, J. Amer. Chem. Soc., 79 (1957) 3062.
3. J. Chatt, L. A. Duncanson, and L. M. Venanzi, J. Chem. Soc.,
   (1955) 4456.
4. Ibid, (1955) 4461.
5. J. Chatt and B. L. Shaw, J. Chem. Soc., (1959) 705.
6. L. E. Orgel, J. Inorg. Nucl. Chem., 2 (1956) 137.
7. D. R. Armstrong, R. Fortune, and P. G. Perkins, Inorg. Chim.
   Acta, 9 (1974) 9.
8. G. E. Coates, M. L. H. Green, and K. Wade, Organometallic
   Compounds: The Transition Elements, Vol. 2, Methuen and
   Company, LTD, London, 1968, p. 32.
9. R. G. Pearson, J. Amer. Chem. Soc., 85 (1963) 3533.
10. M. L. H. Green and P. L. I. Nagy, J. Chem. Soc., (1963), 189.
11. M. L. H. Green and A. N. Stear, J. Organomet. Chem., 1 (1964)
    230.
12. M. Hancock, M. N. Levy, and M. Tsutsui, Organometallic Reactions,
    Vol. 4, E. I. Becker and M. Tsutsui (eds.), John Wiley and
    Sons, New York, 1972, Ch. 1.
13. H. C. Volger and K. Vrieze, J. Organomet. Chem., 13 (1968) 479.
14. M. Tsutsui, M. Hancock, J. Ariyoshi, and M. N. Levy, J. Amer.
    Chem. Soc., 91 (1969) 5233.
15. M. Tsutsui and C. E. Hudman, Chem. Lett., (1972) 777.
16. M. L. H. Green and P. L. I. Nagy, J. Organomet. Chem., 1 (1964)
    58.
17. J. K. P. Ariyaratne and M. L. H. Green, J. Chem. Soc., (1963)
    2926.
18. K. Vrieze, C. MacLean, P. Cossee, and C. W. Hilbers, Rec. Trav.
    Chim. Pays-Bas, 85 (1966) 1077.
19. M. Tsutsui and N. Ely, J. Amer. Chem. Soc., 97 (1975) 3551.
20. T. J. Marks, A. M. Seyam, and J. R. Kolb, J. Amer. Chem. Soc.,
    95 (1973) 5529.
21. G. Winkhaus, L. Pratt, and G. Wilkinson, J. Chem. Soc., (1961)
    3807.
22. J. Chatt and B. L. Shaw, J. Chem. Soc., (1962) 5072.

23. D. R. Armstrong, R. Fortune, and P. G. Perkins, J. Catal., 41 (1976) 51.
24. R. B. Woodward and R. Hoffmann, The Conservation of Orbital Symmetry, Verlag Chemie, Weinheim, Germany, 1971, p.1.
25. K. Fukui and S. Inagaki, J. Amer. Chem. Soc., 97 (1975) 4445.
26. K. Tatsumi, K. Yamaguchi, and T. Fueno, J. Mol. Catal., in press.
27. K. Tatsumi, K. Yamaguchi, and T. Fueno, Tetrahedron, 31 (1975) 2899.
28. D. R. Armstrong, P. G. Perkins, and J. J. P. Stewart, J. Chem. Soc., Dalton, (1972) 1972.
29. R. G. Pearson, Symmetry Rules for Chemical Reactions, John Wiley and Sons, New York, 1976, p. 413.
30. R. G. Pearson, Fortschr. Chem. Forsh., 41 (1973) 75.
31. A. Nakamura and S. Otsuka, J. Amer. Chem. Soc., 94 (1972) 1886.
32. P. M. Henry, Acc. Chem. Res., 6 (1973) 16.
33. A. Seqnitz, P. M. Bailey, and P. M. Maitlais, Chem. Commum., (1973) 698.
34. J. K. Stille, D. E. James, and L. F. Hines, J. Amer. Chem. Soc., 95 (1973) 5062.
35. C. E. Moore, Atomic Energy Levels, Vol. 2 and 3, Natn. Stand. Ref. Data Ser. (1971).
36. J. Smidt, W. Hafner, R. Jira, R. Sieber, J. Sedlmeier, and A. Sabel, Angew. Chem., 74 (1962) 93.
37. W. Hafner, R. Jira, J. Sedlmeier, and J. Smidt, Chem. Ber., 95 (1962) 1575.
38. G. F. Wright, Ann. N. Y. Acad. Sci., 65 (1957) 436.
39. R. R. Grinstead, J. Org. Chem., 26 (1961) 238.
40. R. Criegee, Angew. Chem., 70 (1958) 173.
41. J. Chatt, Chem. Rev., 48 (1951) 7.
42. P. M. Henry, J. Amer. Chem. Soc., 86 (1964) 3246.
43. J. K. P. Ariyaratne and M. L. H. Green, J. Chem. Soc., (1964) 1.
44. Y. Wakatsuki, S. Nozakura, and S. Murahashi, Bull. Chem. Soc. Jap., 42 (1969) 273.
45. H. Thyret, Angew. Chem. Int. Ed. Engl., 11 (1972) 520.
46. M. Tsutsui, M. Ori, and J. Francis, J. Amer. Chem. Soc., 94 (1972) 1414.
47. J. Hillis and M. Tsutsui, J. Amer. Chem. Soc., 95 (1973) 7907.
48. J. Hillis, J. Francis, M. Ori, and M. Tsutsui, J. Amer. Chem. Soc., 96 (1974) 4800.
49. F. A. Cotton, J. N. Francis, B. A. Frenz, and M. Tsutsui, J. Amer. Chem. Soc., 95 (1973) 2483.
50. K. Ziegler, E. Holzkamp, H. Breil, and H. Martin, Angew. Chem., 67 (1955) 541.
51. P. Cossee, J. Catal., 3 (1964) 80.
52. P. Cossee, The Stereochemistry of Macromolecules, Vol. 1, A. D. Ketley (ed), Marcel Dekker, New York, 1967, Ch. 3.
53. O. Novaro, S. Chow, and P. Magnovat, J. Catal., 41 (1976) 91.
54. Ibid., 42 (1976) 131.

55. T. S. Djabiev, R. D. Sabirova, and A. D. Shilov, Kinet. Katal., 5 (1964) 441.
56. G. Henrici-Olivé and S. Olivé, Polymer Lett., 12 (1974) 39.
57. M. Tsutsui, Trans. N. Y. Acad. Sci., Series II, 26 (1964) 423.
58. M. Tsutsui, J. Ariyoshi, T. Koyano, and M. N. Levy, Adv. in Chem. Series, no. 70, American Chemical Society, 1968, p. 266.
59. M. Tsutsui and J. Ariyoshi, Trans. N. Y. Acad. Sci., 26 (1964) 431.
60. B. Gorewit and M. Tsutsui, in press.
61. J. Chatt and R. L. Richards, The Chemistry and Biochemistry of Nitrogen Fixation, J. R. Postgate (ed), Plenum Press, New York, 1971, p. 57.
62. K. Jonas, Angew. Chem., 85 (1973) 1050.
63. K. Jonas, Angew. Chem. Int. Ed. Engl., 12 (1973) 997.
64. K. Jonas, D. J. Braver, C. Krüger, P. J. Roberts, and Y. H. Tsay, J. Amer. Chem. Soc., 98 (1976) 74.
65. K. B. Yatsimirskii and Yv. A. Kruglyak, Dokl. Akad. Nauk SSSR, 186 (1969) 885.
66. J. N. Armor and H. Taube, J. Amer. Chem. Soc., 92 (1970) 2560.
67. M. E. Volpin and V. B. Shur, Nature, 209 (1966) 1236.
68. M. E. Volpin and V. B. Shur, Dokl. Akad. Nauk SSSR, 156 (1964) 1102.
69. G. Henrici-Olivé and S. Olivé, Angew. Chem. Int. Ed. Engl., 8 (1969) 650.
70. G. Henrici-Olivé and S. Olivé, Angew. Chem., 80 (1968) 398.
71. A. E. Shilov, A. K. Shilova, and E. F. Kvashina, Kinet. Katal., 10 (1969) 1402.
72. A. E. Shilov, A. K. Shilova, E. F. Kvashina, and T. A. Vorontsova, Chem. Commun., (1971) 1590.
73. E. E. van Tamelen, Acc. Chem. Res., 3 (1970) 361.
74. H. Brintzinger, J. Amer. Chem. Soc., 89 (1967) 6871.
75. H. Brintzinger, Biochemistry, 5 (1966) 3947.
76. J. E. Bercaw, J. Amer. Chem. Soc., 96 (1974) 5087.
77. J. P. Collman, R. R. Gagne, C. A. Reed, T. R. Halbert, G. Yang, and W. T. Robinson, J. Amer. Chem. Soc., 97 (1975) 1427.
78. R. Guilard, M. Rontesse, and P. Fournari, J. Chem. Soc. Comm., (1976) 161.

CHIRAL FERROCENYLPHOSPHINES AND THEIR USE AS LIGANDS FOR TRANSITION

METAL COMPLEX CATALYZED ASYMMETRIC SYNTHESIS

Tamio Hayashi and Makoto Kumada

Department of Synthetic Chemistry, Kyoto University

Yoshida, Kyoto 606, Japan

I. Introduction

There has been intense interest and activity recently in
asymmetric synthesis catalyzed by transition metal complexes with
chiral ligands [1,2]. This has resulted in so great success in
asymmetric homogeneous hydrogenation by chiral phosphine-rhodium
complexes as to produce α-amino acids with over 90% optical purity
[3]. One of the most significant problems in studies on the
catalytic asymmetric synthesis is how to develop a chiral ligand
which will enable the catalyst for a given reaction to be as efficient in stereoselectivity as possible, and considerable efforts
have been devoted to searching for new chiral phosphine ligands.
These chiral phosphines may be classified into two classes: (A) phosphines bearing an asymmetric center at the phosphorus atom and (B)
phosphines whose chirality is due to asymmetric carbons in groups
bonded to phosphorus.
    Methylphenylpropylphosphine (1) [4] was used by Horner et al.
[5] and Knowles et al. [6] in their early studies on catalytic
asymmetric hydrogenation. Many optically active tertiary phosphines
have been prepared since Mislow et al. developed a new synthetic
method for the chiral phosphines in 1967 [7], and some of them have
been reported to be effective ligands for catalytic asymmetric
reactions. o-Anisylcyclohexylmethylphosphine (ACMP) (2) [8] and
1,2-di(o-anisylphenylphosphino)ethane (DIPAMP) (3) [9] by Knowles
et al. gave, when complexed with rhodium, optical yields of over
90% in the hydrogenation of α-acylaminoacrylic acids.
    Phosphines containing asymmetric alkyl groups were usually
derived from optically active natural compounds. Kagan et al. have
achieved great success with 2,3-O-isopropylidene-2,3-dihydroxy-1,4-

Me
|
Pr—P◀—Ph

1

Me
|
P
MeO

2

P----CH₂)₂
OMe

3

bis(diphenylphosphino)butane (DIOP) (4) [*10*], prepared relatively
easily from tartaric acid. The DIOP has been successfully applied
in various kinds of catalytic asymmetric syntheses, e.g., hydrogenation of olefins [*11-15*] and ketones [*16,17*], hydrosilylation of
ketones [*18-21*] and imines [*11a*], hydroformylation [*22-24*], Grignard
cross-coupling [*25,26*], and allylic alkylation [*27*]. Neomenthyldi-
phenylphosphine (NMDPP) (5) [*28,29*] and its diastereomeric isomer
(MDPP) (6) [*29*], prepared by Morrison et al., have been used for
the hydrogenation of acrylic acids to give up to 62% optical yield
[*3,28*]. They have also studied a bisphosphine ligand (CAMPHOS) (7)

H
O   PPh₂
O   PPh₂
H

4

PPh₂

5

PPh₂

6

PPh₂
PPh₂

7

derived from camphoric acid [*3*]. Dimenthylphosphines (8) have been
synthesized and used as chiral ligands for a nickel catalyzed cool-
igomerization of olefins [*30*]. Optically active amino acids, pro-
line and hydroxyproline, have been employed as chiral sources of
phosphine ligands. They are (S)-2-diphenylphosphinomethylpyrroli-
dines (9) [*31*] and ($2S,4S$)-4-diphenylphosphino-2-diphenylphosphino-
methylpyrrolidines (10) [*32*], respectively, the latter affording
enantiomeric excess of 83-91% in the hydrogenation of α-acylamino-
cinnamic acids. 2,2'-Bis(diphenylphosphinomethyl)-1,1'-binaphthyl
(NAPHOS) (11) [*33*], whose chirality is due to binaphthyl axial
chirality, has been prepared and used as a ligand in several transition metal complex catalyzed asymmetric reactions.

)₂PR

8

N
|
R
CH₂PPh₂

9 R=H, Me

Ph₂P
N
|
R
CH₂PPh₂

10 R=H, COOBu^t

CH₂PPh₂
CH₂PPh₂

11

The primary purpose of this article is to describe the preparation of a new type of phosphines with planar chirality which arises from introducing phosphino groups into the α-ferrocenylethyldimethylamine system [34] and their uses as chiral ligands in some of the transition metal complex catalyzed asymmetric reactions.

II. Preparation of Chiral Ferrocenylphosphines

Chiral ferrocenylphosphines are readily prepared by way of lithiation of optically resolved α-ferrocenylethyldimethylamine (FA). The lithiation of (R)-FA was previously reported by Ugi and coworkers [34] to proceed with high stereoselectivity to give preferentially (R)-α-[(R)-2-lithioferrocenyl]ethyldimethylamine.

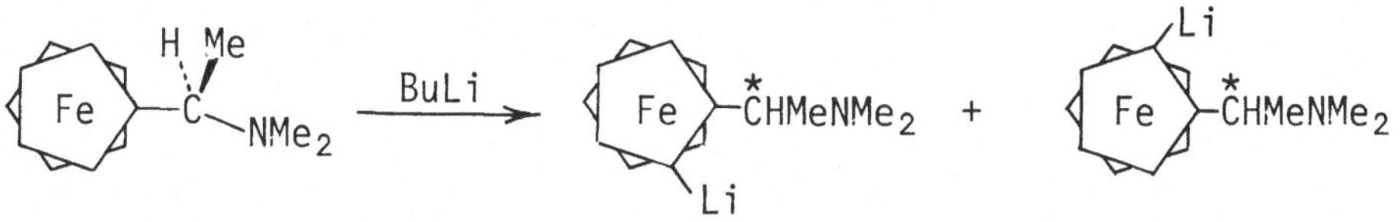

(R)-FA   (R)-(R) (96%)   (R)-(S) (4%)

In our studies, (S)-FA was metalated, after the Ugi's procedures, with a slight excess of butyllithium in ether, and the mixture was then treated with chlorodiphenylphosphine. (S)-α-[(R)-2-diphenylphosphinoferrocenyl]ethyldimethylamine (PPFA) was obtained in 55% yield [35] (eq. 1).

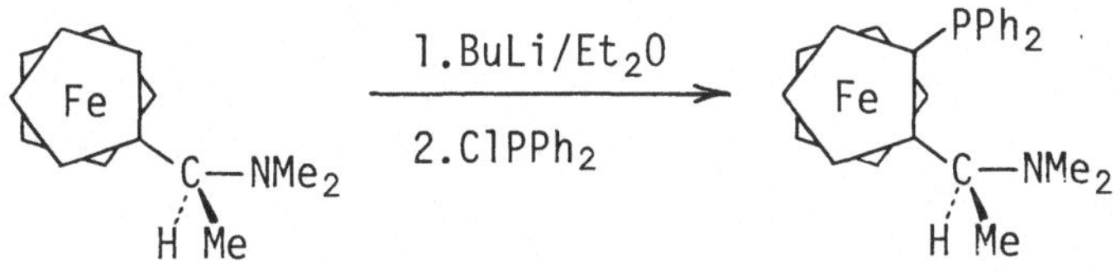

$$\text{(S)-FA} \xrightarrow[\text{2.ClPPh}_2]{\text{1.BuLi/Et}_2\text{O}} \text{(S)-(R)-PPFA} \tag{1}$$

Similarly, (R)-α-[(S)-2-dimethylphosphinoferrocenyl]ethyldimethylamine (MPFA) [35] was prepared in 31% yield starting with (R)-FA and chlorodimethylphosphine (eq. 2).

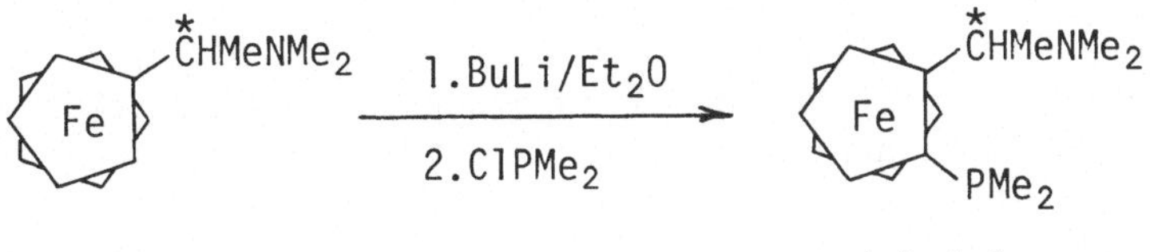

$$\text{(R)-FA} \xrightarrow[\text{2.ClPMe}_2]{\text{1.BuLi/Et}_2\text{O}} \text{(R)-(S)-MPFA} \tag{2}$$

The stepwise lithiation of (S)-FA with butyllithium in ether and then with butyllithium/N,N,N',N'-tetramethylethylenediamine (TMEDA) in ether led to the introduction of two diphenylphosphino groups, one onto each of the cyclopentadienyl rings to give (S)-α-[(R)-1',2-bis(diphenylphosphino)ferrocenyl]ethyldimethylamine (BPPFA) in 40% yield [35] (eq. 3). The analogous bis(dimethylphosphino) derivative (BMPFA) was also prepared.

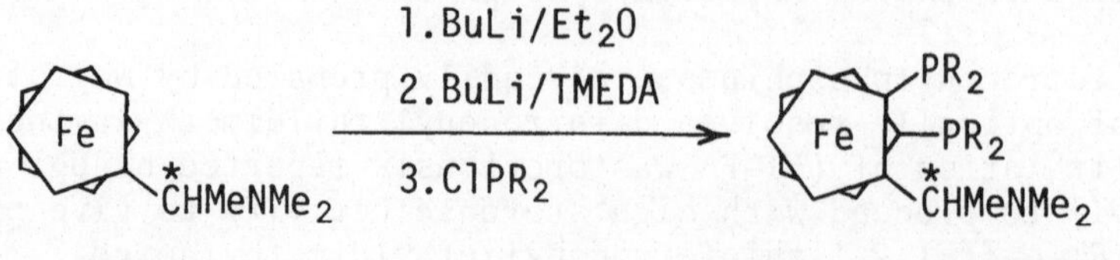

$$\begin{array}{c} \text{1.BuLi/Et}_2\text{O} \\ \text{2.BuLi/TMEDA} \\ \hline \text{3.ClPR}_2 \end{array}$$

(S)-FA

R=Ph, (S)-(R)-BPPFA

R=Me, (S)-(R)-BMPFA

(3)

Chiral ferrocenylphosphines with (R)-(R)-configuration, which are the diastereomeric isomers of (R)-(S)-PPFA and BPPFA, could be prepared, although in low yield, by the sequence of reactions involving trimethylsilylation of (R)-FA followed by diphenylphosphination and finally by desilylation after Price's techniques [36] (eq. 4).

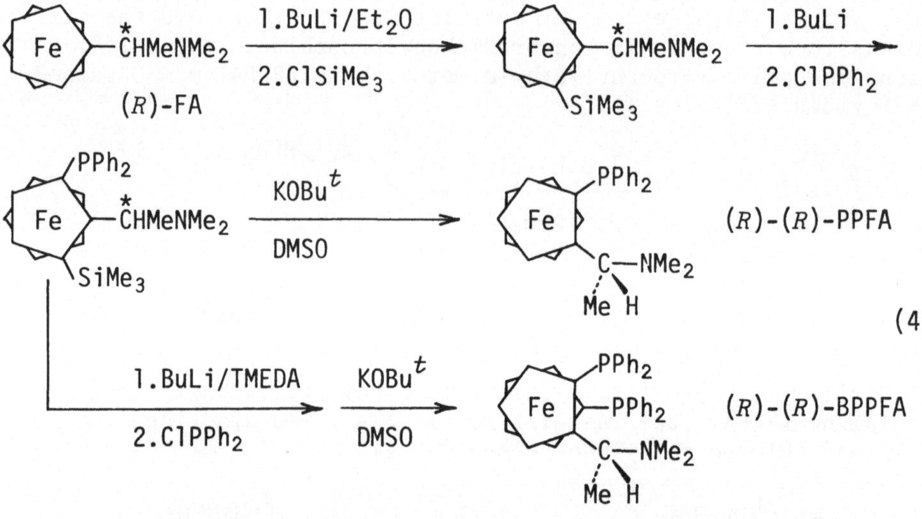

$$\begin{array}{c} \text{1.BuLi/Et}_2\text{O} \\ \hline \text{2.ClSiMe}_3 \end{array}$$

(R)-FA

$$\begin{array}{c} \text{1.BuLi} \\ \hline \text{2.ClPPh}_2 \end{array}$$

$$\begin{array}{c} \text{KOBu}^t \\ \hline \text{DMSO} \end{array}$$

(R)-(R)-PPFA

$$\begin{array}{c} \text{1.BuLi/TMEDA} \\ \hline \text{2.ClPPh}_2 \end{array} \quad \begin{array}{c} \text{KOBu}^t \\ \hline \text{DMSO} \end{array}$$

(R)-(R)-BPPFA

(4)

Replacement of the dimethylamino group in PPFA by other dialkyl-amino groups could be achieved first by converting to the phosphine oxide, then quaternization of the nitrogen atom, treatment with a secondary amine and finally by reduction with alane. The diethyl-amino, piperidino, morpholino, and N-methylpiperazino derivatives have been thus prepared in 50-70% overall yield (eq. 5).

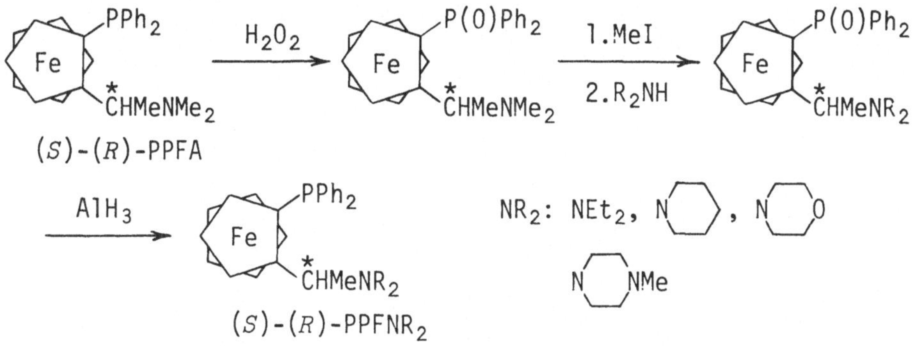

$$(S)-(R)-\text{PPFA} \xrightarrow{H_2O_2} \xrightarrow{\substack{1.\text{MeI} \\ 2.R_2NH}}$$

$$\xrightarrow{AlH_3} \quad (S)-(R)-\text{PPFNR}_2$$

$$NR_2: \quad NEt_2, \quad N\bigcirc, \quad N\bigcirc O \qquad (5)$$

$$N\bigcirc NMe$$

A hydroxy group was also introduced into the side chain of the chiral ferrocenylphosphines by the sequence of reactions shown in eq. 6 [37].

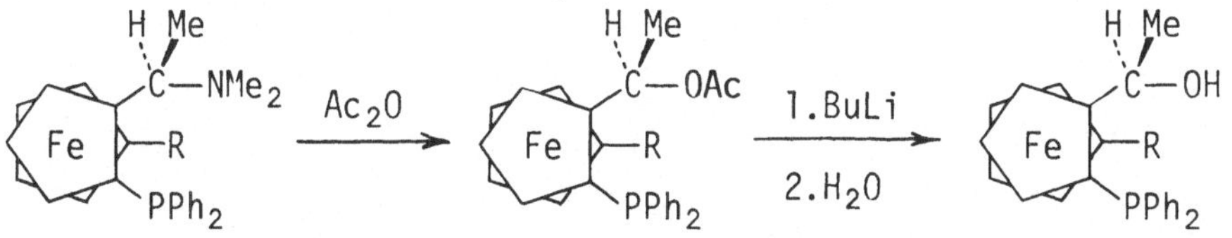

$$\xrightarrow{Ac_2O} \xrightarrow{\substack{1.\text{BuLi} \\ 2.H_2O}} \qquad (6)$$

R=H   , (R)-(S)-PPFA   (R)-(S)-PPFOAc   (R)-(S)-PPFOH

R=PPh₂, (R)-(S)-BPPFA   (R)-(S)-BPPFOAc   (R)-(S)-BPPFOH

These chiral phosphines are quite unique in that they all contain both planar and central elements of chirality and also a functional group such as amino or hydroxy that can interact attractively with an appropriate functional center of the substrate which interacts with the chiral catalyst.

Ferrocenylphosphines, PPEF and BPPEF, having only a planar element of chirality were prepared by a sequence of reactions as shown in eq. 7 and 8.

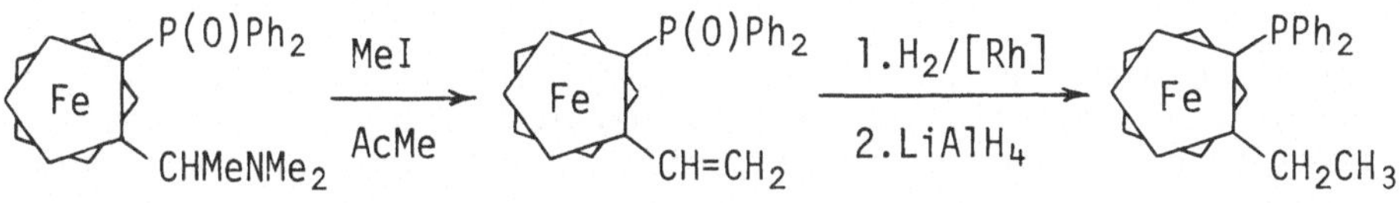

$$\xrightarrow[\text{AcMe}]{\text{MeI}} \xrightarrow{\substack{1.H_2/[\text{Rh}] \\ 2.\text{LiAlH}_4}} \qquad (7)$$

$$(R)-\text{PPEF}$$

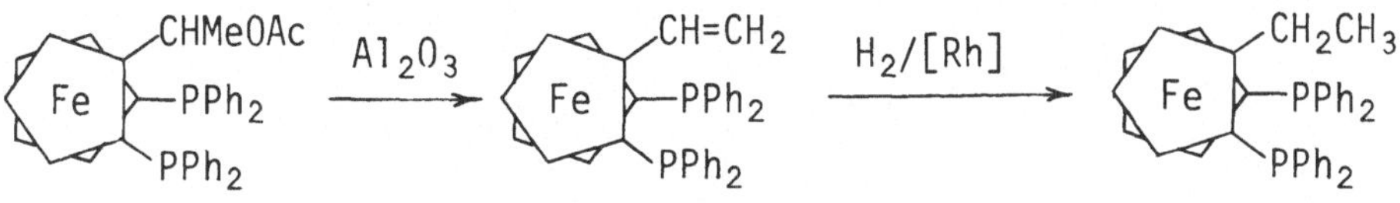

$$\xrightarrow{Al_2O_3} \xrightarrow{H_2/[\text{Rh}]} \qquad (8)$$

$$(S)-\text{BPPFV} \qquad\qquad (S)-\text{BPPEF}$$

1-Dimethylaminomethyl-2-diphenylphosphinoferrocene (FcPN) which lacks carbon central chirality was optically resolved via its phosphine sulfide dibenzoyl-*d*-tartaric acid salt (eq. 9) [38]. The bisphosphine (FcBPN) analogous to FcPN was also prepared from the resolved FcPN (eq. 10).

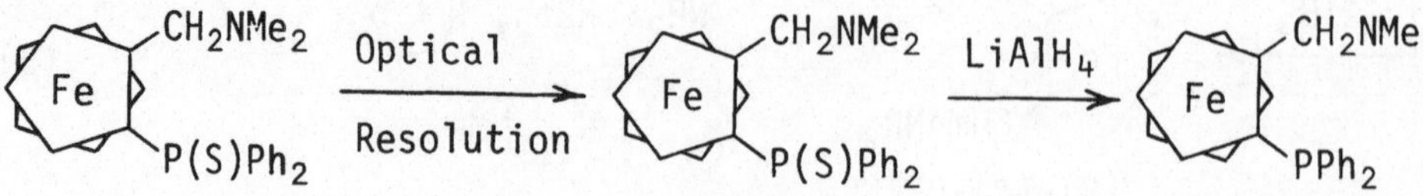

Melting points and specific rotations of these chiral ferrocenylphosphines are summerized in Table 1. The absorption and CD spectra of PPFA, MPFA, BPPFA, and PPEF are shown in Fig. 1 and 2.

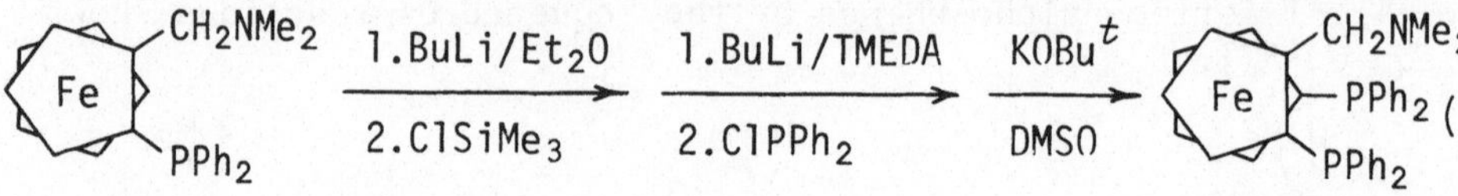

The absorption spectrum of ferrocene has two long wave-length bands at 440 and 325 nm assignable to *d-d* type transition [39]. The CD spectra of chiral ferrocenylphosphines reveal optical activity arising from the planar chirality around these two absorption bands, that is, (S)-(R)-PPFA, (S)-(R)-BPPFA, and (R)-EPPF all have positive Cotton effects around 450-470 and negative ones around 340-350 nm, whereas opposite Cotton effects are observed in the case of (R)-(S)-MPFA.

III. Asymmetric Hydrosilylation of Ketones

First we used the chiral ferrocenylphosphines as chiral ligands for asymmetric catalytic hydrosilylation of ketones [18-21,40,41]. The asymmetric addition of trialkylsilanes and dialkylsilanes readily took place at 50° and 20°C, respectively, in the presence of rhodium complexes containing chiral ferrocenylphosphines (eq. 11).

$$R^1COR^2 + R_3SiH \xrightarrow{[Rh^*]} R^1R^2\overset{*}{C}HOSiR_3 \quad (11)$$

$$[Rh^*] = \frac{1}{2} [(C_6H_{10})RhCl]_2 + 2P^*$$

Table 1. Physical Constants of Chiral Ferrocenylphosphines

Abbr.	Mp (°C)	$[\alpha]_D^{25}$	
(S)-(R)-PPFA	139	+361°	(c 0.6, EtOH)
(R)-(S)-MPFA	oil	-134°	(c 0.3, $CHCl_3$)
(S)-(R)-BPPFA	139-40	+349°	(c 0.5, $CHCl_3$)
(S)-(R)-BMPFA	oil	+14.8°	(c 0.6, $CHCl_3$)
(R)-(R)-PPFA	——	+364°	(c 0.4, $CHCl_3$)
(R)-(R)-BPPFA	——	+335°	(c 0.3, $CHCl_3$)
(S)-(R)-PPFNR$_2$			
NR$_2$ = NEt$_2$		+339°	(c 0.5, $CHCl_3$)
 N⟩	144-45	+331°	(c 0.5, $CHCl_3$)
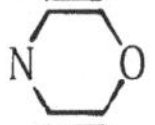 N⟩O	153	+343°	(c 0.5, $CHCl_3$)
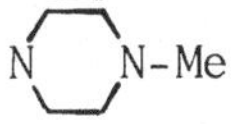 N⟩N-Me	101-2	+381°	(c 0.4, $CHCl_3$)
(S)-(R)-PPFOAc		+320°	(c 0.4, $CHCl_3$)
(S)-(R)-PPFOH		+271°	(c 0.7, $CHCl_3$)
(R)-(S)-BPPFOAc	153.3-55	-311°	(c 0.5, $CHCl_3$)
(R)-(S)-BPPFOH	154-55	-285°	(c 0.5, $CHCl_3$)
(R)-PPEF	95.5	+273°	(c 0.3, $CHCl_3$)
(S)-BPPFV	178.5-82	+83.3°	(c 0.3, $CHCl_3$)
(S)-BPPEF	166-67	-147°	(c 0.3, $CHCl_3$)
(S)-FcPN	91.5	-323°	(c 0.4, $CHCl_3$)
(S)-FcBPN	135-36	-269°	(c 0.3, $CHCl_3$)

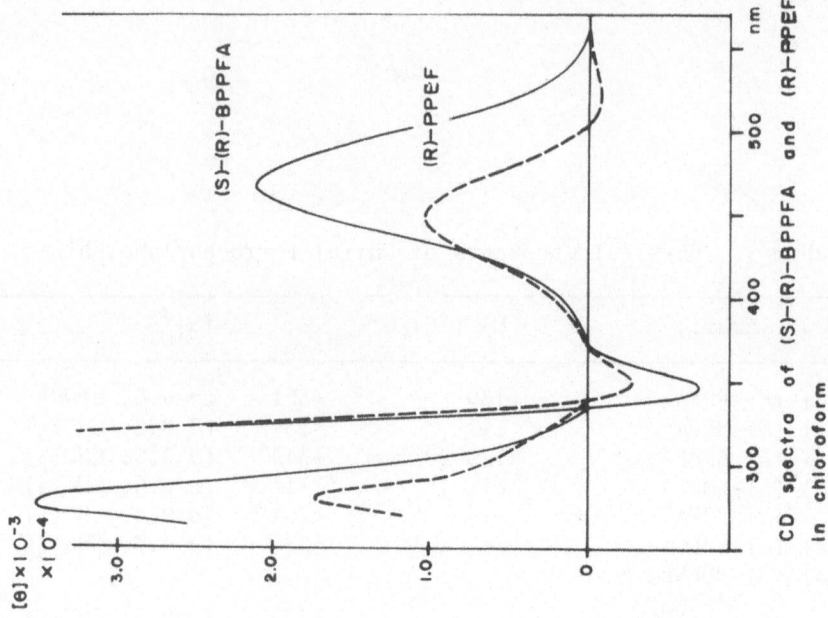

CD spectra of (S)-(R)-BPPFA and (R)-PPEF
in chloroform

Fig. 2

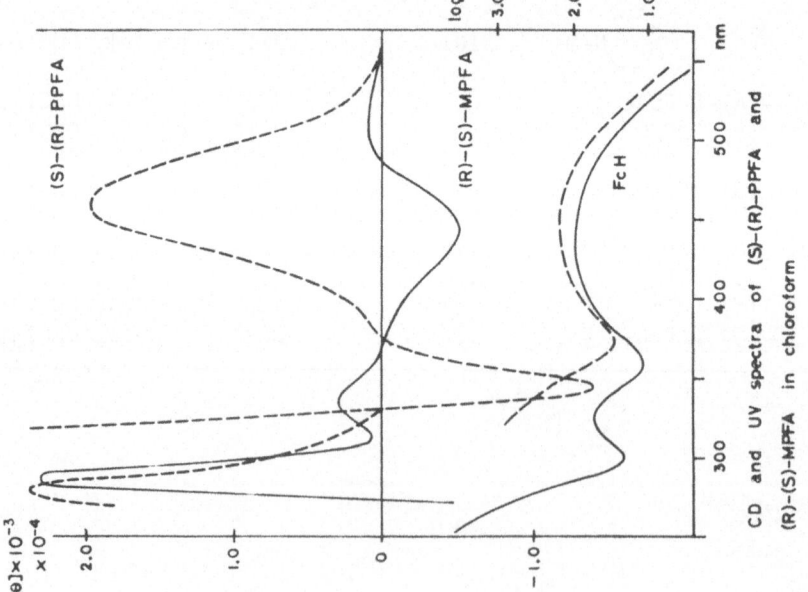

CD and UV spectra of (S)-(R)-PPFA and
(R)-(S)-MPFA in chloroform

Fig. 1

Table 2. Asymmetric Hydrosilylation of Ketones Catalyzed by Chiral Ferrocenylphosphine-Rhodium Complexes[a]

Ketone	Silane	Ligand[b]	Yield[c] (%)	% e.e. (Configuration)	
EtCOPh	PhMe$_2$SiH	(S)-(R)-PPFA	61	10.5	(R)
EtCOPh	PhMe$_2$SiH	(S)-(R)-PPFA[d]	40	9.0	(R)
EtCOPh	Et$_2$SiH$_2$	(S)-(R)-PPFA	82	2.7	(R)
EtCOPh	Et$_2$SiH$_2$	(S)-(R)-PPFA[d]	82	2.7	(R)
t-BuCOMe	PhMe$_2$SiH	(S)-(R)-PPFA	71	19.6	(R)
MeCOPh	Ph$_2$SiH$_2$	(R)-(S)-MPFA	85	49.2	(R)
MeCOPh	α-NpPhSiH$_2$	(R)-(S)-MPFA	74	51.8	(R)
EtCOPh	PhMe$_2$SiH	(R)-(S)-MPFA	58	12.2	(R)
EtCOPh	Ph$_2$SiH$_2$	(R)-(S)-MPFA	83	38.3	(R)
EtCOPh	Me$_3$SiH	(R)-(S)-MPFA	64	7.2	(S)
EtCOPh	Et$_2$SiH$_2$	(R)-(S)-MPFA	88	4.2	(S)
t-BuCOMe	PhMe$_2$SiH	(R)-(S)-MPFA	29	23.5	(S)
t-BuCOPh	Ph$_2$SiH$_2$	(R)-(S)-MPFA	74	41.1	(R)
MeCOPh	Ph$_2$SiH$_2$	(S)-(R)-BPPFA	72	28.6	(R)
EtCOPh	Ph$_2$SiH$_2$	(S)-(R)-BPPFA	73	24.5	(R)
EtCOPh	Et$_2$SiH$_2$	(S)-(R)-BPPFA	84	25.3	(R)
EtCOPh	PhMe$_2$SiH	(R)-PPEF	46	5.4	(R)
EtCOPh	Ph$_2$SiH$_2$	(R)-PPEF	81	0.2	(S)

[a] Catalyst = 0.05 mol%. [b] PPFA, MPFA, or PPEF/Rh = 2, BPPFA/Rh = 1. [c] Based on the amount of the isolated silyl ether. [d] PPFA/Rh = 1.

The results obtained are summerized in Table 2. As is seen from it, the extent of asymmetric induction depends strongly upon not only the structure of chiral phosphine ligands but also that of hydrosilanes employed. The marked effect of the latter on the stereoselectivity is very characteristic of the asymmetric hydrosilylation of ketones, and has been fully discussed by Ojima et al. [18] and by us [41]. Fairly good optical yields were attained when (S)-(R)-PPFA, (R)-(S)-MPFA, or (S)-(R)-BPPFA was used. For example, the reaction of acetophenone with diphenylsilane catalyzed by (R)-(S)-MPFA-rhodium complex resulted in higher optical yield than (R)-benzylmethylphenylphosphine (BMPP) [18,41] or DIOP [19,20] was used as a ligand. The high ability of (R)-(S)-MPFA was also found in the reaction of pinacolone. On the other hand, no significant asymmetric induction was observed when (R)-PPEF-rhodium complex was used as a catalyst. PPFA and MPFA are expected to form a chelate with the rhodium atom using both phosphorus and nitrogen atoms [42]. The results that the hydrosilylation reaction with a PPFA/Rh ratio of 1 and 2 gave almost the same optical yields may support the

chelation mentioned above. BPPFA must coordinate to the rhodium
with two diphenylphosphino groups present at 1- and 1'-positions of
the ferrocene [43]. It seems likely to assume that these chelate
structures with relatively rigid conformation around the rhodium
metal would be of advantage in giveng rise to a higher asymmetric
induction in the present reaction than those with two monodentate
PPEF ligands.

IV. Asymmetric Hydrogenation of Olefins

    Since the Wilkinson catalyst, tris(triphenylphosphine)chloro-
rhodium(I), was found to have a good catalytic activity for homogeneous hydrogenation of olefins [44], studies on asymmetric hydrogenation of olefins have appeared by use of chiral phosphine-rhodium
complexes, some of them achieving remarkable results. Kagan [10,11]
and Knowles [8,9] have reportèd that acylamino acids with 80-96%
enantiomeric purity were obtained by the hydrogenation of α-acyl-
aminocinnamic acids using their phosphines, DIOP, ACMP, and DIPAMP,
as exemplified by the following reaction.

$$PhCH=C(NHCOMe)COOH \xrightarrow[L^*-Rh]{H_2} PhCH_2 \overset{*}{C}H(NHCOMe)COOH \qquad (12)$$

$$
\begin{array}{lll}
L^* = DIOP, & 81\% & e.e. \\
ACMP, & 90\% & e.e. \\
DIPAMP, & 96\% & e.e.
\end{array}
$$

    We have examined similar asymmetric hydrogenation by use of
several chiral ferrocenylphosphines [45]. (S)-(R)-BPPFA was found
to exert an effective chiral influence in the hydrogenation of α-
acetaminoacrylic acid substrates.
    In the presence of [Rh(1,5-hexadiene)Cl]$_2$ and (S)-(R)-BPPFA in
a 1:2.4 ratio, the hydrogenation was completed in 20 hr at 50 atm
initial hydrogen pressure and room temperature (eq. 13).

$$\underset{NHCOMe}{RCH=C-COOH} \xrightarrow[(S)-(R)-BPPFA-Rh]{H_2} \underset{NHCOMe}{RCH_2-\overset{*}{C}HCOOH} \qquad (13)$$

12a: R = H   13a-13e
12b: R = phenyl
12c: R = 4-acetoxyphenyl
12d: R = 3-methoxy-4-acetoxyphenyl
12e: R = 3,4-methylenedioxyphenyl

Table 3 shows the results obtained in a variety of solvents. The
reaction of 12b always gives (S)-N-acetylphenylalanine (13b) with
extremely high stereoselectivity regardless of the solvent employed.
In cases of 12c and 12d, however, the optical yields were dependent

Table 3.   Asymmetric Hydrogenation of Olefins (12) Catalyzed by
           (S)-(R)-BPPFA-Rh Complex

Olefin	Solvent	Isolated yield (%)	$[\alpha]_D$	% e.e. (Configuration)	
12a	EtOH	90	-29.9	69	(S)
	H$_2$O/MeOH(1/1)	99	-26.9	62	(S)
12b	MeOH	94	+42.8	93	(S)
	H$_2$O/EtOH(1/1)	87	+42.1	92	(S)
	H$_2$O/MeOH(1/1)	86	+41.0	89	(S)
12c	MeOH	87	+3.1	8	(S)
	EtOH	92	+15.3	38	(S)
	H$_2$O/MeOH(1/3)	86	+35.1	87	(S)
12d	EtOH	86	+14.7	36	(S)
	H$_2$O/MeOH(1/10)	98	+25.7	63	(S)
	H$_2$O/MeOH(1/2)	94	+35.1	86	(S)
12e	MeOH	96	+18.2	34	(S)
	H$_2$O/MeOH(3/4)	93	+27.7	52	(S)

markedly on the nature of solvents. It can be seen from Table 3
that the (S)-(R)-BPPFA-Rh complex shows much higher catalytic
ability in aqueous solution than in methanol or ethanol. Such a
high ability may be attributed to, in addition to the expected
steric effects, attractive interactions forming an ammonium carboxylate between the amino group in (S)-(R)-BPPFA and the carboxy group
in the olefinic substrates. Relatively low optical yields were
observed in the hydrogenation of methyl α-acetaminocinnamate (S,
21% e.e.) and of 12b with one equivalent of triethylamine added (S,
23% e.e.). This fact may support the attractive interactions mentioned above. The results summarized in Table 4 may permit one to
perceive stereochemical influence of structural factors of the
ferrocenylphosphine ligands on the catalytic asymmetric hydrogenation. The low optical yield obtained with (S)-(R)-BMPFA seems to
indicate that the steric bulkiness of the phosphorus group is
essential for asymmetric induction in the hydrogenation of α-acetaminocinnamic acid (12b). (R)-(R)-BPPFA is a diastereomeric isomer
of (S)-(R)-BPPFA, i.e., they differ only in configuration of the
carbon chirality. (R)-FcBPN does not contain the carbon central
element of chirality on the side chain of the ferrocene nucleus.
The (R)-(R)-BPPFA and (R)-FcBPN gave lower asymmetric bias than the
(S)-(R)-BPPFA, though the hydrogenated product 13b had the same S

Table 4. Asymmetric Hydrogenation of α-acetaminocinnamic acid (12b)

Ligand	Solvent	Isolated yield (%)	$[\alpha]_D$	% e.e. (Configuration)	
(S)-(R)-BPPFA	MeOH	94	+42.8	93	(S)
(S)-(R)-BMPFA	H_2O/MeOH(1/1)	65	+6.5	13	(S)
(R)-(R)-BPPFA	MeOH	96	+25.0	54	(S)
(R)-FcBPN	MeOH	89	+26.7	58	(S)
	H_2O/MeOH(1/1)	95	+32.2	70	(S)

configuration. It follows that the planar chirality of the ferrocenylphosphines plays a major part for the asymmetric induction and
the central chirality a minor one. As pointed out above, attractive
interactions between the dimethylamino group and the carboxy group
on substrate seem to contribute to the asymmetric potential of the
reaction. It may be possible to argue that the amino group in the
(S)-(R)-BPPFA catalyst is fixed to a preferable direction by the
carbon S configuration and the amino group in (R)-(R)-BPPFA or (R)-
FcBPN is not.

V. Asymmetric Hydrogenation of Prochiral Carbonyl Compounds

Recently, an increasing attention has been focused on a catalytic asymmetric hydrogenation of prochiral carbonyl compounds, which
is achieved by means of certain chiral phosphine-rhodium complexes
as catalysts [16,17,46-49]. However, the chiral rhodium catalysts
thus far reported have not afforded so satisfactory results in their
catalytic activity or asymmetry-inducing ability as to make the
asymmetric hydrogenation practically useful. Scorrano et al. [16,
46] have reported the hydrogenation of acetophenone in the presence
of $[Rh(NBD)L_2]^+ClO_4^-$ (L = (R)-BMPP or (-)-DIOP) gave 1-phenylethanol
with R configuration of 8.6% and 8.1%, respectively. Higher optical
yields, up to 51%, have been recorded by Heil and coworkers [17] in
the hydrogenation of acetophenone catalyzed by rhodium complex
generated in situ from $[Rh(1,5-hexadiene)Cl]_2$ and (+)-DIOP, though
the chemical yields were rather low. In the studies on catalytic
asymmetric hydrogenation of acrylic acids, we pointed out that
attractive interactions between functional groups on a substrate
and on a chiral ligand operate effectively in giving rise to a high
asymmetric induction. Here we describe that BPPFOH ligand, which
has a hydroxy group on the side chain of a chiral ferrocenylphos-

phine, brings about a high degree of stereoselectivity in a rhodium complex-catalyzed hydrogenation of prochiral carbonyl compounds.

A cationic rhodium catalyst precursor, [Rh(COD){(R)-(S)-BPPF-OH}]$^+$ClO$_4$$^-$ (abbr. (R)-(S)-BPPFOH-Rh$^+$) was prepared from [Rh(COD)-Cl]$_2$, AgClO$_4$, and (R)-(S)-BPPFOH according to the reported procedure [50]. Hydrogenation of carbonyl compounds was carried out in the presence of (R)-(S)-BPPFOH-Rh$^+$ (0.25-0.50 mol%) in MeOH (2% H$_2$O) solvent at 0-30°C and 50 atm initial hydrogen pressure (eq. 14).

$$R^1COR^2 \xrightarrow[\text{(R)-(S)-BPPFOH-Rh}^+]{H_2} R^1\overset{*}{C}HR^2 \atop \underset{OH}{|} \qquad (14)$$

14a: R^1 = Me, R^2 = Ph
14b: R^1 = Et, R^2 = Ph   15a-15e
14c: R^1 = Me, R^2 = Bu-t
14d: R^1 = Me, R^2 = Bu-n
14e: R^1 = Me, R^2 = COOH

The results are summerized in Table 5, which contains also data obtained with some other rhodium catalysts for comparison. Acetophenone (14a) was hydrogenated quantitatively and rapidly at 23°C in the presence of (R)-(S)-BPPFOH-Rh$^+$ to give (R)-1-phenyl-ethanol (15a) with an optical purity of 40%. An in situ catalyst (abbr. (R)-(S)-BPPFOH-Rh) formed from (R)-(S)-BPPFOH and [Rh(1,5-hexa-diene)Cl] showed almost the same degree of stereoselectivity though the chemical yield was rather low. (R)-(S)-BPPFOH-Rh$^+$ was also fairly effective for the hydrogenation of propiophenone (14b) and pinacolone (14c). Of particular interest are the high optical yields attained with BPPFOH in the hydrogenation of pyruvic acid (14e). These values (59-83% e.e.) are comparable to those reported for asymmetric reduction of α-keto esters via hydrosilylation [18b] and are the highest ever known for asymmetric hydrogenation of carbonyl compounds.

The ability of BPPFOH ligand to cause high asymmetric induction can probably be ascribed to hydrogen bonding possible between the carbonyl group on a substrate and the hydroxy group on BPPFOH ligand, which may increase conformational rigidity in the diastereo-meric transition states or intermediates. The hydrogenated products (15a and 15e) with lower optical purity and reversed configuration S were obtained with (R)-(S)-BPPFA and (S)-BPPEF, which are both analogous to (R)-(S)-BPPFOH but lack the hydroxy group. This fact may well support the above mentioned participation of the hydroxy group in asymmetric hydrogenation of carbonyl compounds.

BPPFOH was also found to be an effective ligand for rhodium complex-catalyzed asymmetric hydrogenation of imines [51]. For example, N-benzyl-1-phenylethylamine (17) of 48% optical purity was obtained from the reaction of N-benzylimine of acetophenone (16) (eq. 15).

Table 5. Asymmetric Hydrogenation of Carbonyl Compounds

Substrate	Catalyst[a]	Reaction conditions	Conversion (%)	$[\alpha]_D$[b]	Optical yield (%) (Configuration)
MeCOPh	(R)-(S)-BPPFOH-Rh$^+$	0°C, 8 hr	96	+18.6	43 (R)
MeCOPh	(R)-(S)-BPPFOH-Rh$^+$	23°C, 3 hr	100	+17.2	40 (R)
MeCOPh	(R)-(S)-BPPFOH-Rh	30°C, 47 hr	71	+15.3	35 (R)
MeCOPh	(R)-(S)-BPPFA-Rh$^+$	20°C, 65 hr	80	-6.7	15 (S)
MeCOPh	$(-)$-DIOP-Rh	30°C, 170 hr	5	—	—
EtCOPh	(R)-(S)-BPPFOH-Rh$^+$	20°C, 4 hr	94	+8.70	31 (R)
MeCOBu-t	(R)-(S)-BPPFOH-Rh$^+$	28°C, 2 hr	100	-3.29	43 (R)
MeCOBu-n	(R)-(S)-BPPFOH-Rh$^+$	20°C, 6 hr	100	+1.70	15 (S)
MeCOCOOH	(R)-(S)-BPPFOH-Rh$^+$	23°C, 1 hr	100	+4.86	59 (R)
MeCOCOOH	(R)-(S)-BPPFOH-Rh$^+$	0°C, 7 hr	100	+5.60	68 (R)
MeCOCOOH^c	(R)-(S)-BPPFOH-Rh$^+$	20°C, 16 hr	100	+6.79	83 (R)
MeCOCOOH	(R)-(S)-BPPFOH-Rh	20°C, 24 hr	100	+4.51	55 (R)
MeCOCOOH	(R)-(S)-BPPFA-Rh$^+$	20°C, 4 hr	100	-1.28	16 (S)
MeCOCOOH	(S)-BPPEF-Rh	20°C, 48 hr	71	-1.28	16 (S)
MeCOCOOH	$(-)$-DIOP-Rh	20°C, 110 hr	44	+0.47	6 (R)

[a] (R)-(S)-BPPFA-Rh$^+$ = [Rh(COD){(R)-(S)-BPPFA}]$^+$ClO$_4^-$, (S)-BPPEF-Rh = (S)-BPPEF + $\frac{1}{2}$[Rh(C$_6$H$_{10}$)Cl]$_2$, $(-)$-DIOP-Rh = $(-)$-DIOP + $\frac{1}{2}$[Rh(C$_6$H$_{10}$)Cl]$_2$. [b] In case of pyruvic acid, produced lactic acid was esterified to methyl lactate and its specific rotation measured. [c] One equiv. of Et$_3$N was added.

$$\text{PhMeC=NCH}_2\text{Ph} \xrightarrow[\text{MeOH}]{\text{H}_2/[\text{Rh*}]} \text{PhMeCHNHCH}_2\text{Ph} \qquad (15)$$

16   17

[Rh*]: (S)-(R)-BPPFOH-Rh, (S) 48% e.e.
       (S)-(R)-BPPFA-Rh, (R) 18% e.e.

VI. Asymmetric Grignard Cross-Coupling

We describe here that chiral (aminoalkylferrocenyl)phosphines are effective ligands for the nickel-phosphine catalyzed asymmetric Grignard cross-coupling to form an optically active hydrocarbon, with the aminoalkyl side chain being essential for the high asymmetric induction [53].

In 1972, Corriu and Masse [54] and we [55] reported the selective cross-coupling reaction of Grignard reagents with aryl and alkenyl halides in the presence of nickel-phosphine complexes as catalysts. For this catalysis we proposed a mechanism involving a labile diorganonickel complex as a key intermediate (Scheme 1). Since that time this catalytic reaction has been employed in synthetic practice for reasons of simple procedures, mild reaction conditions, high yields, high purity of the coupling products, and its wide applicability [56].

Scheme 1

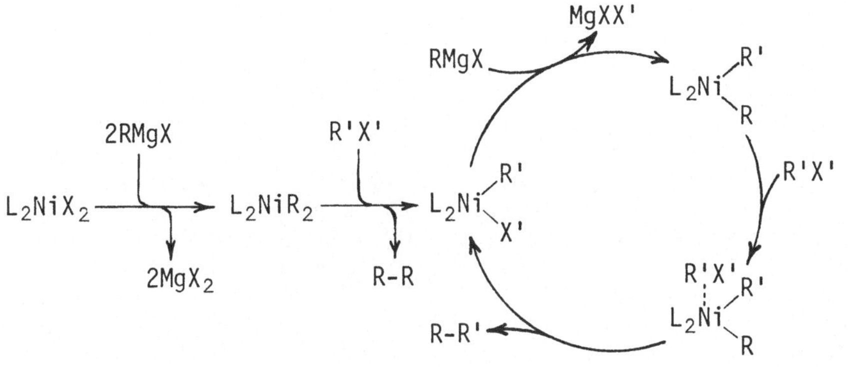

This Scheme suggests that the use of chiral phosphine-nickel complexes as catalysts will provide a new type of asymmetric reaction, giving optically active hydrocarbons. In fact, it has been reported [25,26] that the Ni[(-)-DIOP]Cl$_2$ catalyst gave partially active (7-16% e.e.) coupling products in the reaction of *sec*-alkyl Grignard reagents with phenyl and vinyl halides (eq. 16). This asymmetric reaction is classified as a kinetic resolution of racemic *sec*-alkyl Grignard reagents.

$$R^1R^2CH-MgX \quad + \quad R^3X \quad \xrightarrow{\text{[Ni*]}} \quad R^1R^2\overset{*}{C}H-R^3 \qquad (16)$$

racemic   optically active

We have examined the cross-coupling reaction of the 1-phenyl-ethyl Grignard reagent (18) with vinyl bromide (19) using chiral ferrocenylphosphines, (S)-(R)- and (R)-(R)-PPFA, (S)-FcPN, (R)-PPEF, and (S)-(R)-BPPFA (eq. 17). (S)-(R)-PPFA contains both planar and

$$PhMeCHMgCl \quad + \quad CH_2=CHBr \quad \xrightarrow{\text{[Ni*]}} \quad PhMe\overset{*}{C}HCH=CH_2 \qquad (17)$$

18   19   20

central elements of chirality, and also dimethylamino group on a side chain of the ferrocene. (R)-(R)-PPFA is a diastereomeric isomer of (S)-(R)-PPFA. (S)-FcPN lacks the central chirality and (R)-PPEF possesses the planar chirality only. These phosphines are expected to permit one to estimate separately the role which each element of chirality and the functionality may play in the present Grignard cross-coupling.

A 1:2 mixture of anhydrous $NiCl_2$ and each chiral phosphine was used as a catalyst precursor. Coupling occurred smoothly at -20 to 0°C within several hours to give optically active 3-phenyl-1-butene (20) in good chemical yields. Results summarized in Table 6 contain three significant features. Firstly, the coupling product of high optical purity (50-60%) was obtained with PPFA, FcPN, and BPPFA. Although the present asymmetric reaction is due to kinetic resolution of the *sec*-alkyl Grignard reagent, the optical purity of 20 was not largely affected by the 18/19 ratio, indicating that the inversion of the Grignard reagent 18 is fast as compared with the coupling. The high ability of the ligands to cause asymmetric induction is apparent by comparing these results with 7-16% optical purity [25,26] obtained with (-)-DIOP. Secondly, the (S)-FcPN ligand, which is analogous to (R)-(S)-PPFA but has only planar chirality, and (R)-(R)-PPFA, which has R configuration of carbon central chirality as opposed to its epimer (S)-(R)-PPFA having S configuration, showed the asymmetric induction of comparable efficiency to the (S)-(R)- or (R)-(S)-PPFA ligand. The result demonstrates that ferrocene planar chirality plays an important role rather than the carbon central chirality. Finally, a dramatic decrease in the asymmetric induction was observed with PPEF as a ligand which contains no dimethylamino group. Thus the dimethylamino group on chiral ferrocenylphosphines is the first requisite for the high stereoselectivity in the present asymmetric cross-coupling reaction.

The important role of the amino group may be visualized by its

Table 6. Asymmetric Cross-Coupling of 18 with 19 Catalyzed by
Nickel Complexes with PPFA, BPPFA, FcPN, and PPEF[a]

Chiral ligand	18/19[b]	Reaction temp (°C)	Yield[c] (%)	$[\alpha]_D^{22}$	Optical purity (Configuration)	
(R)-(S)-PPFA	4	-20	99	+4.04	63	(S)
(S)-(R)-PPFA	4	0	99	-3.75	59	(R)
(S)-(R)-PPFA	2	0	97	-3.58	56	(R)
(S)-(R)-PPFA[d]	2	0	98	-3.62	57	(R)
(S)-(R)-PPFA	1	0	83	-3.32	52	(R)
(R)-(R)-PPFA	2	0	99	-3.19	50	(R)
(S)-FcPN	4	0	98	+3.82	60	(S)
(R)-PPEF	4	0	86	+0.72	4	(S)
(S)-(R)-BPPFA[e]	4	0	73	-3.82	60	(R)

[a] Ratio [Ni*]/19 = 5×10^{-3}. [b] Concentration of 18 in ether was 1.5
M unless otherwise noted. [c] Yields based on 19 used were determined
by GLC. [d] 18 of 0.5 M was used. [e] (S)-(R)-BPPFA/Ni = 1.

Table 7. Asymmetric Cross-Coupling of 18 with 19 Catalyzed by
Nickel Complexes with (S)-(R)-PPFNR$_2$

NR_2	NMe_2 (PPFA)	NEt_2 (21)	N⬡ (22)	N⬡O (23)	N⬡NMe (24)
% e.e. of 20	59	32	34	16	60
(Configuration)	R	R	S	R	R

strong ability to coordinate with the magnesium atom in the Grignard
reagent. It is most probable that the configuration of the coupling
product has already been determined before the chiral carbon-nickel
bond is formed. The magnesium atom in the Grignard reagent must be
coordinated with the dimethylamino group and such a complexation
should increase the stereoselectivity via the enhanced steric
interactions.

The data in Table 7 obtained with several other (aminoalkyl)-
ferrocenylphosphines, PPFNR$_2$, (21-24) could provide further insight
into the present asymmetric Grignard cross-coupling reaction. The
steric bulkiness of the amino substituent had a powerful effect on

Table 8.   Asymmetric Cross-Coupling of 25 with 26 Catalyzed by
           Chiral Phosphine-Nickel Complexes[a]

Chiral ligand	3-Phenyl-1-butene (20)		
	Yield (%)	$[\alpha]_D^{22}$	Optical purity (Configuration)
(S)-(R)-PPFA	16	+1.21	18.8 (S)
(S)-(R)-BPPFA	58	+2.14	33.4 (S)
(R)-(S)-BPPFA	58	-2.16	33.8 (R)
$(-)$-DIOP	28	-0.69	5.7 (R)

[a] $[Ni^*]/26/25 = 5 \times 10^{-3}/1/1.2$.   Refluxed for 46 hr.

the stereoselection. Thus, the ferrocenylphosphine with piperidino
group (22) afforded S configuration of 20 with 34% e.e., while the
use of phosphines with dimethylamino group (PPFA) and with diethyl-
amino group (21) resulted in 59% (R) and 32% (R), respectively.
Morpholino and N-methylpiperazino groups are thought to have almost
the same size of steric bulkiness as the piperidino group, but the
phosphines 22, 23, and 24 gave very different stereochemical
results. A chelate coordination is expected in the reaction with
24 or 23, with two nitrogen atoms or both nitrogen and oxygen atoms
coordinating to the magnesium atom in the Grignard reagent. The
piperidino group must coordinate to the magnesium atom with one
nitrogen atom. This difference in the manner of coordination is
most likely the cause of the different stereochemical control.
      Another type of asymmetric induction was also possible in the
coupling reaction of phenylmagnesium bromide (25) with 4-bromo-1-
butene (26) (eq. 18) [57]. This reaction is a true asymmetric

$$PhMgBr \quad + \quad CH_2{=}CHCH_2CH_2Br \xrightarrow{\quad [Ni^*] \quad} PhMe\overset{*}{C}HCH{=}CH_2 \qquad (18)$$

$$25 \qquad\qquad\qquad 26 \qquad\qquad\qquad\qquad 20$$

synthesis in which the achiral halide is converted into a chiral
moiety by the action of catalyst, while the previous reaction (eq.
17) is a catalytic kinetic resolution of the racemic sec-alkyl
Grignard reagent. The results are summarized in Table 8. It seems
of interest to compare these results with those obtained from reaction (17). Firstly, optical yields attained by reaction (18) are
about one half or less of those by reaction (17) for each chiral
phosphine ligand. With respect to stereoselectivity, ferrocenylphosphines are much more effective ligands than DIOP. Secondly,

these two types of asymmetric coupling reaction give the enantiomeric products in the presence of the same chiral phosphine ligand;
e.g., with (*S*)-(*R*)-BPPFA as a ligand (*R*)-(-)-**20** is obtained from
reaction (17), while reaction (18) forms (*S*)-(+)-**20**.

VII. Conclusions

We have shown that the functionalized chiral ferrocenylphosphine ligands give rise to a high asymmetric induction in several
transition metal complex catalyzed asymmetric reactions. The high
ability of the chiral ferrocenylphosphines is ascribed not only to
their highly asymmetric structure but also to the attractive interactions between functional groups on a substrate and on the chiral
ligands coordinated to a transition metal catalyst. Although the
correlations between the structural features of chiral ligands and
the extent of asymmetric induction are not unambiguous at this time,
the results described here may provide a new conception for the
design and synthesis of chiral phosphine ligands of high ability
for asymmetric catalytic reactions [58].

VIII. Acknowledgements

We thank the Ministry of Education, Japan, for Grant-in-Aid
(No. 011006) and the Asahi Glass Foundation for the Contribution to
Industrial Technology for financial support.

References

1 J.D. Morrison and H.S. Mosher, "Asymmetric Organic Reactions"
   Prentice-Hall Englewood Cliffs, New Jersey, 1971.
2 J.W. Scott and D. Valentine, Jr., *Science*, **184**, 943 (1974).
3 For a review: J.D. Morrison, W.F. Masler, and M.K. Neuberg,
   Advan. Catalysis, **25**, 81 (1976).
4 L. Horner, H. Winkler, A. Rapp, A. Mentrup, H. Hoffman, and
   P. Beck, *Tetrahedron Lett.*, 161 (1961).
5 L. Horner, H. Siegel, and H. Büthe, *Angew. Chem.*, **80**, 1034
   (1968).
6 W.S. Knowles and M.J. Sabacky, *Chem. Commun.*, 1445 (1968).
7 (a) O. Korpium and K. Mislow, *J. Amer. Chem. Soc.*, **89**, 4784
   (1967); (b) K. Naumann, G. Zon, and K. Mislow, *J. Amer. Chem.
   Soc.*, **91**, 7012 (1969).
8 (a) W.S. Knowles, M.J. Sabacky, and B.D. Vineyard, *J.C.S. Chem.
   Commun.*, 10 (1972); (b) *Idem, Chem. Tech.*, 591 (1972); (c) *Idem,
   Ann. N.Y. Acad. Sci.*, **214**, 119 (1973).
9 W.S. Knowles, M.J. Sabacky, B.D. Vineyard, and D.J. Weinkauff,
   J. Amer. Chem. Soc., **97**, 2569 (1975).

10 (a) T.P. Dang and H.B. Kagan, *Chem. Commun.*, 481 (1971); (b)
    H.B. Kagan and T.P. Dang, *J. Amer. Chem. Soc.*, **94**, 6429 (1972).
11 (a) H.B. Kagan, N. Langlois, and T.P. Dang, *J. Organometal.*
    Chem., **90**, 353 (1975); (b) G. Gelbard, H.B. Kagan, and S. Stern,
    Tetrahedron, **32**, 233 (1976) and references cited therein.
12 R. Glaser and B. Vainas, *J. Organometal. Chem.*, **121**, 249 (1976)
    and their previous papers cited therein.
13 M. Tanaka, Y. Watanabe, T. Mitsudo, Y. Yasumori, and Y.
    Takegami, *Chem. Lett.*, 137 (1974).
14 A.P. Stoll and R. Süess, *Helv. Chim. Acta*, **57**, 2487 (1974).
15 B.R. James, D.K.W. Wang, and R.F. Voigt, *J.C.S. Chem. Commun.*,
    574 (1975).
16 A. Levi, G. Modena, and G. Scorrano, *J.C.S. Chem. Commun.*, 6
    (1975).
17 B. Heil, S. Törös, S. Vastag, and L. Marko, *J. Organometal.*
    Chem., **94**, C47 (1975).
18 (a) I. Ojima, T. Kogure, M. Kumagai, S. Horiuchi, and T. Sato,
    J. Organometal. Chem., **122**, 83 (1976); (b) I. Ojima, T. Kogure,
    and M. Kumagai, *J. Org. Chem.*, **42**, 1671 (1977) and their
    previous papers cited therein.
19 T. Hayashi, Dissertation, Kyoto University, 1975.
20 W. Dumont, J.C. Poulin, T.P.Dang, and H.B. Kagan, *J. Amer.*
    Chem. Soc., **95**, 8295 (1973).
21 R.J.P. Corriu and J.J.E. Moreau, *J. Organometal. Chem.*, **85**,
    19 (1975).
22 (a) G. Consiglio, C. Botteghi, C. Salomon, and P. Pino, *Angew.*
    Chem., **85**, 663 (1973); (b) *Idem, Chimia*, **27**, 215 (1973); (c) G.
    Congilio and P. Pino, *Helv. Chim. Acta*, **59**, 642 (1976).
23 B. Stern, A. Hirshauer, and L. Sajus, *Tetrahedron Lett.*, 3247
    (1973).
24 M. Tanaka, Y. Ikeda, and I. Ogata, *Chem. Lett.*, 1115 (1975).
25 G. Consiglio and C. Botteghi, *Helv. Chim. Acta*, **56**, 460 (1973).
26 Y. Kiso, K. Tamao, N. Miyake, K. Yamamoto, and M. Kumada,
    Tetrahedron Lett., 3 (1974).
27 B.M. Trost and P.E. Strege, *J. Amer. Chem. Soc.*, **99**, 1649 (1977).
28 J.D. Mirrison, R.E. Burnett, A.M. Aguiar, C.J. Morrow, and C.
    Phillips, *J. Amer. Chem. Soc.*, **93**, 1301 (1971).
29 (a) J.D. Morrison and W.F. Masler, *J. Org. Chem.*, **39**, 270 (1974);
    (b) A.M. Aguiar, C.J. Morrow, J.D. Morrison, R.E. Burnett, W.F.
    Masler, and N.S. Bhacca, *J. Org. Chem.*, **41**, 1545 (1976). See
    also: M. Tanaka and I. Ogata, *Bull. Chem. Soc. Japan*, **48**, 1094
    (1975). Y. Kiso, K. Yamamoto, K. Tamao, and M. Kumada, *J. Amer.*
    Chem. Soc., **94**, 4373 (1972).
30 B. Bogdanović, B. Henc, A. Lösler, B. Meister, H. Pauling, and
    G. Wilke, *Angew. Chem.*, **85**, 1013 (1973).
31 F. Mizukami, Y. Ikeda, M. Tanaka, and I. Ogata, 22nd Symposium
    on Organometallic Chemistry Japan, 311A (1974).
32 K. Achiwa, *J. Amer. Chem. Soc.*, **98**, 8265 (1976).

33 K. Tamao, H. Yamamoto, H. Matsumoto, N. Miyake, T. Hayashi, and M. Kumada, *Tetrahedron Lett.*, 1389 (1977).

34 D. Marquarding, H. Klusacek, G. Gokel, P. Hoffmann, and I. Ugi, *J. Amer. Chem. Soc.*, 92, 5389 (1970).

35 T. Hayashi, K. Yamamoto, and M. Kumada, *Tetrahedron Lett.*, 4405 (1974).

36 C.C. Price and J.R. Sowa, *J. Org. Chem.*, 32, 4126 (1967).

37 T. Hayashi, T. Mise, and M. Kumada, *Tetrahedron Lett.*, 4351 (1976).

38 T. Hayashi, T. Mise, Y. Tamura, and M. Kumada, unpublished results.

39 M. Rosenblum, "Chemistry of the Iron Group Metallocenes," Wiley, New York, N.Y., 1965, p 40.

40 I. Ojima, K. Yamamoto, and M. Kumada, in R. Ugo, Ed., "Aspects of Homogeneous Catalysis," Vol. 3, D. Reidel Pub. Co., Donarecht-Holland, 1977.

41 (a) T. Hayashi, K. Yamamoto, and M. Kumada, *J. Organometal. Chem.*, 112, 253 (1976); (b) T. Hayashi, K. Yamamoto, K. Kasuga, H. Omizu, and M. Kumada, *J. Organometal. Chem.*, 113, 127 (1976).

42 (a) T.B. Rauchfuss and D.M. Roundhill, *J. Amer. Chem. Soc.*, 96, 3098 (1974); (b) J.C. Kotz, C.L. Nivert, and J.M. Lieber, *J. Organometal. Chem.*, 84, 255 (1975).

43 Y. Kiso, M. Kumada, K. Tamao, and M. Umeno, *J. Organometal. Chem.*, 50, 297 (1973).

44 For reviews: (a) B.R. James, "Homogeneous Hydrogenation," Wiley, New York, N.Y., 1973; (b) R.E. Harmon, S.K. Gupta, and D.J. Brown, *Chem. Rev.*, 73, 21 (1973); (c) A.J. Birch and D.H. Williamson, "Organic Reactions," Vol. 24, Wiley, New York, N.Y., 1976, p 1.

45 T. Hayashi, T. Mise, S. Mitachi, K. Yamamoto, and M. Kumada, *Tetrahedron Lett.*, 1133 (1976).

46 P. Bonvicini, A. Levi, G. Modena, and G. Scorrano, *J.C.S. Chem. Commun.*, 1188 (1972).

47 M. Tanaka, Y. Watanabe, T. Mitsudo, H. Iwane, and Y. Takegami, *Chem. Lett.*, 239 (1973).

48 J. Solodar, *Chem. Tech.*, 421 (1975).

49 C.J. Sih, J.B. Heather, G.P. Peruzzotti, P. Price, R. Sood, and L.F.H. Lee, *J. Amer. Chem. Soc.*, 95, 1676 (1973).

50 R.R. Schrock and J.A. Osborn, *J. Amer. Chem. Soc.*, 93, 3089 (1971).

51 T. Hayahsi, M. Nakatsuka, A. Katsumura, and M. Kumada, unpublished results.

52 R.R. Schrock and J.A. Osborn, *Chem. Commun.*, 567 (1970).

53 T. Hayashi, M. Tajika, K. Tamao, and M. Kumada, *J. Amer. Chem. Soc.*, 98, 3718 (1976).

54 R.J.P. Corriu and J.P. Masse, *J.C.S. Chem. Commun.*, 144 (1972).

55 K. Tamao, K. Sumitani, and M. Kumada, *J. Amer. Chem. Soc.*, 94, 4374 (1972).

56 K. Tamao, K. Sumitani, Y. Kiso, M. Zembayashi, A. Fujioka, S.
    Kodama, I. Nakajima, A. Minato, and M. Kumada, *Bull. Chem. Soc.
    Japan*, **49**, 1958 (1976) and references cited therein.
57 M. Zembayashi, K. Tamao, T. Hayashi, T. Mise, and M. Kumada,
    Tetrahedron Lett., 1799 (1977).
58 Since this article was written, a paper concerning a new chiral
    phosphine ligand, (2S,3S)-bis(diphenylphosphino)butane, appeared:
    M.D. Fryzuk and B. Bosnich, *J. Amer. Chem. Soc.*, **99**, 6262 (1977).

HOMOGENEOUS ASYMMETRIC HYDROGENATION AND HYDROSILYLATION OF KETO ESTERS CATALYZED BY CHIRAL RHODIUM COMPLEXES

Iwao Ojima

Sagami Chemical Research Center

Nishi-Ohnuma 4-4-1, Sagamihara, Kanagawa 229, Japan

Although the homogeneous hydrogenation of carbon-carbon multiple bonds catalyzed by various transition metal complexes has been extensively studied,[1] e.g., asymmetric hydrogenation of acetamidocinnamic acid is a great success,[2] that of carbon-oxygen double bonds has received relatively little attention. Thus, there had been only a few papers[3] before Schrock and Osborn reported[4] in 1970 that cationic rhodium complexes with relatively basic phosphines as ligands catalyze the reduction of ketones under mild conditions. This is partly due to the fact that the Wilkinson's type non-cationic rhodium(I) complexes lack activity toward the hydrogenation of carbonyl functionalities[5] and that these complexes also catalyze the decarbonylation of carbonyl compounds, especially of aldehydes.[5,6] On the basis of these findings a catalytic asymmetric hydrogenation of ketones has been achieved with low optical yields in 1972.[7] Along this line, Solodar reported a good result on the asymmetric hydrogenation of methyl acetoacetate.[8] Ohgo and coworkers also reported the asymmetric homogeneous hydrogenation of carbonyl groups with Co(dimethylglyoxymato)$_2$(quinine)(benzylamine) as a catalyst, but except for the rather special case of the reduction of benzil to benzoin, the product alcohols were obtained with a low enantiomeric excess.[9]

A similar trend can be seen in catalytic hydrosilylation. The hydrosilylation of olefins and acetylenes has been extensively studied in the last two decades.[10] However, hydrosilylation of carbon-hetero atom multiple bonds has received less attention. Since a silicon-oxygen bond or silicon-nitrogen bond can be easily hydrolyzed, the hydrosilylation of carbonyl compounds or imines is equivalent to hydrogenation. Thus, the catalytic hydrosilylation of the compounds containing such a double bond may provide a powerful new reduction method if the reaction is achieved effectively. In 1972, we found that tris(triphenylphosphine)chlororhodium was quite effective for the hydrosilylation of carbonyl compounds.[11] Corriu and Moreau also studied[12] the addition of dihydrosilanes to ketones in the presence of tris(triphenylphosphine)dichlororuthenium as well as the rhodium complex, but the ruthenium(II) complex appeared to be less effective than the rhodium(I) complex. Thus, the hydrosilylation of carbonyls catalyzed by the rhodium complex followed by hydrolysis can serve as a powerful new method in organic synthesis and has already been developed in the selective reductions of terpene carbonyl compounds[13] and Schiff bases.[14] Asymmetric reduction of prochiral ketones by hydrosilylation in the presence of a chiral rhodium catalyst followed by hydrolysis has been developed by four groups independently after our first report in 1972 and continues to be active investigation.[15]

Among many asymmetric reactions, the synthesis of optically active α-hydroxycarboxylic acids has gathered much interest for a long time, and a large number of reports has been published on the Grignard reaction and the reduction of α-keto esters.[16] The asymmetric reductions of chiral α-keto esters by catalytic hydrogenation and metal hydride reduction have been extensively studied. Relatively little is known, however, about the asymmetric reduction of α-keto esters by chiral reducing agents, although reductions of phenylglyoxylic acid and its esters by the use of chiral magnesium alkoxides[17] and lithium aluminum hydride-chiral alcohol complexes[18] have been reported. The optical yield attained by the former agents was reported to be 15-33%, and that obtained by the latter systems was 4-17%. Ohgo et al. have reported the catalytic asymmetric hydrogenation of phenylglyoxylates using Co(dimethylglyoxymato)$_2$-quinine complex to give mandelates in 11.5-19.5% optical yield.[9a] As for β-keto esters, an effective asymmetric hydrogenation of methyl acetoacetate (71%ee) has been reported by Solodar using a cationic rhodium complex with chiral phosphines.[8] We would like to describe here our recent advance in the asymmetric reduction of keto esters via homogeneous hydrogenation and hydrosilylation catalyzed by rhodium complexes with chiral phosphine ligands, which has achieved the high optical yield (> 80%) production of hydroxy esters in excellent yields.

HOMOGENEOUS ASYMMETRIC HYDROGENATION OF α-KETO ESTERS

The asymmetric hydrogenation of ketones catalyzed by cationic rhodium complexes is attracting much interest.[7,8,19] However, the optical yields attained in the reaction have been rather low so far for simple prochiral ketones. Recently, Hayashi et al. developed a unique chiral ferrocenyldiphosphine, BPPFOH, which brought about the effective asymmetric hydrogenation of prochiral ketones

when it was employed as ligand in a rhodium complex, and high optical yields were realized for the hydrogenation of pyruvic acid.[20]
Nevertheless, the asymmetric hydrogenation of methyl pyruvate catalyzed by the same rhodium catalyst resulted in only 10% asymmetric induction.[21] We have found that neutral Wilkinson-type catalysts are quite effective for the hydrogenation of α-keto esters
although it is known that the corresponding neutral rhodium complexes exhibit only a low catalytic activity toward the hydrogenation of simple ketones, the Wilkinson catalyst itself, $[(Ph_3P)_3$-
RhCl], being completely ineffective.[19d] We report here the effective homogeneous hydrogenation of pyruvate and phenylglyoxylate
catalyzed by neutral rhodium complexes with phosphine ligand and
its application to the asymmetric hydrogenation of pyruvates and
ketopantoyl lactone.

Homogeneous Hydrogenation of α-Keto Esters Catalyzed by
Neutral Rhodium(I) Complexes with Phosphine Ligands[22]

Typically, the hydrogenation of isobutyl pyruvate was carried
out in dry benzene at 25°C under initial hydrogen pressure of 50
atm in the presence of tris(triphenylphosphine)chlororhodium (0.5
mol%). The reaction completed within 24 hr to give isobutyl lactate in quantitative yield.

As Table 1 shows, the pressure of hydrogen exerts a large influence on the reaction rate, viz., the hydrogenation of methyl,
propyl and isobutyl pyruvate catalyzed by $[(Ph_3P)_3RhCl]$ or
$[(DPPB)Rh(S)Cl]$ [DPPB = 1,4-bis(diphenylphosphino)butane, S = sol-

Table 1
Effects of Hydrogen Pressure on the Rate of
the Hydrogenation of Isobutyl Pyruvate[a]

Pressure of Hydrogen(atm)	Yield (%) of Isobutyl Lactate[b]	
	$(Ph_3P)_3RhCl$	$(DPPB)Rh(S)Cl$
1	44	22
20	98	95
50	100	100

a All reactions were carried out with 15 mmol of isobutyl pyruvate and 7.5×10^{-2} mmol of the catalyst in
5 ml of dry benzene at 20°C for 24 hr. b The yield
was estimated by GLC analysis.

vent], was virtually accomplished within 20-30 hr under an initial
hydrogen pressure of 20-50 atm at ambient temperature, whereas it
required 70-100 hr under atmospheric pressure of hydrogen. The
reaction rate is also markedly dependent upon the structure of the
substrate, e.g., the hydrogenation of ethyl phenylglyoxylate was
fairly slow and 72 hr was necessary to complete the reaction under
a hydrogen pressure of 50 atm at 30°C in tetrahydrofuran (THF).
The solvent effects on the rate of the hydrogenation of isobutyl
pyruvate are illustrated in Figure 1. It is noteworthy that the
effects of solvents on the catalytic activities of $[(Ph_3P)_3RhCl]$
and [(DPPB)Rh(S)Cl] are considerably different from each other.
The difference may principally be due to the <u>trans</u> configuration
in the active species of the former catalyst and <u>cis</u> configuration
in the latter.

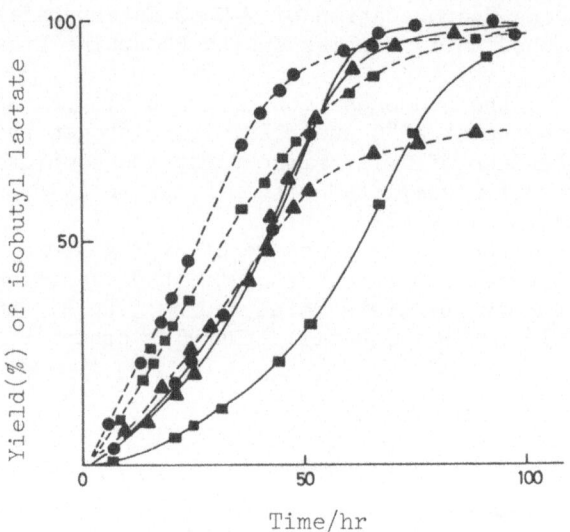

Figure 1. Solvent effects on the rate of the hydrogenation of
isobutyl pyruvate (15 mmol) under atmospheric pressure of
hydrogen at 20°C in 5 ml of dry solvent: ----, with $[(Ph_3P)_3Rh-$
Cl]; ——, with [(DPPB)Rh(S)Cl]. ● in C_6H_6; ■ in THF;
▲ in C_6H_6-MeOH (1:1).

Asymmetric Hydrogenation of Pyruvates Catalyzed by Neutral
Rhodium(I) Complexes with Chiral Phosphines[22]

Asymmetric hydrogenation of methyl, propyl and isobutyl pyruvate was carried out in dry organic solvent at 20°C under an initial hydrogen pressure of 20 atm in the presence of the rhodium

catalyst (0.5 mol%) which was prepared in situ from $[Rh(COD)Cl]_2$
(COD = cycloocta-1,5-diene) and chiral diphosphine, (2S,4S)-N-t-
butoxycarbonyl-4-diphenylphosphino-2-diphenylphosphinomethylpyr-
rolidine [(-)BPPM][23] or (-)-2,3-isopropylidene-2,3-dihydroxy-1,4-
bis(diphenylphosphino)butane [(-)DIOP].[24] The reaction was virtually complete after 24 hr to afford the corresponding (R)-(+)-lac-
tates in almost quantitative yields.

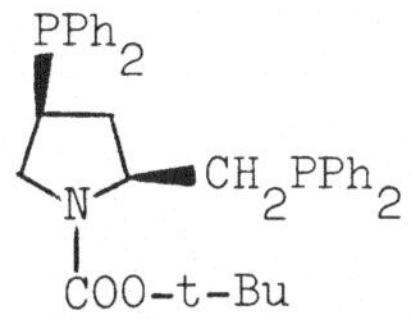

(-)BPPM 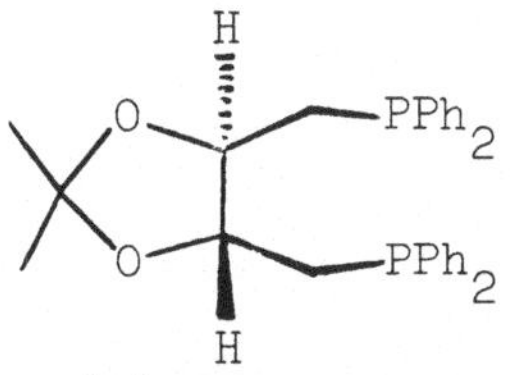   (-)DIOP

The results (Table 2) show that the optical yield is clearly
dependent on the nature of the chiral ligand employed, i.e., BPPM
is much more effective than DIOP in the asymmetric induction. The
remarkable effect of solvent is also illustrated in Table 2 for
the reduction of methyl pyruvate. It should be noted that a dry
aprotic solvent, namely, benzene or THF, gives much better results
than methanol does in the present system although a protic solvent,
e.g., methanol or ethanol, is usually employed[8,19] for good results
in the hydrogenation of ketones catalyzed by cationic rhodium complexes; also addition of small quantities of water (1%) to the
present system does not increase the rate, but it decreases the
extent of asymmetric induction. Therefore, it is strongly suggested that the mechanism of the present reaction is different
from that proposed by Schrock and Osborn for the reactions catalyzed by cationic rhodium complexes.[4]

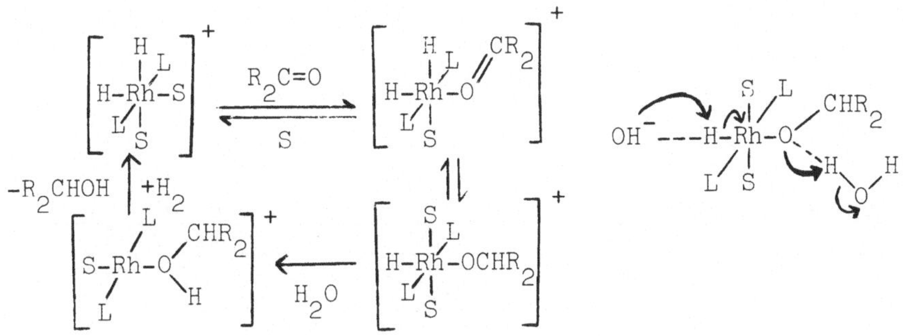

L = phosphine ligand; S = solvent

Scheme 1. Schrock-Osborn Mechanism

Table 2
Asymmetric Hydrogenation of the Esters of Pyruvic Acid[a]

$$CH_3COCOOR \xrightarrow[\text{H}_2]{[Rh]^*} CH_3\overset{*}{C}HCOOR$$
$$\underset{OH}{|}$$

R	Ligand	Solvent[b]	Time (hr)	Conversion (%)[c]	$[\alpha]_D^{18-20}$ (neat)	Optical Purity(%ee)[d]
Me	(−)BPPM	benzene	24	100	+5.47	66.3
		THF	24	97	+5.40	65.4
		MeOH	90	87	+3.49	42.4
	(−)DIOP	benzene	24	100	+2.62	31.7
		THF	20	99	+3.40	41.2
		THF[e]	20	93	+2.89	35.0
		monoglyme	20	99	+3.12	37.8
		MeOH	90	95	+1.50	18.2
n-Pr	(−)BPPM	benzene	24	99	+9.17	75.8
		THF	24	98	+9.17	75.8
	(−)DIOP	THF	24	100	+5.07	41.9
		THF[e]	24	91	+4.85	40.1
i-Bu	(−)BPPM	benzene	24	100	+10.70	70.7
		THF	24	100	+10.76	71.1
	(−)DIOP	THF	24	99	+5.73	37.9

[a] All hydrogenations were run with 15 mmol of pyruvate, 3.8×10^{-2} mmol of $[Rh(COD)Cl]_2$, and 9.0×10^{-2} mmol of diphosphine in 4 ml of solvent at 20°C under initial hydrogen pressure of 20 atm. [b] Dry solvent was used unless otherwise noted. [c] Determined by GLC. [d] Optical yields were calculated on the basis of reported values for the specific rotations of pure enantiomers: (S)-methyl lactate (Ref. 25); (S)-n-propyl lactate (Ref. 26); (S)-i-butyl lactate (Ref. 27). [e] 1% water was added.

Asymmetric hydrogenation of propyl pyruvate catalyzed by cationic rhodium complex with (−)DIOP or (−)BPPM as chiral ligand. In order to clarify the differences between neutral complexes and cationic complexes in the mode of catalysis, we prepared the cationic rhodium catalysts with (−)DIOP and (−)BPPM, and carried out the asymmetric hydrogenation of propyl pyruvate under similar conditions to those employed in the neutral rhodium complex catalyzed reaction. The results are listed in Table 3.[28]

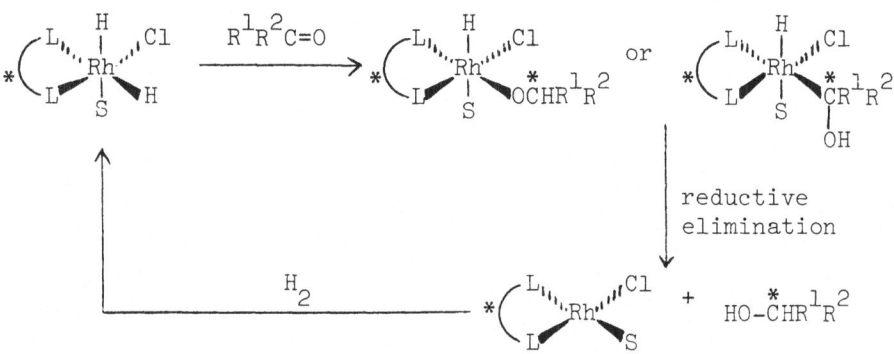

Scheme 2.   Possible Mechanism for the Neutral Rhodium Complex
             Catalyzed Hydrogenation

Table 3
Asymmetric Hydrogenation of Propyl Pyruvate Catalyzed
by Cationic Rhodium Complexes

$$CH_3COCOOC_3H_7 \xrightarrow[\;H_2\;(20\;atm),\;20°C\;]{[Rh]^+} CH_3\overset{*}{C}HCOOC_3H_7$$
$$\underset{OH}{}$$

Catalyst	Solvent	Conversion(%)	Time (hr)	$[\alpha]_D^{18}$ [a]	Optical Purity(%ee)
$[(-)DIOP-Rh(COD)]^+ClO_4^-$	MeOH	100	20	-2.95(S)	24.4
	MeOH[b]	98	6	-2.11(S)	17.5
	benzene	99	20	-1.90(S)	15.7
$[(-)BPPM-Rh(COD)]^+ClO_4^-$	MeOH	100	16	+2.44(R)	20.2

[a] Measured as neat liquid.   [b] 1% water was added.

As Table 3 shows, i) achieved optical yields of the reaction
are considerably lower than those attained in the neutral complex
catalyzed reaction, ii) inversion of the direction of asymmetric
induction is observed in the case of (-)DIOP ligand, iii) methanol
gives better results than benzene does, iv) addition of 1% water
to the solvent (MeOH) increases the reaction rate, but it decreases
the optical yield. The findings, iii) and iv), clearly indicate
that the reaction follows the Schrock-Osborn mechanism.

Accordingly, it is proven that the mechanism of the neutral
complex catalyzed hydrogenation of pyruvate is entirely different
from that of the cationic complex catalyzed one. At this stage,
we postulate that the reaction catalyzed by neutral rhodium complex proceeds in a manner similar to that proposed for the hydrogenation of olefins catalyzed by Wilkinson **catalyst** (Scheme 2).

Asymmetric Hydrogenation of Ketopantoyl Lactone Catalyzed
by Neutral Rhodium Complexes and Its Application to the
Asymmetric Synthesis of D-(+)-Pantothenate[29]

Pantothenic acid is a member of the B complex vitamines and
is an important constitute of Coenzyme A: Pantothenic acid is
converted to pantetheine, which further reacts with ATP to form
Coenzyme A. The biosynthesis of pantothenic acid from valine has
been postulated[30] to involve (a) the oxidative deamination of valine
to α-ketoisovaleric acid, (b) the hydroxymethylation of this acid
to form ketopantoyl lactone, (c) the asymmetric reduction of ketopantoyl lactone to pantoyl lactone, and (d) the coupling of pantoyl lactone with β-alanine to pantothenic acid. Among these processes the step (c) is the most significant process since only D-
(+)-pantothenic acid derived from D-(-)-pantoyl lactone has the
biological activities.[31] Although the biological synthesis of D-
(+)-pantothenic acid has been reported using microbial reduction
of ketopantoyl lactone to pantoyl lactone,[32] no attempts have
been done on the chemical asymmetric synthesis of this substance
following the biosynthetic route. We have found that a rhodium
complex with (-)BPPM displays a high recognition ability comparable
to that of microorganisms, and thus the chiral rhodium complex can
be considered as a functional biomimetic model of the ketopantoyl
lactone reductase. We describe here an effective biomimetic route
to D-(+)-pantothenate using the catalytic asymmetric hydrogenation
in the key step.

One of the key compounds in the biosynthetic route is ketopantoyl lactone since the asymmetric reduction of this compound is
the characteristic process in the biological systems, which forms
a sharp contrast to the optical resolution of racemic pantoyl lactone employed in the commercial synthesis of D-(+)-pantothenic acid
derivatives.[33] As the formation of ketopantoyl lactone is not restricted to enzymatic process but a simple aldol condensation, we
started the asymmetric synthesis from ketopantoyl lactone.

a) transaminase   b) ketopantoaldolase   c) reductase
d) pantothenate synthetase

The asymmetric hydrogenation of ketopantoyl lactone was carried out by means of the rhodium catalyst with BPPM as chiral ligand to afford D-(-)-pantoyl lactone with the optical purity of
86.7%ee in almost quantitative yield under optimum conditions.
The results obtained under a variety of conditions are summarized
in Table 4. The corresponding asymmetric hydrogenation using (-)-
DIOP as chiral ligand in THF resulted in only 35% asymmetric synthesis. As Table 4 shows, i) the optical yield is affected by the
solvent employed and benzene affords the best results as far as we
have examined, and ii) remarkable effect of the reaction temperature on the optical yield is observed: It is of interest that the
extent of asymmetric induction decreases precipitously at temperatures below 20°C, e.g., 10°C in benzene or 0°C in THF. This phenomenon would be interpreted as i) the change of the rate determining step or ii) the exchange of one mechanism for the other
provided that the reaction proceeds via two parallel mechanisms.
The configurational change of the chiral ligand in the rhodium
complex would be also suggested.
     The pantoyl lactone thus obtained was easily purified to the
pure D-isomer by recrystallization from n-hexane-benzene. Accordingly, pure sample of D-(-)-pantoyl lactone was obtained in at
least 70% yield from ketopantoyl lactone. The pure sample of D-
(-)-pantoyl lactone was converted to ethyl ester of D-(+)-panto-
thenic acid by the reaction with β-alanine ethyl ester.[34] The
transformation of ethyl D-(+)-pantothenate to D-(+)-pantothenic
acid is a known process.[31] As the optical yield of the reduction
of ketopantoyl lactone using baker's yeast has been reported[32] to
be ca. 72%, our chiral rhodium catalyst is shown to be superior to
baker's yeast in this reaction. Although Lanzilotta et al. recently have found that specific strains of the ascomycete, Byssochlamys
fulva, can achieve exceedingly high optical yield production of
the D-isomer,[32] the isolation procedure from aqueous reaction
media, i.e., extraction, recovery of raw materials, and purifi-

cation, is very troublesome because of the high solubility of the product in water. Thus, the present process has some advantages in a synthetic point of view, e.g., i) conversion of the reaction is virtually 100%, ii) the isolation of the product is quite simple and convenient since the reaction is carried out in small quantity of non-aqueous media, and iii) the D-(-)-pantoyl lactone with low optical purity, which is remained in the mother liquor in the recrystallization, is easily oxidized to the starting keto-pantoyl lactone.

Table 4

Asymmetric Hydrogenation of Ketopantoyl Lactone to D-(-)-Pantoyl Lactone Catalyzed by the Rhodium(I) Complex with BPPM[a]

Solvent	Initial H_2 Pressure (atm)	Conditions	Conversion (%) [b]	$[\alpha]_D^{25}$ [c]	Optical Purity (%ee) [d]
benzene	50	10°, 48hr	95.4	-23.4	46.2
benzene	50	20°, 48hr	99.2	-43.4	85.5
benzene	50	30°, 48hr	100.0	-44.0	86.7
benzene	50	50°, 24hr	100.0	-43.0	84.8
THF	50	0°, 70hr	46.1	- 9.65	19.0
THF	50	30°, 48hr	99.5	-40.9	80.7
chloro-benzene	50	50°, 48hr	94.5	-32.2	63.5
toluene	50	50°, 48hr	99.6	-39.4	77.7

a) All reactions were carried out with 5 mmol of ketopantoyl lactone in the presence of 1 mol% of catalyst in 5 ml of solvent. The BPPM-rhodium complex catalyst was prepared in situ from $[Rh(COD)Cl]_2$ and BPPM in the solvent. [BPPM]/[Rh] = 1.12-1.17. b) Determined by GLC analysis. c)Measured in water: c = 2.101-2.098. d) Optical purity was calculated on the basis of the maximum rotation of the pure enantiomer (Ref. 31).

HOMOGENEOUS ASYMMETRIC REDUCTION OF KETO ESTERS VIA HYDROSILYLATION[35]

The asymmetric reduction of α-,β- and γ-keto esters is described here as an application of our work on the reduction of

Table 5
Asymmetric Reductions of Pyruvates and Phenylglyoxylates via Hydrosilylation

α-Keto Ester	Hydrosilane	Chiral Ligand	α-Hydroxy Ester	
			Chemical Yield(%)[a]	Optical Purity(%ee)[b]
$CH_3COCOOPr-n$	Et_2SiH_2	(+)BMPP	85	30.3 (R)
	$PhMeSiH_2$	(+)BMPP	80	50.0 (R)
	Ph_2SiH_2	(+)BMPP	84	60.3 (R)
	Ph_2SiH_2	(+)DIOP	82	76.5 (S)
	$\alpha-NpPhSiH_2$	(-)DIOP	90	85.4 (R)
	Ph_2SiH_2	(-)BPPM	78	28.9 (R)
	$\alpha-NpPhSiH_2$	(-)BPPM	85	78.7 (R)
$CH_3COCOOBu-n$	Ph_2SiH_2	(+)DIOP	82	74.2 (S)
	$\alpha-NpPhSiH_2$	(+)DIOP	83	83.1 (S)
$CH_3COCOOBu-i$	Ph_2SiH_2	(-)DIOP	85	63.1 (R)
	$\alpha-NpPhSiH_2$	(-)DIOP	84	72.1 (R)
PhCOCOOEt	Et_2SiH_2	(+)BMPP	80	6.4 (S)
	Ph_2SiH_2	(+)BMPP	82	10.3 (R)
	Ph_2SiH_2	(+)DIOP	80	9.7 (S)
	$\alpha-NpPhSiH_2$	(+)DIOP	87	39.2 (S)
$PhCOCOOC_6H_{11}-c$	Ph_2SiH_2	(+)DIOP	89	42.5 (S)
	$\alpha-NpPhSiH_2$	(+)DIOP	85	47.2 (S)

a GLC yield. b Optical purity for lactates or mandelates is calculated from the specific rotation of the pure enantiomer which
is reported in the literature: for lactates (Ref. 26); for ethyl
mandelate (Ref. 36); for cyclohexyl mandelate (Ref. 35).

carbonyl[11,13]and imino[14]functionalities using hydrosilylation catalyzed by rhodium(I) complexes with phosphine ligands. The mechanism of the induction of asymmetry is also discussed.

Asymmetric Reduction of α-Keto Esters via Hydrosilylation

The asymmetric reduction of α-keto esters, typically alkyl
pyruvate and phenylglyoxylate, was carried out via hydrosilylation
catalyzed by rhodium complexes with (R)-(+)-benzylmethylphenyl-
phosphine [(+)BMPP], (+)DIOP, (-)DIOP and (-)BPPM as chiral ligands. Reactions were performed in accordance with a general procedure as shown in equation 1.

$$R^1COCOOR^2 \ + \ R^3R^4SiH_2 \xrightarrow{\ [Rh]^* \ } \ R^1\overset{*}{C}HCOOR^2$$
$$\underset{OSiHR^3R^4}{|}$$

$$\xrightarrow{\ H^+ \ } \ R^1\overset{*}{C}HCOOR^2 \qquad (eqn.\ 1)$$
$$\underset{OH}{|}$$

Results are summarized in Table 5. As is seen from Table 5,
optical yields depend on the structure of chiral ligand and hydrosilane employed: The combination of DIOP and α-NpPhSiH$_2$ gives the
best result in each case. The optical yields realized for the
asymmetric reduction of pyruvates by this method are much higher
than those obtained by other methods, e.g., n-propyl and n-butyl
pyruvate were effectively converted to n-propyl and n-butyl lactate in 85.4 and 83.1% optical yields, respectively. A similar
asymmetric reduction of ethyl phenylglyoxylate resulted in rather
low optical yield, whereas cyclohexyl ester afforded by far better
results than ethyl ester did. These results clearly indicate that
the bulkiness of the ester group is an essential factor for determining the effectiveness of the asymmetric induction. A possible
mechanism which can accommodate these results will be presented
(vide post).

"Double Asymmetric Reduction" of (-)-Menthyl Phenylglyoxylate
and Pyruvate Using Catalytic Hydrosilylation

The "double asymmetric reduction" of (-)-menthyl ester of
phenylglyoxylic acid and pyruvic acid were carried out using the
DIOP-rhodium complexes. Conceptually, there are several distinct
ways in which the asymmetric reduction of an α-keto ester to give
the corresponding optically active α-hydroxy ester can be achieved:
(a) by the reduction of a chiral ester with an achiral reducing
agent; (b) by the reduction of an achiral ester with a chiral re-

ducing agent; (c) by a combination of chiral ester and chiral reducing agent.

Horeau and coworkers have investigated these possibilities
with (-)-menthyl and ethyl phenylglyoxylate.[18] Either a simple
asymmetric reduction of (-)-menthyl phenylglyoxylate with an achiral agent, $LiAlH_4$-cyclohexanol, process (a), or of ethyl phenylglyoxylate with a chiral reducing agent, $LiAlH_4$-(+)-camphor, process (b), gives (R)-(-)-mandelate (10 and 4%ee, respectively).
However, "double asymmetric reduction" using both chiral ester and
chiral reducing agent, process (c), results in 49% asymmetric synthesis. This "double asymmetric induction" is higher than would
be anticipated on the basis of any simple additive effect.   Although this discussion seems to overlook the effect of the bulkiness of ester substituent on the extent of asymmetric induction
(vide supra), " double asymmetric induction" was found to be an
effective way of asymmetric synthesis. Accordingly, we applied
this concept to a catalytic system.

In the first place, we performed the catalytic asymmetric reduction of (-)-menthyl phenylglyoxylate using Ph_2SiH_2 as reducing
agent in the following ways: (a) a simple asymmetric reduction by
the hydrosilylation catalyzed by a rhodium complex with achiral
phosphine ligands, $(Ph_3P)_3RhCl$; (b) double asymmetric reduction
using a rhodium complex with a chiral phosphine ligand, [(-)DIOP]-
Rh(S)Cl (S = solvent); (c) double asymmetric reduction using a
rhodium complex with (+)DIOP as chiral ligand, [(+)DIOP]Rh(S)Cl.
Results are summarized in Table 6.

Based on the results from experiment (a), we can estimate the
influence of (-)-menthyl group on the induction of asymmetry.
Namely, the stereochemical control by (-)-menthyl group operated
to produce (-)-menthyl (S)-(+)-mandelate predominantly. The attained optical yield (21%) is a little higher than that obtained
with the use of $LiAlH_4$-cyclohexanol. In an experiment (b), it was
shown that the opposing asymmetric induction by the chiral catalyst and (-)-menthyl group afforded (R)-(-)-mandelate with rather
low stereoselectivity (37%ee). Accordingly, the rhodium catalyst
with (-)DIOP as chiral ligand may favor the production of (R)-
mandelate also in this system, and therefore, the direction of
asymmetric induction is opposite to that by (-)-menthyl group. In
an experiment (c), it was demonstrated that the effective double
asymmetric induction was realized (60%ee) when the rhodium complex
with (+)DIOP as chiral ligand was employed as a catalyst, which
favored the production of (S)-mandelate. Thus, in this case, the
direction of asymmetric induction by [(+)DIOP]Rh(S)Cl matched well
to the effect of (-)-menthyl group. As is also shown in Table 6,
α-NpPhSiH$_2$ displayed a considerable effect on the extent of asymmetric induction. Namely, α-NpPhSiH$_2$-[(+)DIOP]Rh(S)Cl combination
afforded (S)-mandelate with 77% optical purity on the α-carbon.
The increase of the optical purity should be due to better matching of α-NpPhSiH$_2$ with the chiral ligand in the coordination sphere

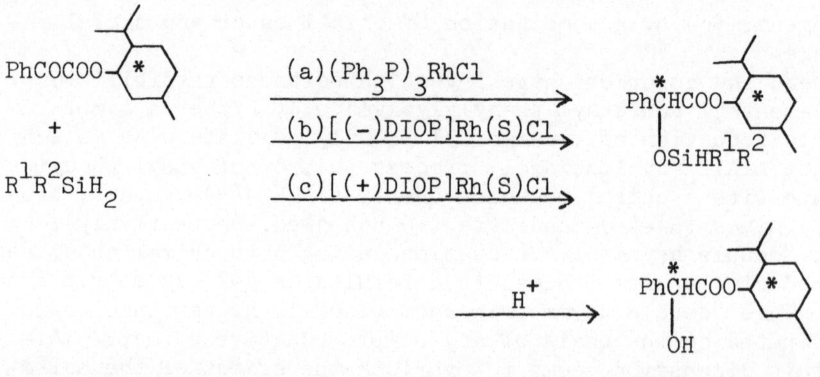

since a simple asymmetric induction using α-NpPhSiH$_2$ and (Ph$_3$P)$_3$-RhCl resulted in the formation of (S)-mandelate with only 17% optical purity.

As for the effect of (-)-menthyl group, the hydrosilylation with the rhodium complex showed an asymmetric induction of opposite direction compared with LiAlH$_4$-cyclohexanol reduction. Namely, the former favored the formation of (S)-mandelate,while the latter did (R)-mandelate.

In a similar manner, (-)-menthyl pyruvate was hydrosilylated using rhodium complexes with (+)DIOP and (-)DIOP as chiral ligand. The optical yield was determined on the basis of the NMR spectrum of the reaction mixture using a shift reagent, Eu(fod)$_3$. Results are listed in Table 7. Optical rotations of the isolated samples of (-)-menthyl lactate are also shown in the table. As Table 7

Table 6
Double Asymmetric Reduction of (-)-Menthyl
Phenylglyoxylate

Method	Hydrosilane	Ligand	Yield(%)[a]	Optical Purity (%ee) [b]
(a)	Ph$_2$SiH$_2$	Ph$_3$P	99	21 (S)
(b)	Ph$_2$SiH$_2$	(-)DIOP	98	37 (R)
(c)	Ph$_2$SiH$_2$	(+)DIOP	98	60 (S)
(a)	α-NpPhSiH$_2$	Ph$_3$P	99	17 (S)
(c)	α-NpPhSiH$_2$	(+)DIOP	99	77 (S)

[a] GLC yield. [b] Optical purity on the α-carbon is estimated on the basis of NMR spectra (100MHz, C$_6$D$_6$).

Table 7
Double Asymmetric Reduction of (-)-Menthyl Pyruvate

$$CH_3COCOO-\langle * \rangle \xrightarrow[\text{[Rh]*}]{R^1R^2SiH_2} \xrightarrow{H^+} CH_3\overset{*}{C}HCOO-\langle * \rangle$$
$$\underset{OH}{|}$$

Hydrosilane	Chiral Ligand	Chemical Yield(%)[a]	$[\alpha]_D^{20}$ [b]	Optical Purity(%ee)[c]
Ph$_2$SiH$_2$	(+)DIOP	85(98)	-79.21	63.5 (S)
Ph$_2$SiH$_2$	(-)DIOP	82(98)	-72.42	62.1 (R)
α-NpPhSiH$_2$	(+)DIOP	81(95)	-80.23	85.6 (S)
α-NpPhSiH$_2$	(-)DIOP	80(96)	-71.23	82.8 (R)

[a] Isolated yield. The values in parentheses are GLC yield.
[b] Measured in EtOH. Optical rotation of (-)-menthyl d,l-lactate
was reported to be -75.9°(EtOH).[37] [c] Estimated on the basis of
NMR spectrum using Eu(fod)$_3$.

shows, the optical yields attained by this system are not so high
as expected as far as Ph$_2$SiH$_2$ is concerned. However, a remarkable
increase of optical yield was observed by the entry of α-NpPhSiH$_2$,
i.e., α-NpPhSiH$_2$-[(+)DIOP]Rh(S)Cl combination afforded the (S)-
lactate with high optical purity such as 85.6%ee, and α-NpPhSiH$_2$-
[(-)DIOP]Rh(S)Cl system gave the (R)-lactate with 82.8%ee. Accordingly, it is concluded that (-)-menthyl which favors the production of (S)-lactate has only a slight effect on asymmetric induction and behaves only as a bulky substituent in the given reaction.

Double asymmetric reduction of (-)-menthyl pyruvate by catalytic hydrogenation. At this point, it is of interest to compare
the results of "double asymmetric reduction" of (-)-menthyl pyruvate by hydrogenation with those by hydrosilylation. As Table 8
shows,[38](-)-menthyl has a considerable effect on the asymmetric
induction and favors the production of (R)-lactate. It is noteworthy that the direction of asymmetric induction by (-)-menthyl
group observed in this reaction is opposite to that observed in
the corresponding hydrosilylation of (-)-menthyl phenylglyoxylate
and (-)-menthyl pyruvate, and is the same to that realized in the
reduction of (-)-menthyl phenylglyoxylate by LiAlH$_4$-cyclohexanol.

Table 8
Double Asymmetric Hydrogenation of (-)-Menthyl Pyruvate[a]

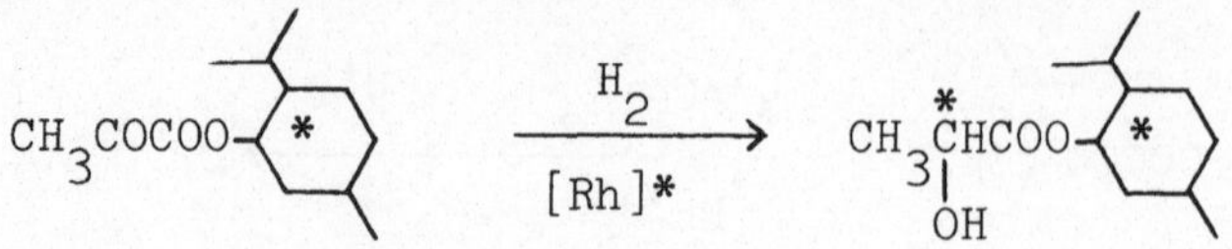

$$CH_3COCOO-\boxed{*} \xrightarrow[\text{[Rh]*}]{H_2} CH_3\overset{*}{C}HCOO-\boxed{*}$$
$$\underset{OH}{\overset{|}{}}$$

Chiral Ligand	Chemical Yield(%)[b]	$[\alpha]_D^{20}$[c]	Optical Purity(%ee)[d]
(+)DIOP	99	-76.13	19.2 (S)
(-)DIOP	99	-73.09	35.5 (R)
(-)BPPM	97	-71.77	73.4 (R)

a All reactions were carried out at 20°C under the initial hydrogen pressure of 20 atm using 0.5 mol% of catalyst in THF for 48 hr. b Isolated yield. c Measured in EtOH. d Estimated based on NMR spectrum using $Eu(fod)_3$.

The results may be accommodated by the fact that the stereochemical course of the hydrosilylation catalyzed by rhodium complexes does not follow apparent steric approach control,[39] and that the reaction proceeds via α-silyloxyalkyl-rhodium complex, as shown previously.[40]

Asymmetric Reduction of Acetoacetates and Levulinates via Hydrosilylation

The optical yields attained in the asymmetric reduction of α-keto esters are much higher than those obtained for simple prochiral ketones. The marked increase of the optical yield for the reaction may be due to a ligand effect, i.e., an attractive interaction, of the ester moiety in the transition state. In this point of view, we have investigated the effects of the ester moiety on the optical yield of the asymmetric hydrosilylation of other keto esters catalyzed by the rhodium complex with DIOP as chiral ligand. We chose acetoacetates and levulinates as substrates since these compounds seemed to be appropriate to estimate such an effect of the ester moiety. In the transition state, an acetoacetate is assumed to form a five membered ring chelate complex I, while a levulinate may afford a six membered ring chelate II.

The catalytic asymmetric hydrosilylation followed by hydrolysis of acetoacetates afforded the corresponding optically active

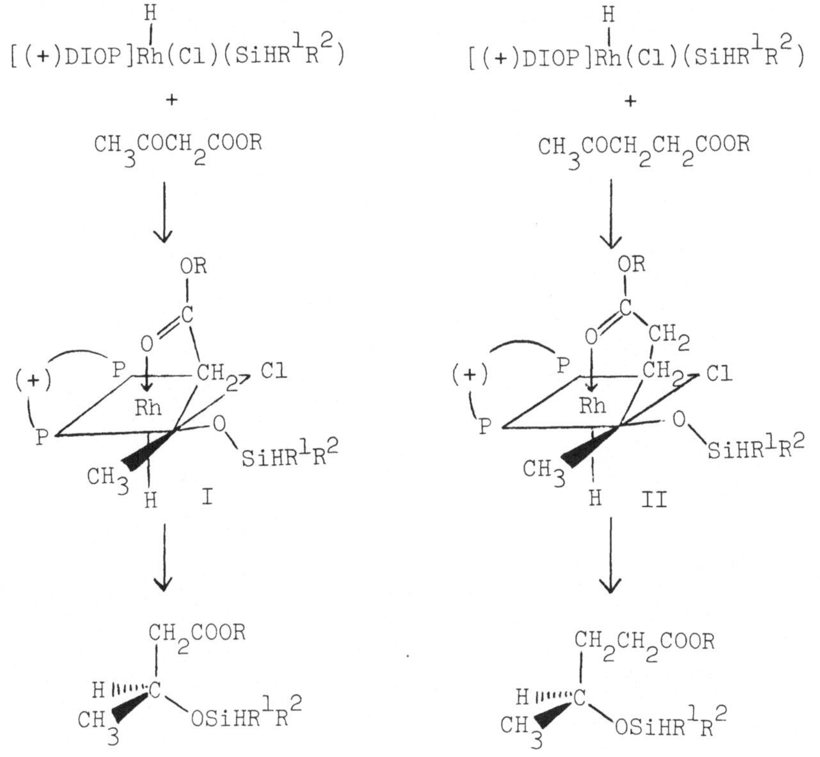

$[(+)DIOP]\overset{H}{Rh}(Cl)(SiHR^1R^2)$

+

CH_3COCH_2COOR

$[(+)DIOP]\overset{H}{Rh}(Cl)(SiHR^1R^2)$

+

$CH_3COCH_2CH_2COOR$

I

II

3-hydroxybutyrates (eqn. 2), while that of levulinates gave optically active 4-methyl-γ-butyrolactone through the silyl ether of
4-hydroxypentanoates (eqn. 3). Results of the asymmetric reduction of acetoacetates are listed in Table 9, and those of levulinates are summarized in Table 10.

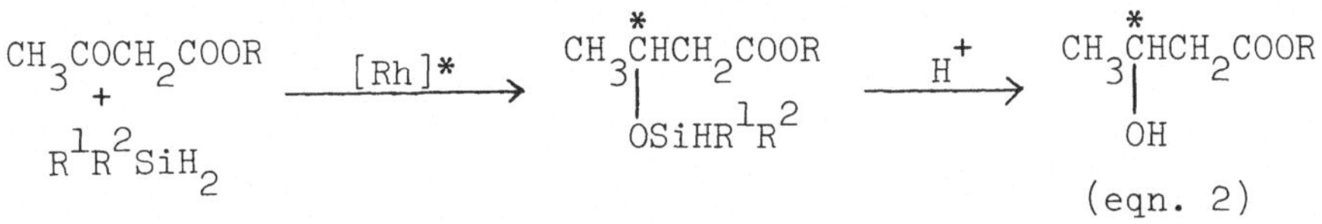

$$CH_3COCH_2COOR + R^1R^2SiH_2 \xrightarrow{[Rh]^*} CH_3\overset{*}{C}HCH_2COOR\underset{OSiHR^1R^2}{|} \xrightarrow{H^+} CH_3\overset{*}{C}HCH_2COOR\underset{OH}{|}$$

(eqn. 2)

$$CH_3COCH_2CH_2COOR + R^1R^2SiH_2 \xrightarrow{[Rh]^*} CH_3\overset{*}{C}HCH_2CH_2COOR\underset{OSiHR^1R^2}{|} \xrightarrow{H^+}$$

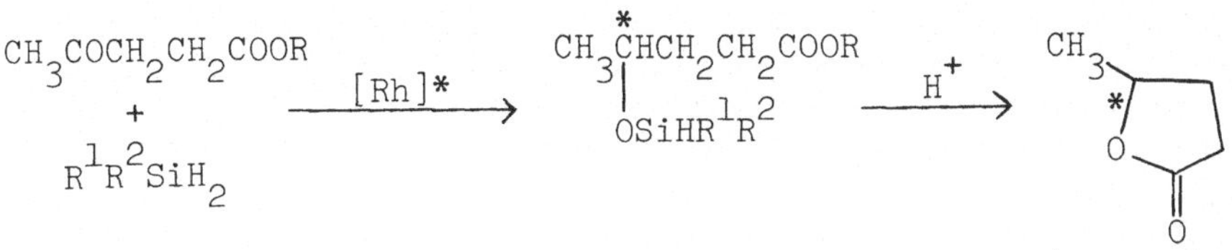

(eqn. 3)

Table 9
Asymmetric Reduction of Acetoacetates via Hydrosilylation
Catalyzed by [(+)DIOP]Rh(S)Cl (S = Solvent)

R	Hydrosilane	Chemical Yield(%)[a]	$[\alpha]_D^{20}$ [b]	Optical Purity(%ee)[c]
Me	Ph_2SiH_2	84	+2.89	12(13.7)(S)
	α-NpPhSiH$_2$	89	+4.96	21(23.5)(S)
Et	Ph_2SiH_2	80	+2.16	11 (S)
	α-NpPhSiH$_2$	86	+5.03	26 (S)
n-Bu	Ph_2SiH_2	83	+1.77	12 (S)
	α-NpPhSiH$_2$	92	+3.66	24 (S)

a GLC yield. b Optical rotations are for the neat liquid. c
Optical purity was determined on the basis of NMR measurement
using Eu(facam)$_3$ in CCl$_4$. The values in parentheses are optical
purity calculated from the reported specific rotation of the
pure enantiomer (Ref. 41).

In order to estimate the effect of the ester moiety, the asymmetric reduction of hexan-2-one was carried out under similar conditions for comparison: The optical purity of (S)-hexan-2-ol thus
obtained was 15.1%ee when Ph_2SiH_2 was employed and 26.6%ee when α-
NpPhSiH$_2$ was used.

$$CH_3CO(CH_2)_3CH_3 \; + \; R^1R^2SiH_2 \; \xrightarrow{[Rh]^*} \; \xrightarrow{H^+} \; CH_3\overset{*}{C}H(CH_2)_3CH_3$$
$$\underset{OH}{|}$$

As is immediately seen from Table 10, a remarkable increase
of optical yield is observed in the case of levulinates. Thus,
the results may suggest the existence of the postulated ligand effect of the ester moiety in the transition state. However, such
an effect cannot be observed in the case of acetoacetates. Accordingly, the asymmetric hydrosilylation of ethyl benzoylacetate followed by hydrolysis using α-NpPhSiH$_2$ and (+)DIOP-rhodium catalyst
was performed for comparison. The optical purity, determined by
NMR using Eu(facam)$_3$, of the obtained ethyl (S)-3-hydroxy-3-phe-

nylpropionate was 62.8%ee,[35] which was a little higher than that realized for simple alkyl phenyl ketones.[15] At any rate, the expected large attractive interaction cannot be observed also in this case.

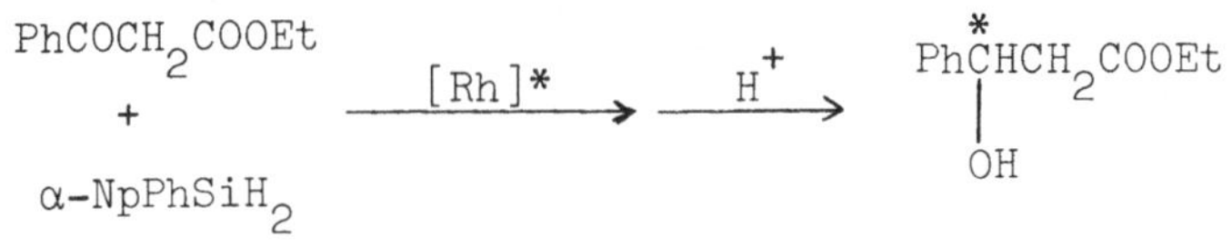

$$\begin{array}{c} PhCOCH_2COOEt \\ + \\ \alpha\text{-NpPhSiH}_2 \end{array} \xrightarrow{\ [Rh]^*\ } \xrightarrow{\ H^+\ } \begin{array}{c} \overset{*}{Ph\overset{|}{C}HCH_2COOEt} \\ | \\ OH \end{array}$$

For a possible explanation of these results, we suppose at this stage that transfer hydrogenation of the preformed silyl enol ether of acetoacetate or benzoylacetate, which would be formed by dehydrogenative coupling of the enol with the hydrosilane and would be a mixture of E and Z isomer, may take place (Scheme 3) together with hydrosilylation. Namely, if the asymmetric induction by transfer hydrogenation is considerably lower than that by hydrosilylation, or if the direction of asymmetric induction by the former reaction is opposite to that by the latter, the apparently obtained optical yield of the given reaction may be unexpectedly low.

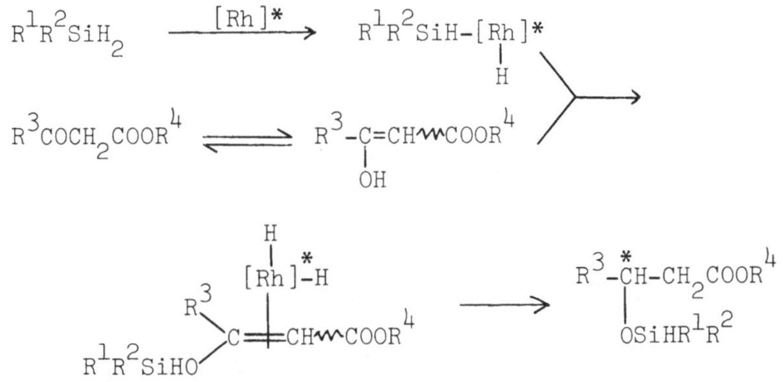

$$R^1R^2SiH_2 \xrightarrow{\ [Rh]^*\ } \underset{\underset{H}{|}}{R^1R^2SiH\text{-}[Rh]^*}$$

$$R^3COCH_2COOR^4 \rightleftharpoons \underset{\underset{OH}{|}}{R^3\text{-}C=CH\text{-}COOR^4}$$

$$\underset{R^1R^2SiHO}{\overset{R^3}{>}}C\!=\!\underset{\underset{[Rh]\text{-}H}{\overset{H}{|}}}{\overset{}{C}}H\text{-}COOR^4 \longrightarrow \underset{\underset{OSiHR^1R^2}{|}}{R^3\text{-}\overset{*}{C}H\text{-}CH_2COOR^4}$$

Scheme 3. Possible Mechanism for the Transfer Hydrogenation in the Reaction of Acylacetates

From a synthetic point of view, a preparation of optically active γ-valerolactone is important for the synthesis of optically active terpenes or alkaloides via optically active aminoalcohols or diols. However, little is known about the effective route to the optically active γ-valerolactone except an optical resolu-

tion. Thus, the facile asymmetric synthesis of 4-methyl-γ-butyro-
lactone via hydrosilylation has a significant value. Recently,
Meyers and Mihelich reported an effective asymmetric synthesis of
2-substituted γ-butyrolactones using optically active oxazolines
in 64-73% optical yields.[42] It should be noted, as shown in Table
10, that 4-methyl-γ-butyrolactone with high optical purity such as
76-84%ee was easily obtained via catalytic hydrosilylation of levulinates followed by hydrolysis in excellent yields. As is seen
from Table 10, the effect of the bulkiness of ester substituent on
the asymmetric induction is not so remarkable, but a relatively
bulky substituent afforded a better result. On the other hand,
the bulkiness of dihydrosilane exerts a large influence on the extent of asymmetric induction, i.e., α-NpPhSiH$_2$ is by far better
reagent than Ph$_2$SiH$_2$. These results clearly indicate that α-NpPh-
SiH$_2$ has an appropriate bulkiness which satisfies the steric re-

Table 10
Asymmetric Reduction of Levulinates via Hydrosilylation
Catalyzed by [(+)DIOP]Rh(S)Cl (S = Solvent)

R	Hydrosilane	4-methyl-γ-butyrolactone		
		Chemical Yield(%)[a]	$[\alpha]_D^{22}$(neat)	Optical Purity(%ee)[b]
Me	Ph$_2$SiH$_2$	99	-11.01	39.6 (S)
	α-NpPhSiH$_2$	99	-21.14	76.2 (S)
Et	Ph$_2$SiH$_2$	100	-10.56	38.1 (S)
	α-NpPhSiH$_2$	99	-22.34	80.5 (S)
n-Pr	Ph$_2$SiH$_2$	96	- 9.61	34.6 (S)
	α-NpPhSiH$_2$	94	-22.96	82.7 (S)
i-Bu	Ph$_2$SiH$_2$	96	-10.88	39.2 (S)
	α-NpPhSiH$_2$	96	-23.43	84.4 (S)
c-C$_6$H$_{11}$	Ph$_2$SiH$_2$	94	-11.16	43.1 (S)
	α-NpPhSiH$_2$	98	-23.17	83.5 (S)
PhCH$_2$	Ph$_2$SiH$_2$	95	-10.56	38.1 (S)
	α-NpPhSiH$_2$	100	-20.86	75.1 (S)

a GLC yield. b Optical purity is calculated from the reported
specific rotation of the pure enantiomer (Ref. 43).

quirements for the effective asymmetric induction in the chiral coordination sphere. In all cases, S configuration of the lactone
was preferred when (+)DIOP was employed as a chiral ligand.
     (-)-Menthyl and (-)-bornyl levulinate were also chosen as
substrate. As Table 11 shows, the effect of (-)-menthyl group on
the extent of asymmetric induction was not large. However, a considerable effect was observed when Ph_2SiH_2 was employed: (-)Men-
thyl levulinate-[(-)DIOP]Rh(S)Cl combination gave a better result
(ca. 10%ee difference) than the ester-[(+)DIOP]Rh(S)Cl combination
did. The bulkier (-)-bornyl group has a negative effect on the
extent of asymmetric induction, especially when α-NpPhSiH$_2$ was employed. The results may be due to the fact that the coordination
sphere is much too crowded by the entry of the bulky silane and
the ester group. A similar phenomenon was observed when the (-)-
menthyl ester was reduced by using α-NpPhSiH$_2$-[(-)DIOP]Rh(S)Cl
system.

Table 11

Asymmetric Reduction of (-)-Menthyl and (-)-Bornyl Levulinate
via Hydrosilylation Catalyzed by Rhodium(I) Complex with (+)-
DIOP or (-)DIOP

R	Hydrosilane	Chiral Ligand	4-methyl-γ-butyrolactone	
			Chemical Yield(%)[a]	Optical Purity(%ee) [b]
(-)-menthyl	Ph_2SiH_2	(+)DIOP	93	37.1 (S)
	Ph_2SiH_2	(-)DIOP	90	47.8 (R)
	α-NpPhSiH$_2$	(+)DIOP	92	82.9 (S)
	α-NpPhSiH$_2$	(-)DIOP	98	79.8 (R)
(-)-bornyl	Ph_2SiH_2	(+)DIOP	97	41.7 (S)
	Ph_2SiH_2	(-)DIOP	98	38.7 (R)
	α-NpPhSiH$_2$	(+)DIOP	99	72.5 (S)
	α-NpPhSiH$_2$	(-)DIOP	98	67.9 (R)

a GLC yield. b see Table 10.

On the Mechanism of Asymmetric Induction

     As for the induction of asymmetry, we already proposed pre-
viously[40] a probable mechanism for the BMPP-rhodium(I) complex
system based on the stereochemical inspection of the relationship

between the configuration of the chiral phosphine and that of the resulting alcohol. According to the proposed mechanism, the inter- mediate α-silyloxyalkyl-rhodium complex plays a key-role for the asymmetric induction, and the following steps are involved as shown in Scheme 4: (a) oxidative addition of the hydrosilane to the rhodium(I) complex; (b) insertion of the ketone carbonyl into the resulting silicon-rhodium bond to form diastereomeric α-silyl- oxyalkyl-rhodium intermediate 4 in accordance with product devel- opment control,[39] (c) formation of an optically active silyl ether of secondary alcohol by reductive elimination. Among these steps, step (b) must play the most important role in inducing asymmetry at the carbonyl carbon because this step determines a predominant configuration and the extent of enantiomeric excess of the product.

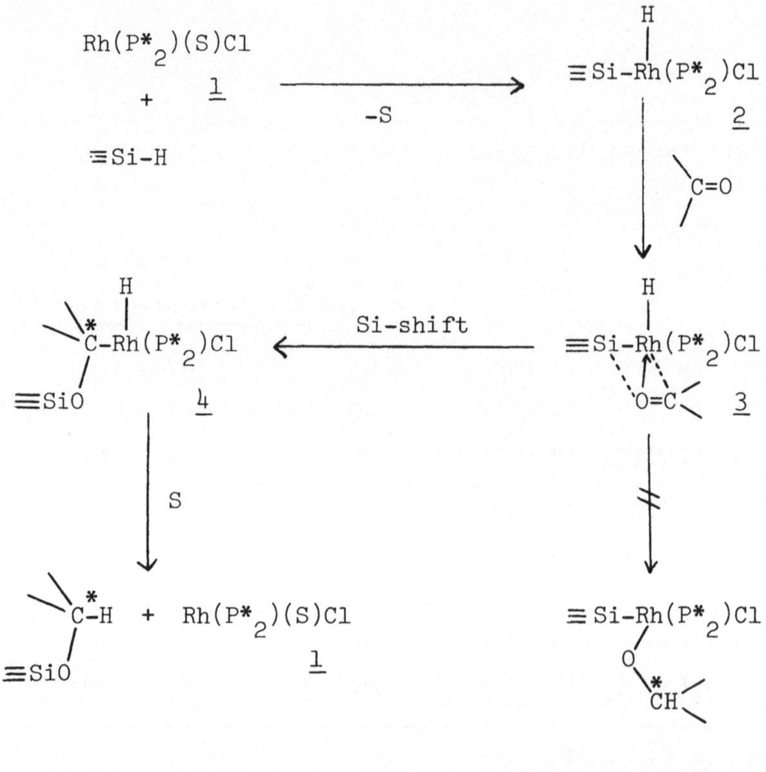

Scheme 4. Proposed Mechanism of Asymmetric Hydrosilylation

We also reported previously that the predominant configura- tion of the produced secondary alcohol was satisfactorily predict- ed by the inspection of the most preferable conformation of the intermediate complex 4 bearing (+)BMPP as chiral ligand using the

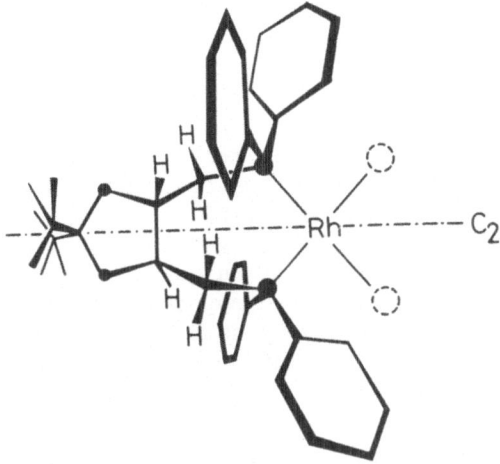

Figure 2.   Illustration of the most preferable
conformation of (+)DIOP-rhodium complex in
solution based on the Dreiding model.

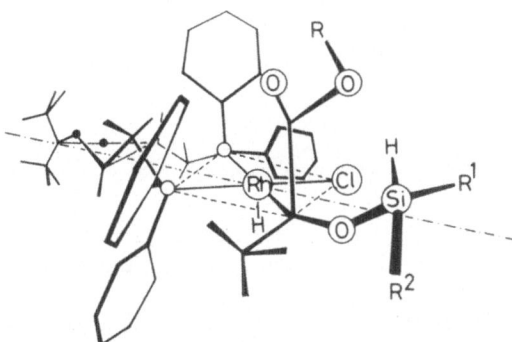

Figure 3.   Illustration of the most favorable
structure of the key intermediate complex 4
when pyruvate is chosen as substrate for the
reaction.

Dreiding models.[41] This prediction is also successfully applied
to the reaction of α-keto esters, where certain electronic attractive interaction between the ester moiety and the central rhodium metal is taken into account. Although the steric requirements in the coordination sphere of the DIOP-rhodium complex are
quite different from those of the BMPP-rhodium complex, the intermediate α-silyloxyalkyl-rhodium complex 4 may also play a key role
for the induction of asymmetry. The coordinated DIOP ligand can
take several conformations; nevertheless the most preferable conformation in solution based on the Dreiding model is the one depicted in Figure 2 where steric repulsion between substituents
concerning especially four phenyl groups is the smallest. In this
conformation DIOP-rhodium moiety has a C_2 axis as shown in Figure
2. It should be noted that the proposed conformation corresponds
exactly to the averaged one of possible conformations. Inspection
of the Dreiding model of the intermediate complex 4 indicates that
the most preferable structure of the complex 4 is the one depicted
in Figure 3 on account of the attractive interaction mentioned
above. Accordingly, the ester group takes a quasi-apical position,
methyl occupies the most congested site, and then bulkier silyl-
oxy group occupies the least hindered site. The hydride shift in
this conformation should afford (S)-lactate. This model explains
well the observed results and is also successfully applied to the
case of levulinates.

ACKNOWLEDGEMENT

The author is grateful to Mr. Tetsuo Kogure and Mrs. Miyoko
Kumagai for their valuable contribution to this work. He is also
indebted to Dr. Kazuo Achiwa of University of Tokyo for a gift of
a chiral ligand, BPPM.

REFERENCES

1. B. R. James, "Homogeneous Hydrogenation" Wiley Interscience,
   New York, 1973; R. E. Harmon, S. K. Gupta and D. J. Brown,
   Chem. Rev., 73, 35 (1973).
2. e.g., B. D. Vineyard, W. S. Knowles, M. J. Sabacky, G. L.
   Bachman and D. J. Weinkauff, J. Amer. Chem. Soc., 99, 5946
   (1977) and references therein.
3. H. B. Henbest and T. R. B. Mitchell, J. Chem. Soc., (C) 785
   (1970).
4. R. R. Schrock and J. A. Osborn, Chem. Commun., 567 (1970);
   idem, J. Amer. Chem. Soc., 93, 2398 (1971).
5. A. J. Birch and K. A. M. Walker, J. Chem. Soc., (C) 1894
   (1966).
6. K. Ohno and J. Tsuji, J. Amer. Chem. Soc., 90, 99 (1968);

M. C. Baird, C. J. Nyman and G. Wilkinson, J. Chem. Soc., (A) 348 (1968); J. Blum, E. Oppenheimer, and E. D. Bergmann, J. Amer. Chem. Soc., 89, 2338 (1967).

7. P. Bonvicini, A. Levi, G. Modena and G. Sccorano, JCS Chem. Commun., 1188 (1972).

8. J. Solodar, CHEMTECH, 421 (1975).

9. (a) Y. Ohgo, Y. Natori, S. Takeuchi and J. Yoshimura, Chem. Lett., 709 (1974); (b) idem, ibid., 1377 (1974); (c) Y. Ohgo, S. Takeuchi and J. Yoshimura, Bull. Chem. Soc. Japan, 44, 583 (1971).

10. C. Eaborn and R. W. Bott in A. G. MacDiarmid (Ed.), "Organometallic Compounds of the Group IV Elements," Vol. 1, Marcel Dekker, 1968, pp. 213-279 and references therein.

11. I. Ojima, M. Nihonyanagi and Y. Nagai, JCS Chem. Commun., 938 (1972); I. Ojima, M. Nihonyanagi, T. Kogure and Y. Nagai, Bull. Chem. Soc. Japan, 45, 3506 (1972).

12. R. J. P. Corriu and J. J. E. Moreau, JCS Chem. Commun., 38 (1973).

13. I. Ojima, M. Nihonyanagi and Y. Nagai, Bull. Chem. Soc. Japan, 45, 3722 (1972); I. Ojima, T. Kogure and Y. Nagai, Tetrahedron Lett., 5035 (1972).

14. I. Ojima, T. Kogure and Y. Nagai, Tetrahedron Lett., 2475 (1973); N. Langlois, T.-P. Dang and H. B. Kagan, ibid., 4865 (1973).

15. I. Ojima, K. Yamamoto and M. Kumada in R. Ugo (Ed.) "Aspects of Homogeneous Catalysis," Vol. 3, Reidel Publishing Co., 1977, pp. 186-228 ar¯ references therein.

16. J. D. Morrison and H. S. Mosher, "Asymmetric Organic Reactions," Prentice-Hall Inc., Englewood Cliffs, N. J., 1971, Section 2 and 5, and references therein.

17. G. Vagon and A. Antonini, C. R. Acad. Sci., 230, 1870 (1950); idem, ibid., 232, 1120 (1951).

18. A. Horeau, H. B. Kagan and J. P. Vigneron, Bull. Soc. Chim. France, 3795 (1968).

19. (a) M. Tanaka, Y. Watanabe, T. Mitsudo, H. Iwane and Y. Takegami, Chem. Lett., 239 (1973); (b) C. J. Sih, J. B. Heather, G. P. Perzzotti, P. Price, R. Sood and L. F. H. Lee, J. Amer. Chem. Soc., 95, 1676 (1973); (c) A. Levi, G. Modena and G. Scorrano, JCS Chem. Commun., 6 (1975); (d) B. Heil, S. Törös, S. Vestag and L. Markó, J. Organometal. Chem., 94, C47 (1975).

20. T. Hayashi, T. Mise and M. Kumada, Tetrahedron Lett., 4351 (1976).

21. T. Mise, T. Hayashi and M. Kumada, 35th Annual Meeting of the Chemical Society of Japan 1976, Abstr., 1K20.

22. I. Ojima, T. Kogure and K. Achiwa, JCS Chem. Commun., 428 (1977).

23. K. Achiwa, J. Amer. Chem. Soc., 98, 8265 (1976).

24. H. B. Kagan and T.-P. Dang, J. Amer. Chem. Soc., 94, 6429 (1972).

25. T. Purdie and J. C. Irvine, J. Chem. Soc., 483 (1899).
26. E. Wassmer and P. A. Guye, Chem. Zentralbl., 2, 1419 (1903).
27. C. E. Wood, J. E. Such and F. Scarf, J. Chem. Soc., 1928 (1926).
28. I. Ojima and T. Kogure, to be published.
29. I. Ojima, T. Kogure and K. Achiwa, to be published.
30. M. Purko, W. O. Nelson and W. A. Wood, J. Biol. Chem., 207, 51 (1954); G. M. Brown and J. J. Reynolds, Annual Rev. Biochem., 32, 419 (1963).
31. e.g., E. T. Stiller, S. A. Harris, J. Finkelstein, J. C. Keresztesy and K. Folkers, J. Amer. Chem. Soc., 62, 1785 (1940).
32. R. P. Lanzilotta, D. G. Bradley and K. M. McDonald, Appl. Microbiology, 27, 130 (1974).
33. e.g., F. Kagan, R. V. Heinzelman, D. I. Weisblat and W. Greiner, J. Amer. Chem. Soc., 79, 3545 (1957) and references therein.
34. cf. A. Güssner, M. Gätzi-Fischer and T. Reichstein, Helv. Chim. Acta, 23, 1276 (1940).
35. I. Ojima, T. Kogure and M. Kumagai, J. Org. Chem., 42, 1671 (1977).
36. R. Roger, J. Chem. Soc., 2178 (1932).
37. A. McKenzie and H. B. Thompson, J. Chem. Soc., 87, 1016 (1905).
38. I. Ojima and T. Kogure, unpublished results.
39. W. G. Dauben, G. J. Fonken and D. S. Noyce, J. Amer. Chem. Soc., 78, 2579 (1956).
40. I. Ojima, T. Kogure, M. Kumagai, S. Horiuchi and T. Sato, J. Organometal. Chem., 122, 83 (1976); I. Ojima and Y. Nagai, Chem. Lett., 223 (1974).
41. E. Fischer and H. Scheibler, Chem. Ber., 42, 1221 (1909).
42. A. I. Meyers and E. D. Mihelich, J. Org. Chem., 40, 1186 (1975).
43. P. A. Levene and L. Haller, J. Biol. Chem., 69, 165 (1926).

METALLACYCLES IN ORGANOTRANSITION METAL CHEMISTRY

R.H. Grubbs and A. Miyashita

Michigan State University

East Lansing, Michigan   48824

Heterocyclic compounds have played a major role in the development of organic chemistry. It is becoming apparent that heterocycles containing a transition metal play an important role in the transition metal catalyzed reactions of olefins and acetylenes.

Metallacycles appear to be intermediates in the olefin metathesis reaction,[1] the cyclopropanation of olefins,[2] and other reactions of metal carbenoid species with olefins.[3]

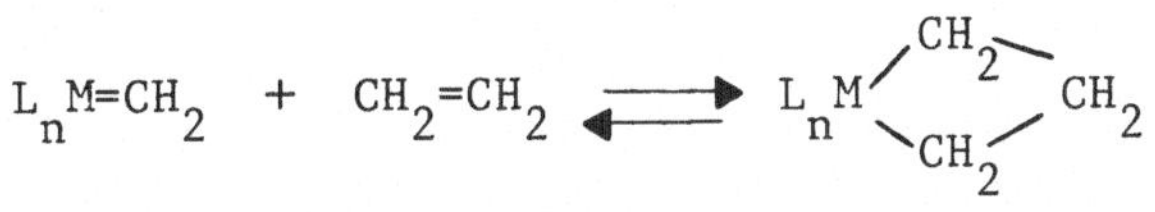

$$L_nM=CH_2 \;+\; CH_2=CH_2 \;\rightleftharpoons\; L_nM\underset{CH_2}{\overset{CH_2}{\diagdown\diagup}}CH_2$$

Stable metallacyclobutanes can be prepared by the interaction of cyclopropanes with platinum salts.[4]

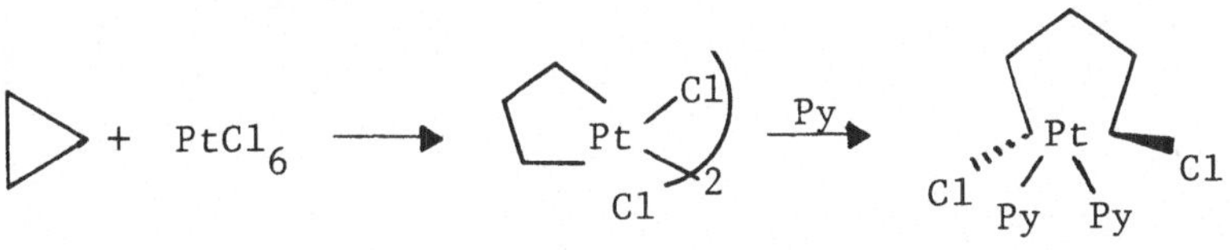

Structure I

A second method which produces an early transition element complexes is the reduction of cationic π-allyl complexes.[5]

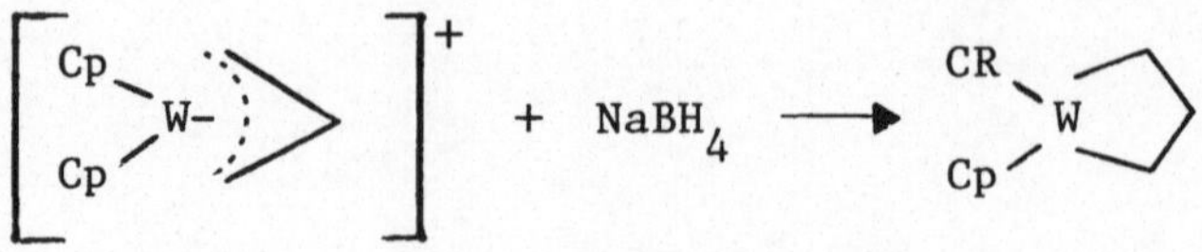

The chemistry of these complexes as well as the study of the reactions of olefins with metal carbene complexes are beginning to deliniate the reaction pathways open to this metallacyclic system. Stable metallacyclopentadienes have been known for some time and appear to be the major intermediate in many acetylene trimerization reactions.[6,7]

Metallacyclopentanes have been proposed or demonstrated to be intermediates in a number of metal catalyzed cycloadditions and cycloreversions of olefins.

The first systems in which metallacycles were isolated involved the addition of an activated olefin to a metal of reduced coordination number. The first such system was reported by Green[8] from the photolysis of an iron carbonyl complex with trifluorethylene. A stable metallacycle (III) resulted from this reaction.

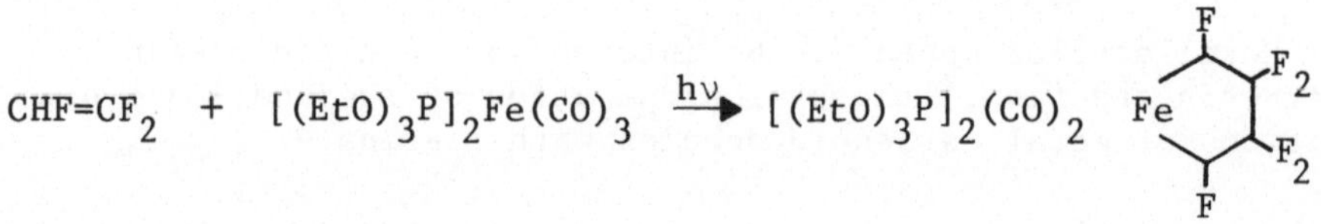

III

Osborn, et al,[9] prepared one of the first such systems from a strained olefin.

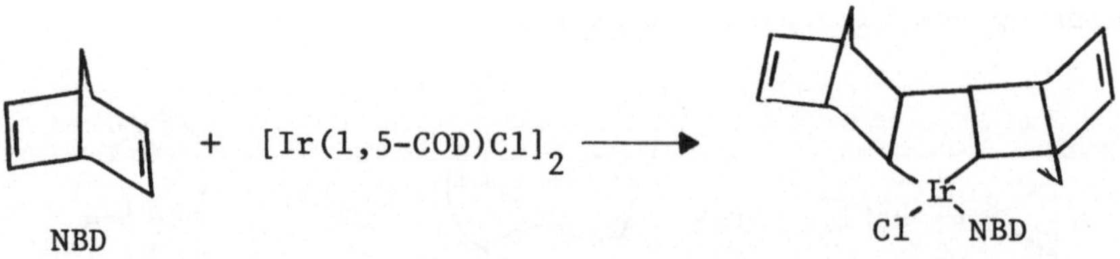

Binger and coworkers reported that bis(cyclooctadiene)nickel (V) catalyzed the cyclodimerization of methylenecyclopropanes.[10] Recently, they have trapped a metallacyclopentane intermediate (VI) in this reaction.[11]

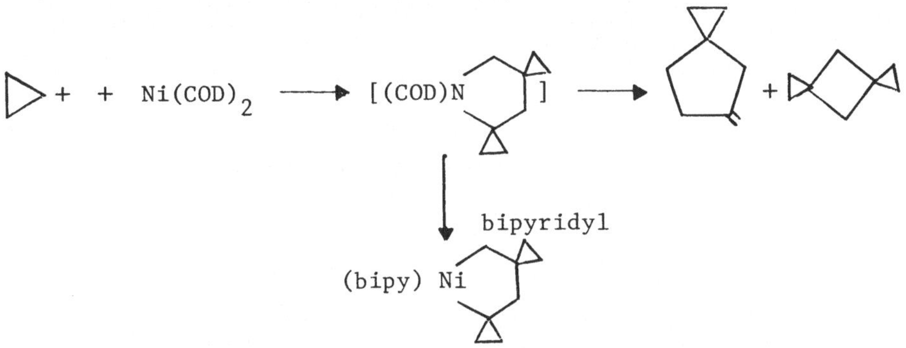

Similar intermediates could be trapped from the reaction of Ni atoms with norbornadiene.[12,13]

Whitesides reported that metallacycles could be prepared in low yields by the reaction of ethylene and other simple olefins with reduced titanocene species. (VII)[14]

Very recently Schrock has reported the preparation of stable Tantalum metallacycles from ethylene. (VIII)[3]

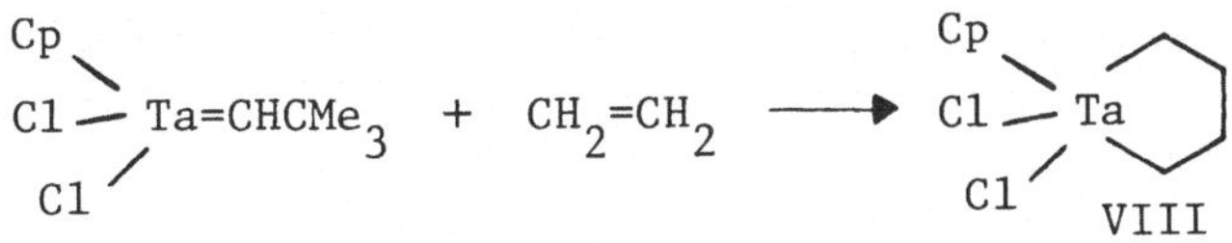

These last two observations as well as the extensions of the following work, suggest that the investigation of metallacycles may open up new catalytic processes for olefin dimerization and oligiomerization.

The majority of evidence for the intermediacy of metallacycles in cycloreversion reactions has resulted from work by the groups of Halpern and Eaton.[15] They found that rhodium (I) complexes catalyzed the ring opening of cubane derivaties and that by using stoichiometric amounts of rhodium (I) carbonyl complex the intermediate metallacycle (IX) could be trapped.

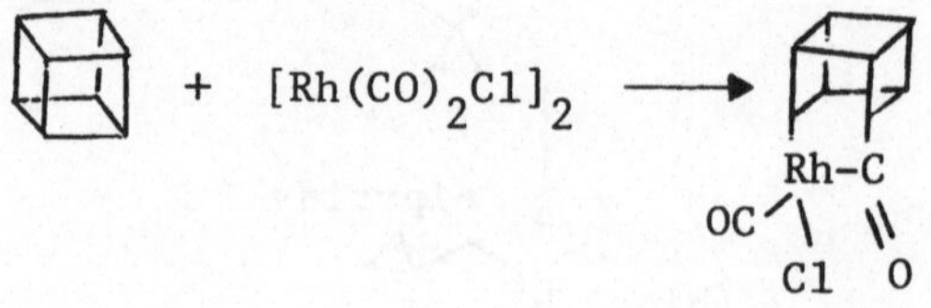
+ [Rh(CO)$_2$Cl]$_2$
Rh–C
OC
Cl O
IX

Stable derivatives of the parent complex have been prepared by the reaction of 1,4-dilithiobutane or the Grignard analog with a variety of transition metal dihalides. Tetramethylene metallacycles containing platinum,[16,17] nickel,[18] titanium,[14] and rhodium[19] have been prepared by this method. The platinum[17] and nickel[20] systems have been characterized by x-ray analysis. Whitesides demonstrated one of the key features of these systems. He found that platinum metallacycles were much more stable toward β-hydride elimination reactions than acyclic analogs.[16] Changing the ligands changed the reaction pathway from β-hydride elimination to reductive elimination.

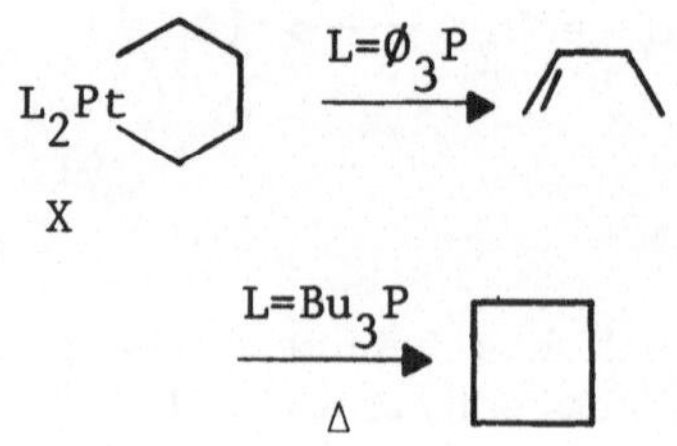
L$_2$Pt
X
L=Ø$_3$P
L=Bu$_3$P
Δ

The most important reaction for olefin reactions, carbon-carbon bond cleavage to produce ethylene, has only been observed to occur to a small extent in a titanocene (VII)[14] and a rhodium system. (XI)[19]

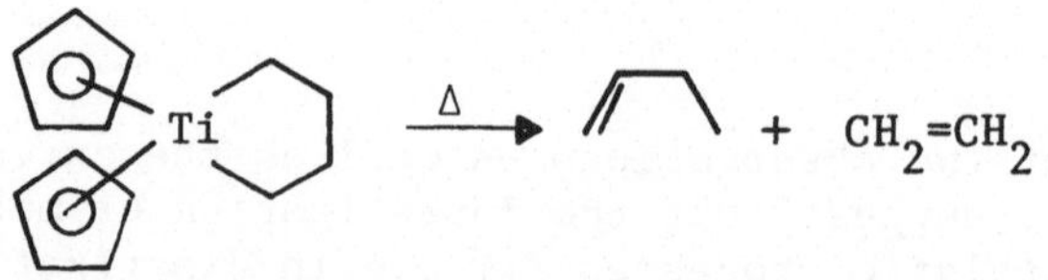
Ti
Δ
+ CH$_2$=CH$_2$
VII

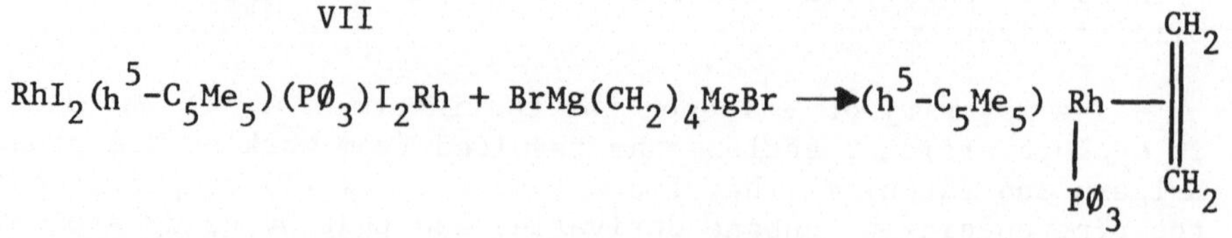
RhI$_2$(h^5-C$_5$Me$_5$)(PØ$_3$)I$_2$Rh + BrMg(CH$_2$)$_4$MgBr → (h^5-C$_5$Me$_5$) Rh
PØ$_3$
CH$_2$
CH$_2$

We have examined a series of phosphinenickel and titanium metallacyclopentanes which undergo three major reactions previously observed for metallacycles;[21],[22] a) reductive elimination b) β-hydride elimination and c) carbon-carbon bond cleavage to ethylene. In order to explore the factors which control the relative rates of these reactions, a series of complexes containing different ligands, and having different coordination numbers have been prepared, character-ized and the products of decomposition examined.

$$P_2Ni \bigcirc \longrightarrow \square + CH_2=CH_2 + \text{butene}$$

Stable metallacycles can be prepared from α,-dilithioreagents and transition metal dihalides.[14],[16],[21]

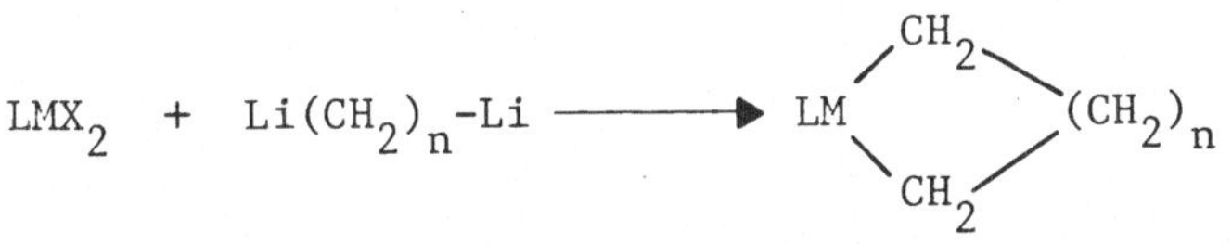

$$LMX_2 + Li(CH_2)_n-Li \longrightarrow LM \overset{CH_2}{\underset{CH_2}{\diagdown}} (CH_2)_n$$

The system prepared from metals other than platinum are very air and temperature sensitive. The majority of our studies have involved metallacyclopentanes of nickel and titanium. Recent preliminary evidence on large rings suggest these systems undergo very interesting reactions.

METALLACYCLOPENTANES

Preparation and Structures

Bisphosphine nickelacyclopentane and titancenacyclopentanes were prepared from 1,4-dililliobutane and either a bisphosphine-dihalonickel(II) or titanocene dichloride.

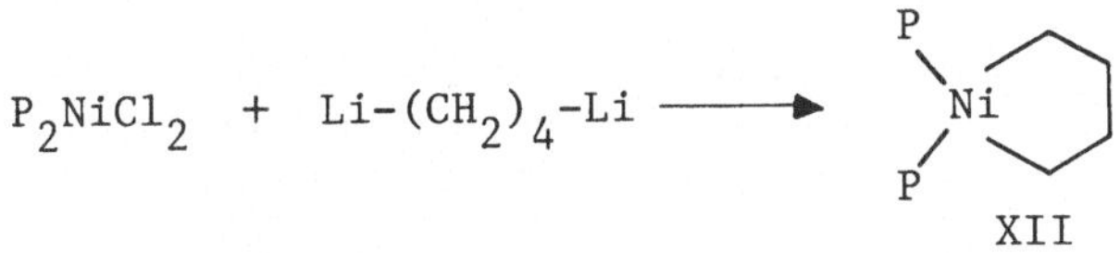

$$P_2NiCl_2 + Li-(CH_2)_4-Li \longrightarrow \quad \text{XII}$$

$P=\emptyset_3P$, $(cyclohexyl)_3P$, $\emptyset_2PCH_2\overset{a}{-}CH_2P\emptyset.\overset{b}{\underset{c}{}}$

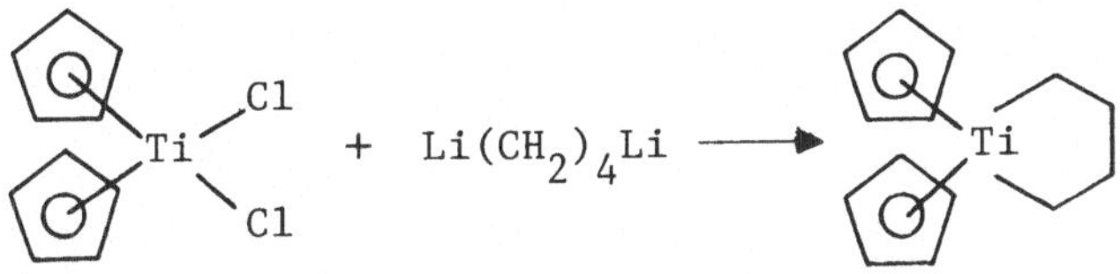

$$\text{(titanocene dichloride)} + Li(CH_2)_4Li \longrightarrow \quad \text{VII}$$

Although the phosphine complexes crystallize as bis-phosphine complexes, in solution the coordination number changes. The triphenylphosphine complex (XIIa) in aromatic solvents disassociates to a small extent. Addition of an excess of triphenylphosphine results in the production of a stable crystalline triphosphine complex.

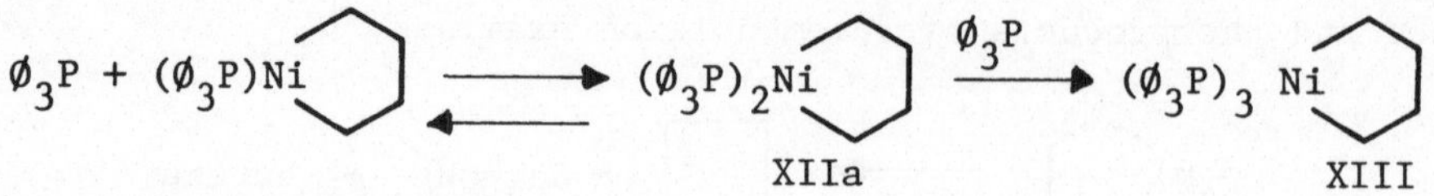

$$\emptyset_3P + (\emptyset_3P)Ni \mathhexbox \rightleftharpoons (\emptyset_3P)_2Ni \xrightarrow{\emptyset_3P} (\emptyset_3P)_3 Ni$$

$$\text{XIIa} \qquad\qquad\qquad \text{XIII}$$

The structures were determined by ^{1}H and ^{31}P NMR, molecular weight measurements and chemical analysis.

In solution the tricyclohexylphosphine complex (XIIb) disassociates to form a monophosphine species.

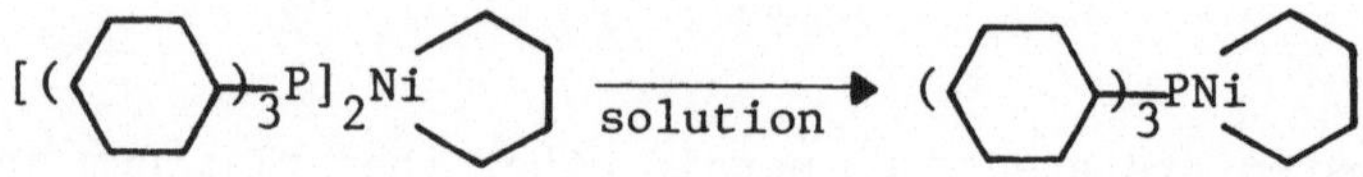

$$[(\hspace{1cm})_3P]_2Ni \hspace{1cm} \xrightarrow{\text{solution}} (\hspace{1cm})_3PNi$$

XIIb

All the evidence suggests that the diphos complex (XIIc) remains four coordinate in solution and showed no tendancy to produce a 5 coordinate complex when a large excess of trialkylphosphine was added.

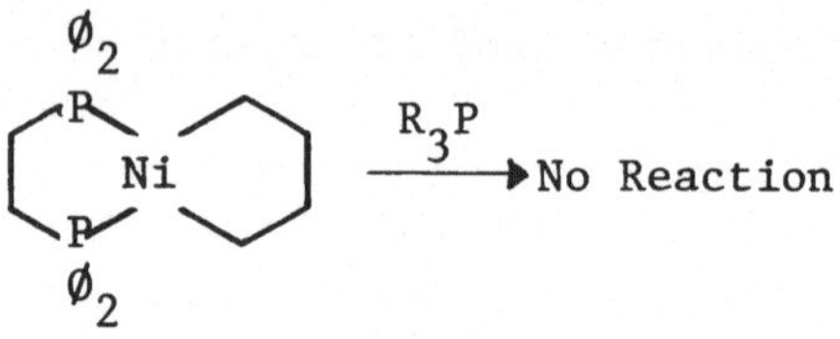

$$\underset{\emptyset_2}{\overset{\emptyset_2}{\underset{P}{\overset{P}{\bigg]}}} Ni \hspace{1cm} \xrightarrow{R_3P} \text{No Reaction}$$

The titanocenemetallacyclopentane VII was unstable above $-30°$ and was difficult to purify. Analytically pure complex could be obtained after chromatography and three recrystallizations.

Decompositions

The metallacyclopentanes were decomposed at $9°C$ in toluene to give near quantitative yields of the products. Product compositions were determined by gas chromatography. The products of each compound are given in Table I.

TABLE I

	$CH_2=CH_2$	☐	linear C_4H_8
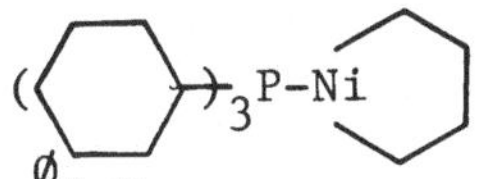 $(\langle\rangle)_3 P-Ni\langle\rangle$	0	0	99%
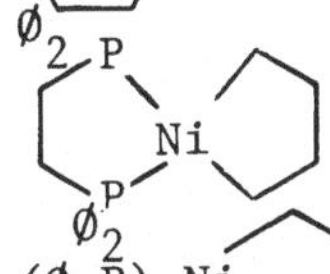 $\emptyset_2 P \diagdown Ni \langle\rangle$ / $\emptyset_2 P \diagup$	2	90	8
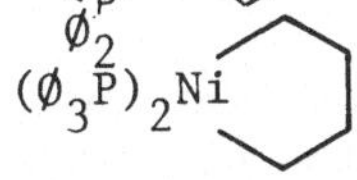 $(\emptyset_3 P)_2 Ni\langle\rangle$	5	68	27
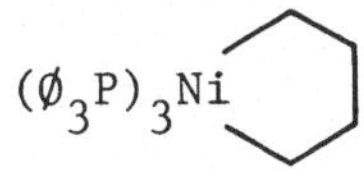 $(\emptyset_3 P)_3 Ni\langle\rangle$	90	10	0

These results suggest that the three coordinate complexes produces linear butenes as the major products, while 4 coordination gives cyclobutane, the product of reductive elimination and the 5-coordinate complex produced ethylene. An examination of the product composition as a function of added free phosphine suggested that this is general.

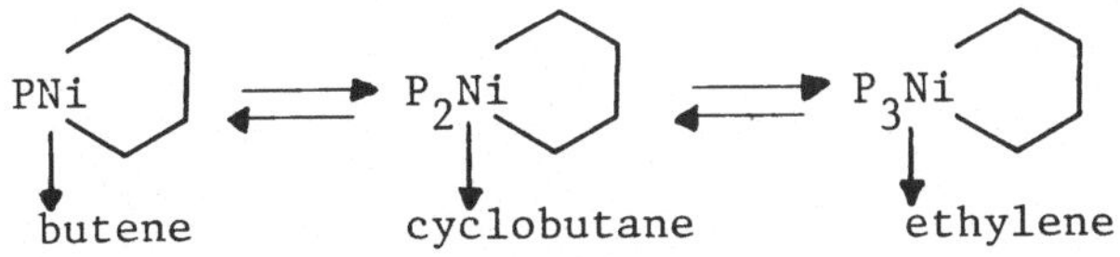

$$PNi\langle\rangle \rightleftharpoons P_2Ni\langle\rangle \rightleftharpoons P_3Ni\langle\rangle$$

$$\downarrow \qquad\qquad \downarrow \qquad\qquad \downarrow$$

butene   cyclobutane   ethylene

Decomposition of the titanocene metallacycle produced high yields of ethylene. The only minor product was butene (no cyclobutane). Addition of tributylphosphine to the system supressed the formation of the minor product.

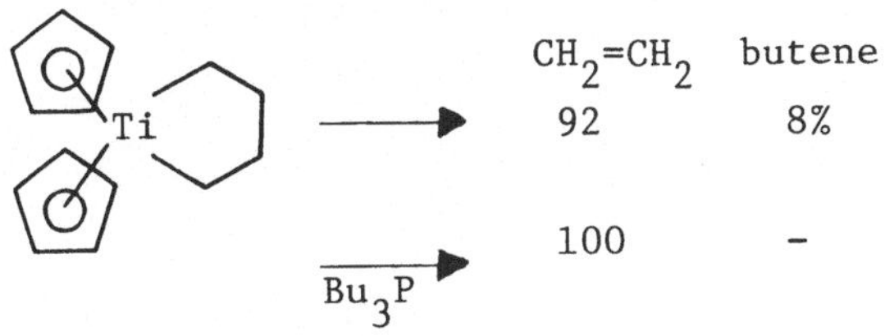

	$CH_2=CH_2$	butene
Ti →	92	8%
$\overline{Bu_3 P}$ →	100	–

Since the equilibrium between the metallacycle and bisolefin complex is the most important reaction if metallacycles are to be involved in catalytic systems, those complexes which produce ethylene on decomposition have been examined in the most detail to determine the facility of such an equilibrium.

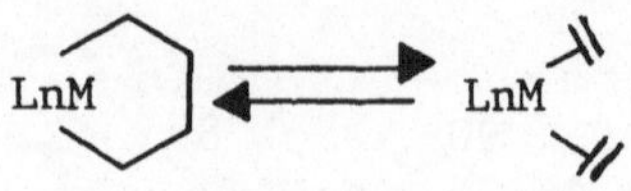

This process can be detected by labeling two of the equivalent carbons of the metallacycle since the two carbons of the olefin should become equivalent in the intermediate.

Consequently, 2,2,5,5-tetradeuterometallacycles of nickel and titanacene were prepared. Equilibration of the metallacycle with the diolefin complex should result in the production of the 2,2,4,4- and 3,3,4,4-d_4 isomers. Hydride rearrangements would result in the production of isomers containing only one deuterium per carbon.[5]

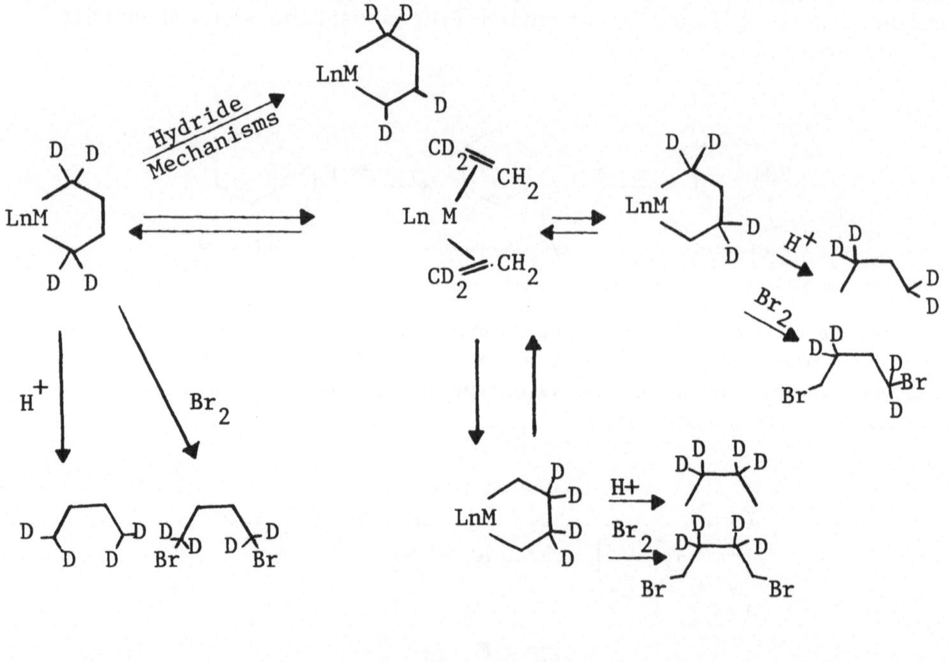

Since the bromination and protonolysis of metallacycle produces quantitative yields of 1,4-dibromobutane or butane,[14,16] the labeling of the metallacycle can be analyzed by determining the labeling pattern of the butanes produced by these reactions.

The labeling pattern of the butane-d_4 was determined by analysis of the mass spectral cracking pattern and that of the 1,4-dibromo-butanes by ^{1}H NMR.

Bis(triphenylphosphine)tetramethylenenickel(11)-2,2,5,5-d_4 (XII-a-d_4) was prepared from the corresponding nickel dihalide[8] and 1,4-dilithiobutane-1,1,4, 4-d_4 (98.0% isotopic purity)[8] and converted into the crystaline trisphosphine complex at -50°C. (XIII-d_4)[22]

$(\emptyset_3P)_2NiCl_2$ + Li-$CD_2CH_2CH_2CD_2$-Li   (O P)   XII-a-d_4 XIII-d_4

The same dilithium reagent[4] was used to prepare the labeled titana-metallacycle.

Solutions of the complexes in toluene were maintained at low temperature for time periods which resulted in low percentage of decomposition. The solutions were then mixed with dry hydrogen chloride or bromine at -30∿40° and the C_4 products were purified by gas chromatography and then analyzed. The amount of isomerization, decomposition and scrambling (mono-hydride rearrangements) are listed in Table II and III.

These results demonstrate a number of general features of metall-acyclopentanes of both early and late transition elements.

1.   The rate of isomerization of both complexes was faster than the decomposition to produce ethylene. This is most consistent with an equilibrium between the metallacycle and bisolefin complex with the decomposition of the bisolefin complex being rate determining.

2.   The bis(triphenylphosphine)nickel complex (III-d_4) which does not decompose to yield ethylene as a major product did not produce isomerized metallacycles. This complex under conditions similar to the analogous trisphosphine complex (I-d_4) only produced a low percentage of mono-hydride scrambled products. For those complexes which produced ethylene as the major decomposition product, iso-merization (D_2 units move pairwise) was faster than scrambling.

3.   Since the ratio of the 2,2,4,4 to 3,3,4,4-isomer was near 2:1 (the equilibrium ratio) even at low conversions, the intermediate must be symmetrical, or more likely the rotation[23] of the olefins in the intermediate was faster than recyclization.

4.   If the trisphosphine nickelacycle decomposed to a bisolefin
complex, a 20 electron system would have been produced (5-coordinate
nickel (0)). Consequently, it is reasonable to assume that a phosphine
is lost prior to or concurrent with the carbon-carbon bond cleavage.
The observation (Table I) that the rate of isomerization decreased
on addition of excess triphenylphosphine shows this to be the case.

$$(\emptyset_3 P)_3 \; Ni \; \bigcirc \; \rightleftharpoons \; \emptyset_3 P \; + \; (\emptyset_3 P)_2 \; Ni \underset{CH_2}{\overset{CH_2}{\diagdown}}\underset{CH_2}{\overset{CH_2}{\parallel}} \xrightarrow[\text{determining}]{\text{rate}} \; C_2 H_4$$

The equilibrium in the equation may involve one or more intermediates.

      Confirmation of the intramolecular nature of the reaction was
obtained by mixing equal amounts of I and I-d_4 in toluene at $0°$ for
2 hrs (sufficient time for extensive isomerization) before protonolysis.   Only butane-d_4 and d_0 was produced.   Intermolecular exchange
and isomerication would have produced butane d_2.

      The above results and those beginning to appear in the literature
suggest that the equilibrium between bisolefin complexes and metallacyclopentanes[3] is a much more general reaction than thought in the
past and may be a key reaction in a number of metal catalyzed
reactions of olefins.

METALLACYCLOHEXANES

Preparation and Structures

      Metallacyclohexanes can be prepared by the same methods used
in the metallacyclopentane synthesis.[14]   The following two complexes
have been prepared and characterized by chemical analysis and [13]Cnmr.

$$Cp_2 TiCl_2 \; + \; Li(CH_2)_5 Li \; \longrightarrow \; Cp_2 Ti \bigcirc$$

XIV

$$(\emptyset_3 P)_2 NiCl_2 \; + \; Li(CH_2)_5 Li \; \longrightarrow \; (\emptyset_3 P)_2 Ni \bigcirc$$

XV

Table II. Isomerization of Phosphine Nickelacyclopentane in Toluene

Compd.	Temp,°C	Time,h	Butane-d_4 (mol %)				% Isom.	% Dec.
			$D_2 \underline{\hspace{0.3cm}} D_2$	$D_2 \underline{\hspace{0.3cm}} D_2$	$D_2 \underline{\hspace{0.3cm}} D_2$	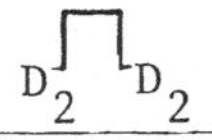 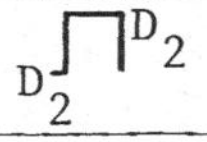 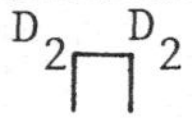 others		
1. XII-a-d_4	10	1	98.3	0	0	1.7	0	--
2. XIII-d_4	-30	0	95.2	2.4	1.2	1.2	4.8[b] (3.6)	0
3. XIII-d_4	0	0.5	84.3	7.0	2.6	6.0	13.2[b] (13.3)	7.4
4. XIII-d_4	0	1.0	47.0	29.7	16.1	7.2	66.7[b] (65.1)	17.0
5. XIII-d_4 + 4.5PØ$_3$	0	1.0	70.9	15.3	8.1	5.7	33.4	14.8
6. XIII-d_4 + 6.9PØ$_3$	0	1.0	84.7	7.4	3.6	4.3	15.2	13.6
7. XIII-d_4 + 8.0PØ$_3$	0	1.0	89.2	5.9	2.9	2.1	12.0[b] (13.7)	13.0
8. XIII-d_4	25	2.0	9.6	16.7	8.7	65.0	97.4	--

a) These isomers include deuterium scrambled butane-d_4 isomers such as 1,1,3,4-, 1,2,3,3- and 1,2,3,4-butane-d_4.

b) These values were obtained by ^{1}H-NMR analysis of 1,4-dibromobutane-d_4 afforded on the reaction of the complex with Br$_2$.

Table III.   Isomerization of Titanacenacyclopentane-d_4 in Toluene

Compd.	Temp,°C	Time,h	Butane-d_4 (mol %)				% Isom.	% Dec.
			D_2 D_2 D_2	D_2 D_2 D_2		others		
VII-d_4	−60	0	84.7	6.8	2.5	6.0	12.6	0
VII-d_4	−35	0.5	79.6	9.7	5.1	5.6	21.1	3
VII-d_4	−35	1.0	73.0	10.7	5.9	10.4	25.1	∿3
VII-d_4	−45	3.0	68.0	14.4	7.5	10.0	32.7 (31.2)	∿5

Decompositions

These complexes were thermally decomposed in solution to determine if this class of compounds also gave products resulting from
carbon–carbon bond cleavage. In the metallacyclopentane systems,
carbon–carbon bond cleavage resulted in the formation of stable
olefin complexes. This is not the case with metallacyclohexanes and
C–C bond cleavage would produce new alkyl complexes. The following
data indicates that extensive C–C bond cleavage will occur even
in these systems which would produce unstable intermediates.

$$CH_4 + C_2H_4 + C_3 + C_4(\ell) + C_5(\ell) + \text{cyclopentane}$$

XIV $\xrightarrow[\text{decalin}]{25°}$ 1   1   0   1   34   63 %

$\xrightarrow[\text{P(n-bu)}_3]{25°}$ 44   15   3.5   19.5   10   5 %

decalin

Preliminary studies of nickel analog (XV) indicate that it also
undergoes C–C bond cleavage reactions. Further studies involving
deuterium labeling, changes in the ligand, and trapping studies are
underway which should define the mechanistic pathways open to these
systems.

REFERENCES

1. N. Calderon, E.A. Ofstead and W.A. Judy, Angew. Chem. Int. Ed.
    Engl., 15, 401 (1976). R.H. Grubbs, D.D. Carr, C. Hoppin and
    P.L. Burk, J. Am. Chem. Soc., 98, 3478 (1976), J. McGinnis,
    T.J. Katz, and S. Iturwitz, ibid. 98, 605 (1976).

2. C.P. Casey, H.E. Tuinastra and M.C. Saeman, ibid., 98, 608 (1976).

3. S.J. McLain, C.D. Wood and R.R. Schrock, J. Am. Chem. Soc., 99, 3519 (1977).

4. R.J. Puddephatt, M.A. Quyser and C.F.H. Tipper, J. Chem. Soc. Chem. Commun., 626 (1976).

5. M. Ephritikhine, M.L.H. Green, R.E. MacKenzie, ibid., 619 (1976).

6. J.P. Collman, J.W. Kang, W.F. Little, M.F. Sullivan, J. Inorg. Chem., 7, 1298 (1968).

7. J.J. Eisch, G.A. Damasavitz, J. Organomet. Chem., 96, C19 (1975).

8. M.L.H. Green, Adv. Organometallic Chem., 8, 29 (1970).

9. A.R. Fraser, et al., J. Am. Chem. Soc., 95, 597 (1973).

10. P. Binger, Angew. Chem., Int. Ed., 11, 309 (1972).

11. Private communication.

12. M.J. Doyle, J. McMeeking and P. Binger, JCS Chem. Comm., 376, 1976.

13. R. Blackborrow, A. Miyashita, R. Grubbs, unpublished work.

14. G. Whitesides and J. McDermott, J. Am. Chem. Soc., 96, 947 (1974); M. Wilson, and G. Whitesides, ibid., 98, 6259 (1976).

15. J. Halpern, P. Eaton, and L. Cassar, J. Am. Chem. Soc., 91, 2405, 1969.

16. G. Whitesides, J. White and J. McDermott, J. Am. Chem. Soc., 95, 4451 (1973); 98, 6521 (1976).

17. R. Grubbs, H. Eick, and C. Biefield, Inorg. Chem., 12, 2166 (1973).

18. M.J. Doyle and P. Binger, unpublished data.

19. P. Diversi, G. Ingrosso, A. Lucherini, J.C.S., Chem. Comm., 52 (1977).

20. C. Krüger and Y.-H. Tsay, unpublished data.

21. R. Grubbs, D. Carr, and P. Burk, Organotransition-Metal Chemistry, 135 (1974).

22.   R. Grubbs, A. Miyashita, M.-I.M. Liu, P.L. Burk, J. Am. Chem.
      Soc., 99, 3863 (1977).

23.   (a) R. Cramer, J.B. Kline, and J.D. Roberts, J. Am. Chem. Soc.,
      9, 2519 (1969).
      (b) C.E. Holloway, G. Hulley, B.F.G. Johnson and J. Lewis,
      J. Chem. Soc., (A), 53 (1969).

NICKEL(0) CATALYZED REACTION OF BICYCLO[1.1.0]BUTANES WITH ELECTRON-DEFICIENT OLEFINS

H. Takaya,[1a] M. Yamakawa,[1b] and R. Noyori[1b]

The Institute for Molecular Science, Myodaiji,
Okazaki, 444, and Department of Chemistry,
Nagoya University, Chikusa, Nagoya, 464, Japan

Among interesting bicyclic hydrocarbons, bicyclo[1.1.0]-butanes are of particular interest since they are the most fundamental and probably the most highly strained. The first authentic bicyclo[1.1.0]butane derivative was reported by Wiberg in 1959.[2] Since then a number of synthesis of this class of compounds have appeared and the abundance of the hydrocarbons allowed extensive studies of their physical and chemical properties.[3] In spite of its excessively high strain energy (66 kcal/mol[4]), bicyclo[1.1.0]-butane (1), the parent hydrocarbon, can exist rather stable and heating to around 200 °C is necessary for a considerable rate of isomerization. The thermal rearrangements of bicyclo[1.1.0]butanes mostly proceed in such a fashion to preserve the central bond giving 1,3-butadiene derivatives. These products are readily rationalized in terms of an orbital-symmetry allowed $[_\sigma 2_s + _\sigma 2_a]$ process. In contrast to such thermal rearrangement, many reactions

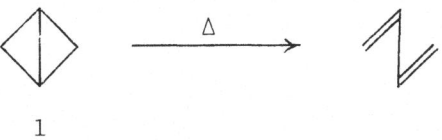

1

of bicyclobutanes with electrophilic reagents such as electron-deficient olefins, alcohols, carbonyl compounds, iodine, etc. involve cleavage of the central bond.[5] Another reaction which breaks the central bond is the radical polymerization.[6] The extraordinary ability of transition metal ions and complexes to effect isomerizations is also worth noting.[7] Historically, transition metal-promoted reaction of bicyclo[1.1.0]butanes appears to come from the hydrogenolysis in the presence of platinum or palladium catalysts.[8] The common feature of the reaction is the cleavage of the central and one of the four neighboring σ bonds and concurrent absorption of two moles of hydrogen, giving linear C_4 products (eq 1). Now various transition metals such as Cu(I),[9a] Ag(I),[9b] Pd(II),[9c] Pt(IV),[9d] Rh(I),[9e] Ir(I),[9a] Fe(0),[9f] and Ru(II)[9d] are known to be active catalysts for the isomerization. In most of the reactions, the geminal two-bond breakage is occurring as had been observed for their hydrogenolysis to give 1,3-dienes (eq 2). The unique catalysis is most plausibly interpreted as proceeding via an allylcarbene—transition metal complexes, which collapse into 1,3-dienes through 1,2-hydrogen transfer with ejection of the metal catalyst. The confirmatory evidence of the intermediacy of carbene —metal complexes has been presented by Dauben and Masamune by detecting the carbene—Pd(II) complexes by NMR using a high catalyst concentration.[10]

$$\text{(1)}$$

$$\text{(2)}$$

As a result of our interest in gaining a better understanding of the nature of the interaction of transition metals and strained carbon—carbon σ bonds, we have made a study of Ni(0) catalyzed reaction of bicyclobutanes and electron-deficient olefins.[11]

RESULTS

Nickel(0) Catalyzed Reaction of Bicyclo[1.1.0]butane ($\underset{\sim}{1}$)
and Methyl Acrylate

When a mixture of $\underset{\sim}{1}$ and methyl acrylate (ten-fold excess) was exposed to a catalytic quantity (ca. 5 mol%) of bis(acrylonitrile)-nickel(0) [Ni(an)$_2$] under nitrogen (60 °C, 36 h), methyl 2-allyl-cyclopropanecarboxylate ($\underset{\sim}{2}$) was obtained in 95% yield as a mixture of $\underline{Z}$ and $\underline{E}$ stereoisomers (65:35 ratio). The reaction with soluble bis(1,5-cyclooctadiene)nickel(0) [Ni(cod)$_2$] proceeded smoothly at 0 °C, affording $\underset{\sim}{2}$ ($\underline{Z}$:$\underline{E}$ = 65:35) in 92% yield.

No coupling products were obtained with simple alkene substrates used in place of electron-deficient olefins. In the absence of electron-deficient olefins, Ni(cod)$_2$ or Ni(CO)$_4$ in toluene did not cause any skeletal change of $\underset{\sim}{1}$, which indicates that the

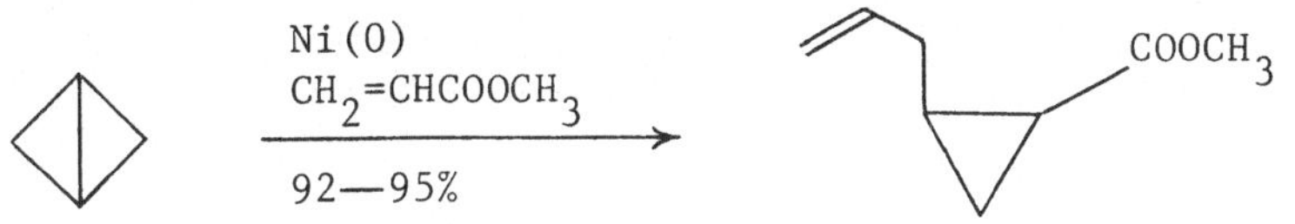

$$\text{1} \xrightarrow[\substack{92-95\%}]{\substack{\text{Ni}(0)\\\text{CH}_2=\text{CHCOOCH}_3}} \text{2 (}\underline{Z}:\underline{E} = 65:35)$$

(3)

coordinating ability of 1 to nickel atom is not high enough to displace the cyclooctadiene or carbonyl ligand. The toluene-insoluble complex, Ni(an)$_2$, was not an effective isomerization catalyst, either. These results led us to conclude that soluble, coordinatively unsaturated Ni(0) complexes having electron-poor olefinic ligand(s) is the true active catalyst.

The Mode of Two-Bond Fission and the Reactive Intermediate

It is clear that the present Ni(0) catalyzed reaction of 1 and methyl acrylate involves cleavage of two σ bonds of 1. However, the simple hydrocarbon 1 having the high symmetry is not an adequate substrate for elucidation of the reaction mode. This ambiguity in reaction mode was made clear by carrying out the reactions of methyl substituted bicyclobutanes 3, 4, and a deuterated derivative 5. The catalysis of 3 and 4 also proceeded quite facilely to give the corresponding cyclopropane carboxylate adducts 6 and 7. The deuterio derivative 5 exhibits analogous behavior and afforded 8. Thus the two-bond cleavage of bicyclo[1.1.0]butanes has revealed to occur specifically at the C-1—C-3 and C-2—C-3 bonds.
The present coupling reaction can be formally viewed as a

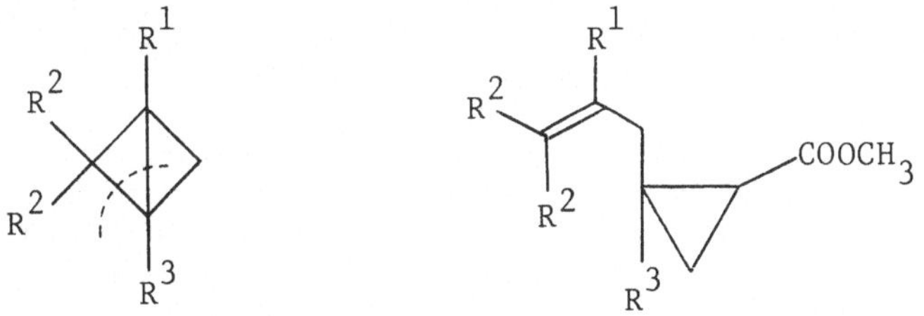

3, R^1 = CH$_3$; R^2 = R^3 = H   6, R^1 = CH$_3$; R^2 = R^3 = H

4, R^1 = R^2 = CH$_3$; R^3 = H   7, R^1 = R^2 = CH$_3$; R^3 = H

5, R^1 = R^2 = CH$_3$; R^3 = D   8, R^1 = R^2 = CH$_3$; R^3 = D

metal-assisted intramolecular retro-carbene addition of the bicyclobutanes followed by intermolecular cycloaddition of the resulting
allylcarbene to an olefin (eq 4). These catalyzed cycloadditions
are in dramatic contrast to the corresponding purely thermal reaction, in which a $[_\sigma 2 + _\pi 2]$-type cycloadduct is formed (eq 5).[5]
It is also interesting to compare these Ni(0)-catalyzed reactions
of bicyclo[1.1.0]butanes and olefins with the behavior of the higher
homolog, bicyclo[2.1.0]pentane which yields exclusively the $[_\sigma 2 +
_\pi 2]$ adduct 9 through the central bond cleavage.[12]

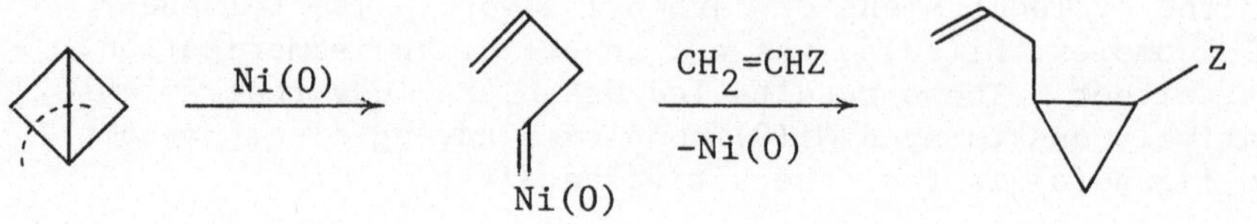

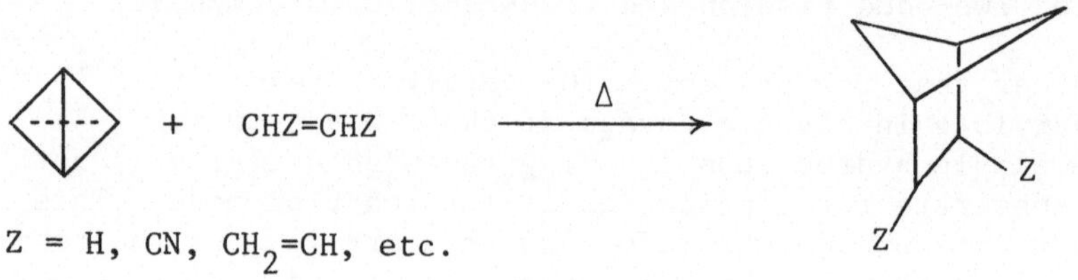

$$Z = H, CN, CH_2=CH, etc.$$

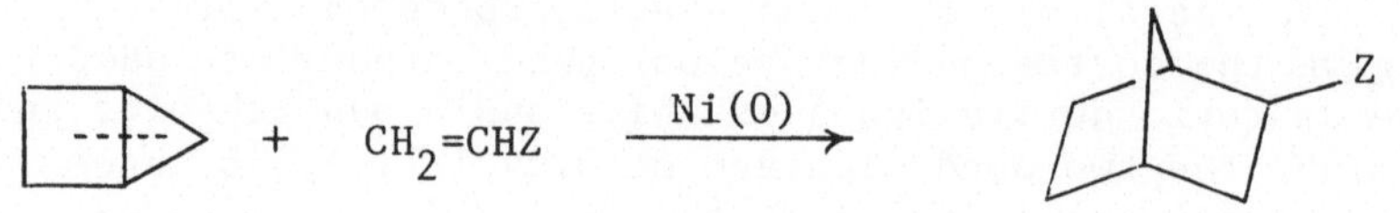

$$Z = COOCH_3, CN, etc.$$

Stereochemistry of the Reaction of the Nickel(0)
Carbenoid with Olefins

An interest in the mechanism of transformation involving allylcarbene—Ni(0) complexes has prompted us to determine the stereochemistry of the coupling reactions. Reaction of 1 and methyl
(Z)-β-deuterioacrylate (10) in the presence of a small amount of
Ni(cod)₂ yielded the cycloadducts 11 and 12 in good yields.[11c]
Comparison of the well-resolved NMR spectra of 11 and 12 taken with
the shift reagent, Eu(fod)₃ with those of undeuterated ones revealed
that the Z configuration of the starting olefin 10 is virtually
retained in the cyclopropane products. This indicates that the
cycloaddition of the allylcarbene—Ni(0) carbenoid to an olefin

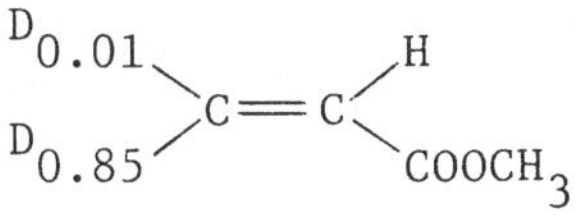

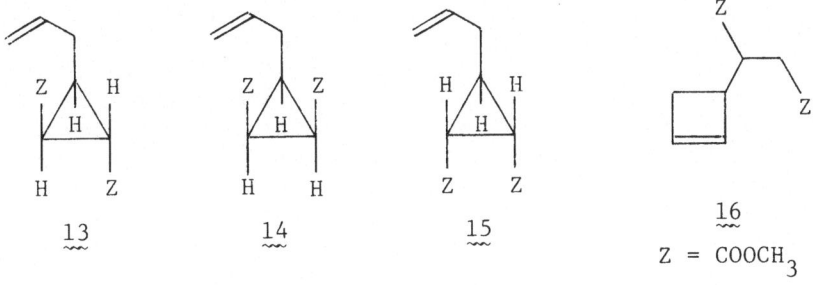

$$D_{0.01} \quad D_{0.85} \quad C=C \quad H \quad COOCH_3$$

10

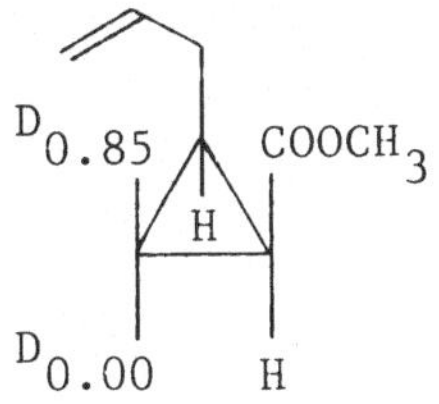

11

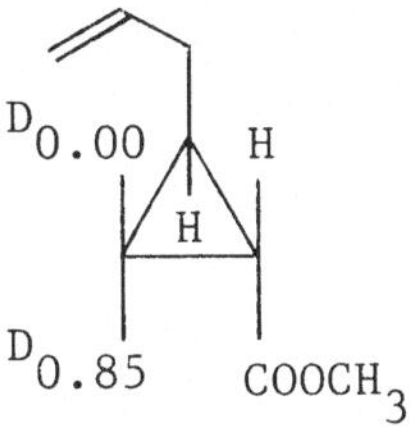

12

proceeds with very high level of stereospecificity.

The reaction of 1 and E- or Z-1,2-disubstituted olefins was also carried out.[11a] The reaction of 1 and dimethyl fumarate in benzene (40 °C, 60 h) in the presence of Ni(an)$_2$ gave 13 as a sole 1:1 cycloadduct in 19% yield, in which the stereochemistry of the fumarate is preserved. In addition, the ene-type reaction product 16 was obtained in 16% yield as a mixture of diastereoisomers. The reaction of 1 and dimethyl maleate gave the stereospecific addition products 14 and 15, and a small amount of the nonstereospecific addition product 13 (31% combined yield) accompanied by a very small amount of 16. Thus complete stereospecific cis addition was attained in the cyclopropanation involving fumarate, while the stereospecificity concerning maleate was >91%. The loss of stereo-chemical integrity in the latter reaction is probably due to the partial isomerization of the starting maleate. In fact, the recov-ered unsaturated esters contained variable amounts (1—3%) of dimethyl fumarate. In addition, control experiment showed that diethyl fumarate reacts about six-times faster than diethyl maleate. Dimethyl fumarate did not isomerize to maleate under the catalytic conditions.

It is worthy to note that the intervenient allylcarbene—Ni(0) complexes have the strong nucleophilicity. This interesting nature could be ascribed to the electropositive property of Ni(0) atom.[13] According to the valence bond method, allylcarbene—transition metal complexes can be expressed as the resonance hybrid as shown

13

14

15

16

Z = COOCH$_3$

below. Relative importance of each resonance contributor should
depend on the nature of the metal M. When M is electron-donating
Ni(0) atom, the ylene (17a) and ylide (17c) structures would be
given much weight (vide infra). Consequently, the carbenic carbon
reveals nucleophilic character and easily reacts with electron-
deficient olefins. Electron-rich simple olefins necessarily
exhibits no reactivity. One of the principal features of the Ni(0)
carbenoid that it has little trend of collapsing into a 1,3-diene
and Ni(0) residue is also reconcile with this view. It is well
recognized that alkylcarbenes are prone to afford olefins in such
a way that the β-hydrogen migrates to the electron-deficient
carbenic carbon.[14] The hydrogen migration is also quite common
to alkylcarbenes complexed with electronegative metals or ions.[14]
This is due to the fact that the inverse ylide structure of the
type 17b is of major importance in the resonance hybrid. The
hitherto reported experimental findings that the allylcarbene
complexes of such metal ions or complexes isomerize to 1,3-dienes
are rationalized in these terms.

L = ligand

        It is also noteworthy that the cycloaddition of the inter-
mediary allylcarbene—Ni(0) complexes to an electron-deficient
olefin, though they have considerable ylide character, proceeds
with a high degree of stereospecificity. This seems rather peculiar
because certain ylides, particularly sulfur ylides, are known to
react with electron-deficient olefins in a nonstereospecific manner
to give cyclopropanes.[15] The present Ni(0)-carbenoid reaction could
be considered to pass according to the stereochemical course
outlined in Scheme I. The carbenic ligand in 18 reacts with the
olefinic substrate coordinated to the same metal atom, giving the
nickelacyclobutane intermediate 19. In such intermediate 19, the
presence of a carbon—metal σ bond or the related bonding interactions prevents the inversion of stereochemistry. Subsequent
reductive elimination of the alkyl ligands give rise to the cyclopropane adduct 20. From the stand point of microscopic reversi-
bility, the latter process can be rationalized by the presence of
the well-established cyclopropane to metallocyclobutane conversion.[16]

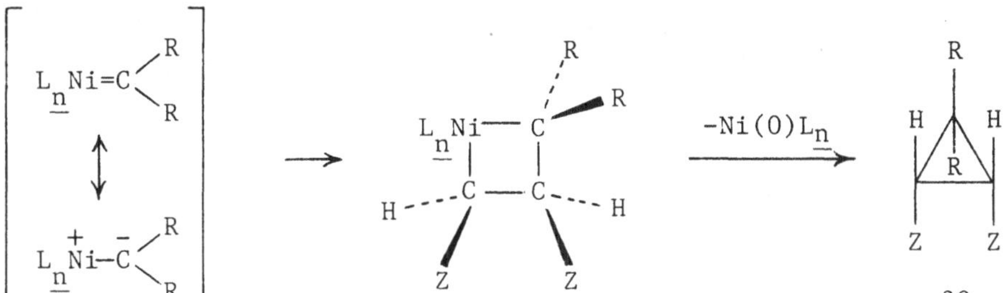

L = 1,2-disubstituted olefin
R,R = H and $CH_2CH=CH_2$

Scheme I

Regioselectivity of the Nickel(0) Promoted Two-Bond Cleavage of Bicyclo[1.1.0]butanes

The reaction course of unsymmetrically substituted bicyclo-butane derivatives is highly dependent on the nature of both substituent at the angular site and transition metal catalysts. Therefore, examination of the regioselectivity of the reaction in more details should offer us informations for understanding the mechanism of the bicyclobutane to allylcarbene conversion.

Gassman examined mainly the Rh(I) catalysis of alkyl substituted bicyclobutanes and proposed a two-step mechanism which involves a cyclopropylcarbinyl cation intermediate 21 (eq 7).[9e] Paquette, after pursuing extensive study on the catalysis with Ag^+ ion, concluded that the geminal two-bond cleavage proceeds so as to generate the most stable "argento carbonium ion"(eq 8).[17] He inter-

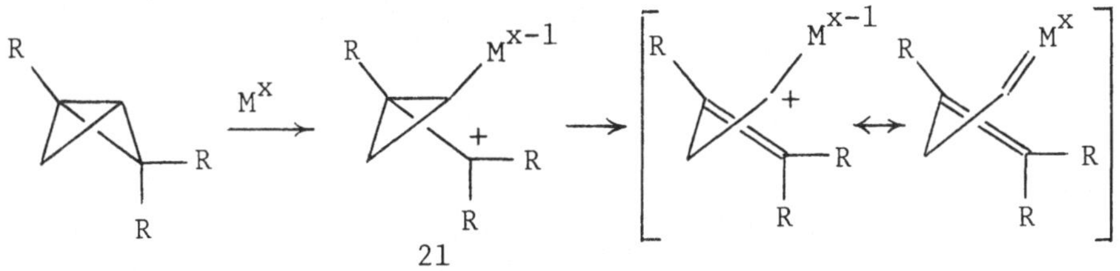

(7)

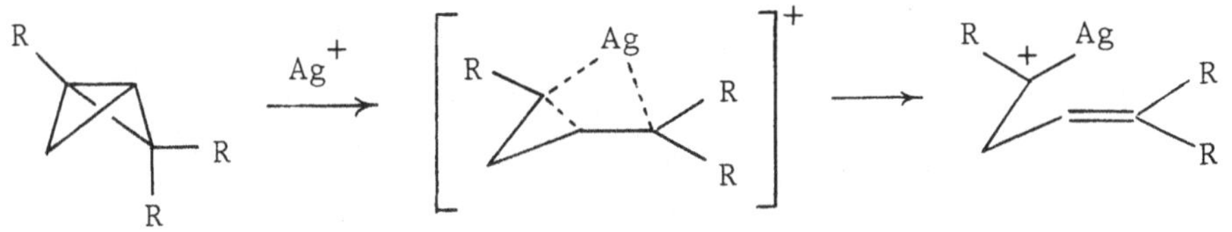

(8)

preted the regioselectivity of the reaction with so-called bulky
metal catalysts in terms of steric factors. Such mechanisms seem
entirely compatible with their own findings but can not adequately
account for the Ni(0) catalyzed reactions. In hopes of obtaining
more insight into the mechanism of the present Ni(0) catalysis, we
examined the catalytic process using bicyclobutanes bearing
electron-donating, electron-withdrawing, and conjugative groups at
the angular position(s).

First, as have been described above, the reactions of methylated derivatives 3 and 4 proceeded quite smoothly to give 6 and 7,
respectively.

With the ester substituted bicyclobutanes 22 and 23, the two
types of ring-opening reactions, path a and path b, occurred,
giving the corresponding cycloadducts in 96—100% yields (50—75%
conversion). In each case, path a which involves an allylcarbo-
alkoxycarbene intermediate was preferred over the alternative path
b.

Introduction of a powerfully electron-attracting cyano group
produced complete regioselectivity. Thus, the catalytic ring
opening of 1-cyanobicyclo[1.1.0]butane (24) and 1-cyano-3-methyl-
bicyclo[1.1.0]butane (25) occurred exclusively at the C-1—C-2 and
C-1—C-3 bonds, as shown by the carbene trapping experiments with
methyl acrylate. In these cases, the catalytic reaction proceeded
slowly and considerable amount of the starting material was
recovered (20—25% conversion). The yields of the carbene adducts
were 60—62%. Control experiments and analysis of the reaction
mixture by GC-MS measurement revealed that the volatile byproducts
formed in the presence of the Ni(0) catalyst were not 1:1 adducts
of the bicyclobutanes and methyl acrylate.

Phenylbicyclo[1.1.0]butane (26) is a thermally unstable hydrocarbon. Without catalyst, it isomerizes easily to 1-phenylcyclo-
butene or forms a dimerization product. However, under the influence of Ni(cod)$_2$ and methyl acrylate, 26 suffers two-bond cleavage
via path a to give only one kind of carbene adduct. The product
yield of this reaction was not high (48%) due to the accompanying

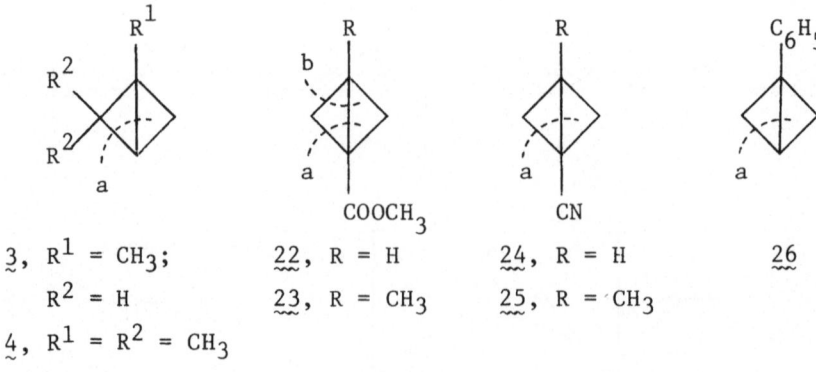

3, R^1 = CH$_3$;   22, R = H   24, R = H   26
   R^2 = H   23, R = CH$_3$ 25, R = CH$_3$

4, R^1 = R^2 = CH$_3$

purely thermal reactions. Direct NMR analysis of the reaction
mixture gave no evidence for formation of other regioisomers.

MO CONSIDERATION ON THE ORIGIN OF THE REGIOSELECTIVITY

General Consideration on the Mechanism of Formation and
Stability of the Intermediary Metal Complexes

Kinetic measurements relating to the initial rates of disappearance of 1 were carried out in methyl acrylate at 0 °C in the
presence of Ni(cod)$_2$. The result revealed that the catalytic
reaction is a second-order process depending upon the concentration
of both 1 and Ni(0) complex. The relative rates of the reaction
(25 °C) of 1, 3, 4, 22, and 23 were 1.0, 2.4, 19, 0.04, and 0.03,
respectively. Methyl substituent at the angular position enhances
the reaction rate. Bicyclobutanes with an electron-withdrawing
group such as CO$_2$CH$_3$, though considered to have stronger affinity
to Ni(0) than 1 (vide infra), undergo the two-bond cleavage much
more slowly. When cyano or carbomethoxy derivatives were mixed
with Ni(cod)$_2$ in methyl acrylate, a dark-orange to wine-red color
developed without any significant consumption of the starting
bicyclobutanes. These phenomena coupled with the kinetic results
lead us to conclude the presence of a preequilibrium complex formation as is shown in Scheme II. In the first place, reversible
π-type complex formation occurs by interaction of a coordinatively
unsaturated metal species ML$_n$ and 1. The central bond of bicyclo-
butane is known to have a 96% π character and hence metals would
coordinate to this molecule from the exo side in an edge-on
manner.[18] Such initial π-complex formation seems requisite for the
subsequent simultaneous two-bond cleavage, since only transition
metal ions or complexes can effect this type of reaction.
Subsequent oxidative addition of the central σ bond to the metal

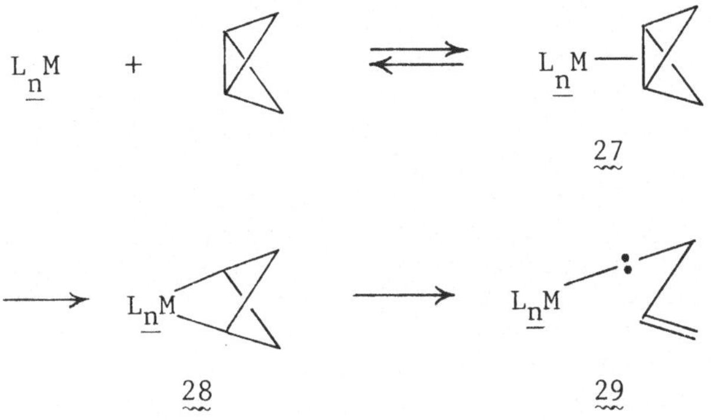

Scheme II

affords the metallocyclic intermediate 28.[19] The complex 28 has
unique structure in that it bears fused two metallocyclobutane
rings. Such perturbation makes the hydrocarbon skeleton so flexible
that it changes into the allylcarbene—metal complex 29. A number
of evidence have been reported for transformation of metallocyclobutanes into the metal carbene complexes.[20] Here it is not clear
whether the terminal carbon—carbon double bond in the carbene
moiety can contribute to the stabilization of the carbenoids by
coordination to the central metal. The electronic configuration
of the complexes 27 and 29 were shown in Figure 1.

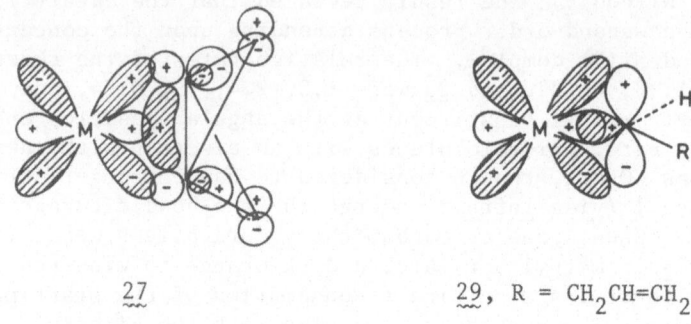

27   29, R = CH₂CH=CH₂

Figure 1. The electronic configuration of the complexes 27 and 29.

Here we would like to consider the stabilization factors for
the intermediary metal complexes. Figure 2 shows the schematical
energy level diagram of organic ligand and Ni(0) atom coordinated
with certain ligands. According to the perturbation molecular
orbital (MO) theory, the stabilization energies of metal complexes
can be estimated from the degree of energy gap between the ligand
highest occupied MO (HOMO) and metal lowest unoccupied MO (LUMO)
(σ interaction) and that between the ligand LUMO and metal HOMO
(π interaction). The relative importance of the contribution
of such pair of interaction to the stability of the complex
is remarkably influenced by both electronic properties of metal M
and ligands. The ionization potential or promotion energy concerning to the $(n - 1)d \rightarrow np$ transition is regarded as a good measure
of the π-donor ability of metals, and metal σ-accepting property
is related to the electron affinity.[13] Accordingly, Ni(0) and
Rh(I), which have low promotion energy (1.72 and ~1.6 eV, respectively) can facilely donate electrons from their HOMOs to ligand
LUMOs. Therefore, π-interaction becomes important. On the other
hand, Ag⁺ and Cu⁺ have promotion energies of 9.94 and 8.25 eV, and
ionization potentials of 21.48 and 20.29 eV, respectively. These
metal ions can thus be regarded as poor π-donors and hence σ-inter-
actions are the major factor in stabilizing the complexes. Other

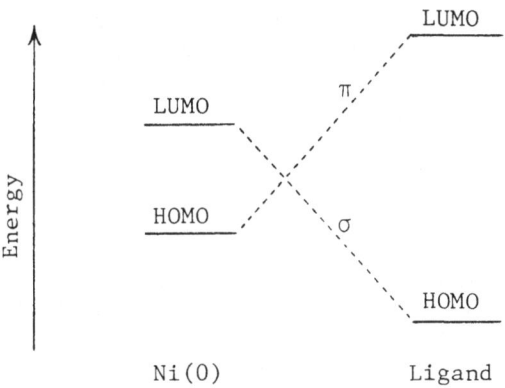

Figure 2. Frontier orbital interactions of Ni(0) and organic ligand.

metals in a high oxidation state are likewise poor π-donors but good σ-acceptors. Pd(II) and Pt(II) are situated between these two extreme cases and their electronic properties are controlled by various secondary factors such as nature of ligands coordinated to the metal.[21]

Substituents in bicyclobutanes exert an influence on their electron-donating and -accepting abilities. As shown in Table I,

Table I

Frontier MO Levels of Certain Bicyclo[1.1.0]butanes[a]

Bicyclo[1.1.0]butane	HOMO, eV	LUMO, eV
1	−12.07	−5.47
3	−11.84	−5.33
22	−12.34	−8.83[b]
24	−12.27	−8.23[b]
26	−11.73	−8.27[b]

[a] Calculated by the extended Hückel method.[22] The geometry of the carbocyclic skeleton was based on the reported data.[23] [b] The lowest unoccupied MO of this compound has large coefficient around the substituent. Therefore we selected low-lying unoccupied orbitals spreading over angular carbons.

bicyclobutane ($\underset{\sim}{1}$) and its alkyl derivatives have a high-lying
HOMO, and therefore σ-interactions appear to be facilitated. For
these substrates, the π back-bonding is less significant than the
σ donation. On the other hand, when an electron-withdrawing group
is introduced to these strained molecules, the π back-bonding
becomes important. As can be seen from Figure 3, HOMOs of bicyclobutanes spread sufficiently in the exo region of the central σ
bond. Moreover, each substrate has antisymmetric LUMO extending
around the angular carbons. Interestingly, the eminent π character
of the central bond is not affected by introduction of a substituent at the angular position, nor the symmetrical shape of the
frontier MOs is deformed to any great extent.

The intermediate $\underset{\sim}{29}$ is considered to be a σ^2 singlet carbene—
Ni(0) complex (Figure 1).[7] The bonding in this complex $\underset{\sim}{29}$ can be
described in terms of the donation of a pair of electrons from an
approximately sp[2] hybridized orbital (HOMO) of the carbenic carbon
to an empty orbital of M (LUMO) (σ interaction) and back-donation
of electrons from an appropriate d or hybridized orbital of M
(HOMO) to a vacant $2p_z$ orbital (LUMO) on carbenic carbon (π-inter-
action). Here again, Ni(0) or Rh(I), having a great π-donor
ability, facilitate the back-donation from M to the carbene LUMO.
In fact, the presence of a great degree (0.56 electron) of charge
transfer from Ni to CH_2 has been indicated by certain MO calculation.[24] Numerous experiments are also available to support
this.[25] By contrast, a carbenoid in which M has little π-donor
ability (Ag^+, Cu^+, etc) is considered to form a coordinate covalent
bond through the interaction of the metal LUMO and carbene HOMO.
Incidentally, frontier orbital energies of carbenes are highly
influenced by the substituents. Electronic properties of certain
σ^2 singlet carbenes obtained by MINDO/3 calculation are shown in
Table II. As can be easily understood from the values of ionization
potential and electron affinity, introduction of an electron-
withdrawing group reduces LUMO level of carbene and thereby
decreases the energy gap between the metal HOMO and the carbene
LUMO. As the result, the carbenoids of π-donating metals such as
Ni(0) and Rh(I) are stabilized. On the other hand, electron-
donating groups elevate the carbene LUMO level and accordingly
destabilize the carbenoids of such type of metals. In contrast,
virtually inverse substituent effects are expected for the carbenoids of σ-accepting type metals in which a large energy separation
is present between metal HOMO and carbene LUMO. In such cases,
electron releasing alkyl groups stabilize the metal—carbene
complexes. MO calculation does support the plausibility of this
consideration on the stability of the Ni(0) carbenoids. Figure 4
outlines the correlation diagram of methylene—Ni(0) complex
calculated by the extended Hückel method.[27] The results indicated
that the stabilization energy of the complex arises from the σ
interaction between the $(3d_{z^2})(4s)$-hybridized orbital of Ni(0) and
sp[2]-like orbital of methylene and the π interaction between the

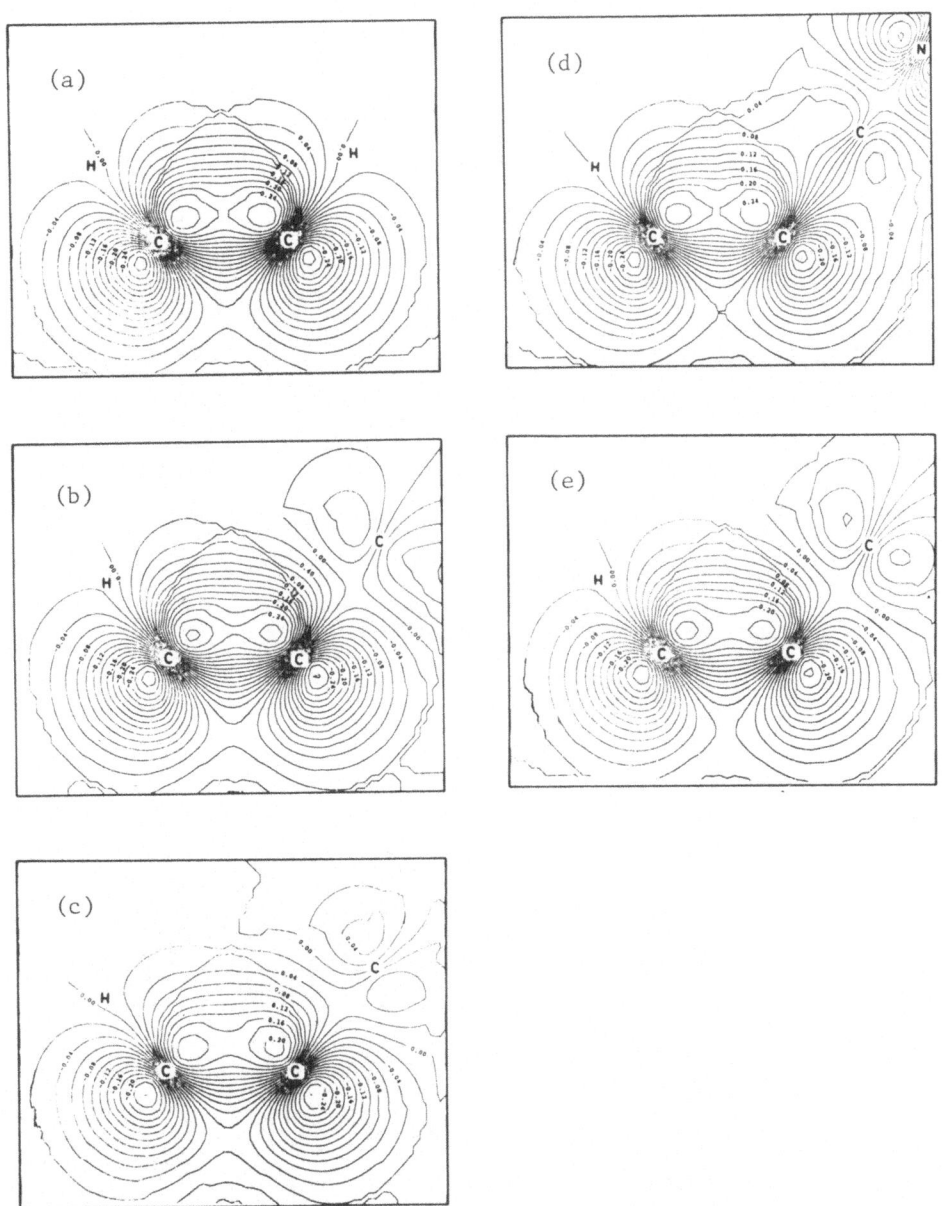

Figure 3. Contour maps of HOMO of (a) 1, (b) 3, (c) 22, (d) 24, and (e) 26 around C-1—C-3 bond in the plane bisecting C-2—C-1—C-4 angle.

Table II

Electronic Properties of Certain Carbenes[a]

Carbene	Ionization potential eV	Electron affinity eV
CH_2	9.94	0.09
$CHCH_3$	8.86	0.83
CHCOOH	9.89	2.20
CHCN	8.92	1.37
CHC_6H_5	8.22	1.55

[a] Calculated by MINDO/3 method. For details of the calculation, see ref. 26.

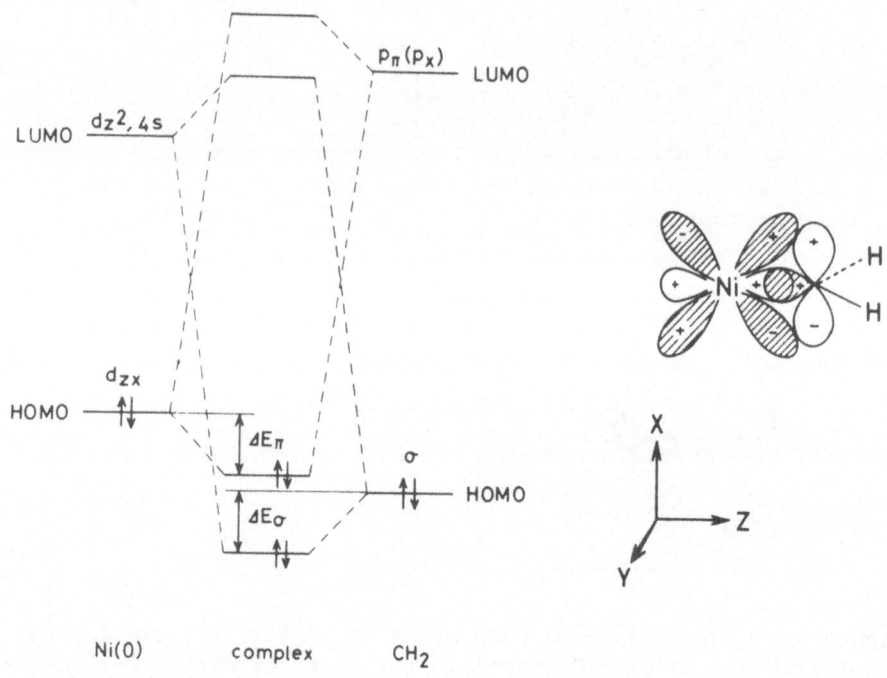

Figure 4.   Correlation diagram of methylene—Ni(0) complex.

d_{zx} orbital of Ni(0) and methylene $2p_x$ orbital. The stabilization
energy, term ΔE_σ, related to the σ donation suffers a change to a
little or moderate extent by introduction of substituents into the
carbenic carbon. On the other hand, the stabilization energy,
ΔE_π, gained through π-back donation greatly varies depending on
the nature of the substituents. In general, electron-withdrawing
groups such as carbomethoxy and cyano groups increase ΔE_π value to
a large extent, as compared with that of the parent methylene
complex, whereas methyl or phenyl substituent decreases this value.
Thus, the calculations suggested that the stabilities of the Ni(0)
complexes are mainly governed by ΔE_π term. The stability decreases
in the order of $CHCOOCH_3 \sim CHCN > CH_2 > CHCH_3 \sim CHC_6H_5$.

Regioselectivity of the Nickel(0) Catalyzed
Two-Bond Cleavage Reaction

As has been discussed above, both HOMO and LUMO of bicyclobutanes were not so distorted by introduction of substituents at
angular positions that Ni(0) atom will not able to discriminate
between the two angular carbons. Thus the regioselectivity could
not be determined by the initial attacking of the Ni(0) complex
on bicyclobutanes. So we must seek other factors which control the
observed clean selectivity. Regioselectivity of a kinetically
controlled reaction is of course determined by relative stabilities
of possible regioisomeric transition states, leading to products.
In principle, the transition states possess the characters of both
starting and product systems, and the relative weight of both
systems depends on where the transition state lies along the
reaction coordinate. As a matter of fact, however, determination
of the exact position of the transition state on reaction coordinate is quite difficult. Some years ago we postulated that the
Ni(0) catalyzed reactions proceed along reaction coordinate by way
of a rather late transition state, and that electronic structure
of the transition state could be approximated by that of the
product allylcarbene—metal complex. If this assumption is valid,
the geminal two-bond cleavage of the unsymmetrically substituted
bicyclobutanes would occur so as to produce the most favorable
allylcarbene—metal complex. Our experimental findings indeed
validate this hypothesis. The retro-carbene additions of 4 and 5
whose angular carbon was substituted by an electron-donating methyl
group occurred so as to escape the methylated carbon from becoming
the carbenic center in the resulting Ni(0) carbenoids. While the
carbon having an electron-withdrawing cyano group was incorporated
as a carbenic center in the reactions of 24 and 25. These observations should be compared with the effect of alkyl group observed
in the rearrangement of bicyclobutanes by Ag^+ ion which is in the
striking contrast to Ni(0) with respect to electronic properties
of transition metals (cf. eq 8). The conjugative phenyl substi-

tuent on a carbenic carbon destabilizes carbene—Ni(0) complex
and this fact let the non-phenylated carbene have preference over
the phenylated one, as was observed in the catalysis of 26.

The results obtained with 1-carbomethoxybicyclo[1.1.0]butanes
22 and 23 had somewhat different features, because two pathways
are competing. Predominance of path a over path b is consistent
with the initial expectation. However, the existence of competitive
two pathways seems peculiar. These facts tell us that for full
understanding of the observations, more precise consideration
regarding the transition state is required. The primary factor of
the difference in degree of regioselectivity found in the reaction
of carbomethoxy and cyano derivatives is not yet clear, but it may
be ascribed to difference in relative positions of the transition
states on reaction coordinate. The reaction of bicyclobutanes
having cyano group which has eminently electron-attracting nature
could be considered to proceed via the later transition state
whose electronic structure resembles closely to the Ni(0) carbenoid.

The regioselectivity observed with the cyano-, carbomethoxy-,
and methylbicyclobutanes can also be predicted based on the
resonance stabilization (Scheme III). The principal polar contributor
in the resonance hybrid of Ni(0) carbenoid is the one which places
significant degree of negative charge on the carbenic carbon.

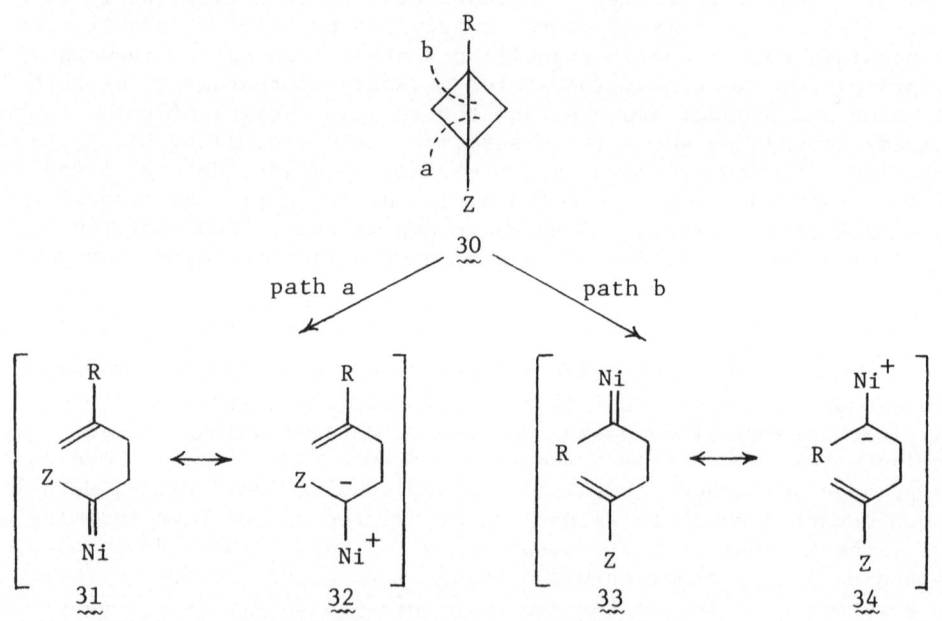

R = electron-donating group or H
Z = electron-withdrawing
       group or H

Scheme III

Therefore, the two-bond cleavage occurs so as to more electronegative angular carbon combines as carbenic center to nickel(0) atom. This can reasonably interpret that path a occurs predominantly in the Ni(0) catalysis. Contrary to the experimental findings, however, conjugation effect of phenyl group implies that incorporation of this group into the carbenic carbon should stabilize the Ni(0) carbenoid. Thus, such consideration based on valence bond method should be done more carefully in certain cases.

It seems important to note that the regiochemistry of the Ni(0) catalyzed reaction of bicyclo[1.1.0]butanes bears many similarities to selectivity in the tungsten catalyzed metathesis of olefins.[28] The degeneracy of terminal olefin metathesis had been discussed in terms of steric interactions. Recently, however, Gassman demonstrated that the degenerate metathesis of terminal olefins by certain tungsten-based catalysts is due to the polarization of the intermediary metal—carbene complex.[29] He examined the regioselectivity of the metathesis reaction shown in eq 9 and concluded that alkylcarbene—tungsten complex $\underline{35}$ has ylide structure $\underline{38}$ as the major resonance contributor.

$$L_n W = CHR \rightleftharpoons L_n W \begin{array}{c} R \\ \\ R' \end{array} \longrightarrow \begin{array}{c} L_n W = CH_2 \\ + \\ RCH = CHR' \end{array} \qquad (9)$$

$\underline{35}$ R,R' = alkyl

$$\overset{-}{L_n} \overset{+}{M} - CHR \quad \longleftrightarrow \quad L_n M = CHR \quad \longleftrightarrow \quad \overset{+}{L_n} \overset{-}{M} - CHR$$

$$\underline{36} \qquad\qquad\qquad \underline{37} \qquad\qquad\qquad \underline{38}$$

ACKNOWLEDGMENTS

We wish to thank Messrs. T. Suzuki, Y. Kumagai, M. Hosoya, N. Hayashi, and H. Kawauchi for their valuable contribution to this work. This work was supported in part by the Matsunaga Science Foundation and the Kurata Science Foundation.

REFERENCES AND NOTES

1. (a) The Institute for Molecular Science; (b) Nagoya University.
2. K. B. Wiberg and R. P. Ciula, J. Am. Chem. Soc., 81, 5261 (1959)
3. K. B. Wiberg, Advan. Alicyclic Chem., 2, 185 (1968).
4. P. v. R. Schleyer, J. E. Williams, and K. R. Blanchard, J. Am. Chem. Soc., 92, 2377 (1970); S. Chang, D. McNally, S. Shary-

Tehrany, M. J. Hickey, and R. H. Boyd, ibid., 92, 3109 (1970).

5. K. B. Wiberg, G. M. Lampman, R. P. Ciula, D. S. Connor, P. Schertler, and J. Lavanish, Tetrahedron, 21, 2749 (1965).

6. H. K. Hall, Jr., C. D. Smith, E. P. Blanchard, Jr., S. C. Cherkofsky, and J. B. Sieja, J. Am. Chem. Soc., 93, 121 (1971).

7. D. J. Cardin, B. Cetinkaya, M. J. Doyle, and M. F. Lappert, Chem. Soc. Rev., 2, 99 (1973); K. C. Bishop III, Chem. Rev., 76, 461 (1976).

8. W. R. Moore, H. R. Ward, and R. F. Merritt, J. Am. Chem. Soc., 83, 2019 (1961); J. Meinwald, C. Switherbank, and A. Lewis, ibid., 85, 1880 (1963); D. M. Lemal and K. S. Shim, Tetrahedron Lett., 3231 (1964); W. v. E. Doering and J. F. Coburn, ibid., 994 (1965); E. P. Blanchard, Jr. and A. Cairncross, J. Am. Chem. Soc., 88, 487 (1966). See also ref. 5.

9. (a) P. G. Gassman, G. R. Meyer, and F. J. Williams, J. Am. Chem. Soc., 94, 7741 (1972); (b) L. A. Paquette, Acc. Chem. Res., 4, 280 (1971); L. A. Paquette and G. Zon, J. Am. Chem. Soc., 96, 224 (1974); (c) M. Sakai, H. Yamaguchi, and S. Masamune, Chem. Commun., 486 (1971). See also ref. 9a. (d) P. G. Gassman and T. J. Atkins, J. Am. Chem. Soc., 93, 1042 (1971); 93, 4597 (1971); (e) P. G. Gassman and F. J. Williams, J. Am. Chem. Soc. 94, 7733 (1972); P. G. Gassman and T. J. Atkins, ibid., 94, 7748 (1972); P. G. Gassman and R. R. Reitz, ibid., 95, 3057 (1973); P. G. Gassman and T. Nakai, ibid., 94, 2877 (1972); (f) R. M. Moriarty, K. -N. Chen, and J. L. Flippen, J. Am. Chem. Soc., 95, 6489 (1973).

10. W. G. Dauben and A. J. Kielbania, Jr., J. Am. Chem. Soc., 94, 3669 (1972); S. Masamune, M. Sakai, and N. Darby, J. Chem. Soc. Chem. Commun., 471 (1972).

11. Part of this work was reported in preliminary form: (a) R. Noyori, T. Suzuki, Y. Kumagai, and H. Takaya, J. Am. Chem. Soc., 93, 5894 (1971); (b) R. Noyori, Tetrahedron Lett., 1691 (1973); (c) R. Noyori, H. Kawauchi, and H. Takaya, ibid., 1749 (1974).

12. R. Noyori, T. Suzuki, and H. Takaya, J. Am. Chem. Soc., 93, 5896 (1971); R. Noyori, Y. Kumagai, and H. Takaya, ibid., 96, 634 (1974).

13. R. S. Nyholm, Proc. Chem. Soc., 273 (1961).

14. W. Kirmse, "Carbene Chemistry", 2nd Ed., Academic Press, New York, N.Y., 1971, Chapter 12.

15. C. R. Johnson and C. W. Schroeck, J. Am. Chem. Soc., 93, 5303 (1971), and references cited therein.

16. S. E. Binns, R. H. Cragg, R. D. Gillard, B. T. Heaton, and M. F. Pilbrow, J. Chem. Soc. A, 1227 (1969); M. Lenarda, R. Ros, M. Granziani, and U. Belluco, J. Organometal. Chem., 65, 407 (1974). Recently metallocyclobutanes were shown to be intermediates of olefin metathesis reactions and cyclopropanes are formed from such intermediates: C. P. Casey and T. J. Burkhardt, J. Am. Chem. Soc., 96, 7808 (1974); C. P. Casey, H. E. Tuinster, and M. C. Saeman, ibid., 98, 608 (1976); P. G. Gassman and T.

H. Johnson, ibid., 98, 6055 (1976).

17. L. A. Paquette, R. P. Henzel, and S. E. Wilson, J. Am. Chem. Soc., 94, 7789 (1972); L. A. Paquette, S. E. Wilson, and R. P. Henzel, ibid., 94, 7771 (1972). See also 9b.

18. M. D. Newton and J. M. Schulman, ibid., 94, 767 (1972); M. Pomerantz and D. F. Hillenbrand, J. Am. Chem. Soc., 95, 5809 (1973).

19. R. D. Gillard, M. Keeton, R. Mason, M. F. Pilbrow, and D. R. Russel, J. Organometal. Chem., 33, 247 (1971); R. M. Moriarty, K. -N. Chen, C. -L. Yeh, J. L. Flippen, and J. Karle, J. Am. Chem. Soc., 94, 8944 (1972); M. Lenarda, R. Ros, M. Graziani, and U. Belluco, J. Organometal. Chem., 65, 407 (1974), and references cited therein.

20. For a review, see C. P. Casey, "Transition Metal Organometallics in Organic Synthesis", Vol. 1, H. Alper Ed., Academic Press, New York, N.Y., 1976, Chapter 3.

21. G. Distefano, G. Innorta, S. Pignataro, and A. Foffani, J. Organometal. Chem., 14, 165 (1968); W. G. Dauben and A. J. Kielbania, Jr., J. Am. Chem. Soc., 93, 7345 (1971); R. Noyori, M. Yamakawa, and H. Takaya, ibid., 98, 1471 (1976).

22. R. Hoffmann, J. Chem. Phys., 39, 1397 (1963).

23. N. C. Baird and M. J. S. Dewar, J. Chem. Phys., 50, 1262 (1969); N. C. Baird, M. J. S. Dewar, and R. Sustmann, ibid., 50, 1275 (1969).

24. A. K. Rappe and W. A. Goddard III, J. Am. Chem. Soc., 99, 3966 (1977).

25. R. M. Kirchner, J. A. Ibers, M. S. Saran, and R. B. King, J. Am. Chem. Soc., 95, 5775 (1973); R. R. Schrock, ibid., 97, 6577 (1975); L. J. Guggenberger and R. R. Schrock, ibid., 97, 6578 (1975).

26. R. Noyori, M. Yamakawa, and W. Ando, Bull. Chem. Soc. Jpn., 51, 811 (1978).

27. The reported value (1.87 Å) was employed for the Ni—C bond length.[24] The bond angle for ∠H-C-X of carbene was set at 120°. The reported values were used for other data on geometry of carbenes.[23]

28. N. Calderon, E. A. Ofstead, and W. A. Judy, Angew. Chem. Int. Ed. Engl., 15, 401 (1976); W. J. Kelly and N. Calderon, J. Macromol. Sci. Chem., 9, 911 (1975); J. McGinnis, T. J. Katz, and S. Hurwitz, J. Am. Chem. Soc., 98, 605 (1976).

29. P. G. Gassman and T. H. Johnson, J. Am. Chem. Soc., 99, 622 (1977).

SIGNIFICANCE OF DIOXYGEN ACTIVATION ON MODEL CATALYTIC

OXYGENATION WITH COBALT-SCHIFF'S BASE COMPLEXES

Akira Nishinaga, Haruo Tomita, Tadashi Shimizu,
and Teruo Matsuura

Department of Synthetic Chemistry
Faculty of Engineering
Kyoto University, Yoshida, Kyoto, Japan

There are many attempts to provide experimental examples for
catalytic oxygenation of organic molecules in connection with biological oxygenations. These attempts have, however, been focused
mostly on monooxygenation, namely the incorporation of monooxygen
from molecular oxygen into organic molecules including the hydroxylation of aliphatic and aromatic substrates and the epoxidation of
olefinic compounds (1). Little attention has, however, been paied
so far to the dioxygenation of organic molecules by homogeneous
catalysis although the dioxygenation of organic substrates is also
an important metabolic path in biological systems. Dioxygenases
catalyze the dioxygenation in living systems as well known. In many
dioxygenase reactions, it has been revealed that metal complexes
are involved in the reaction center (1). For examples, non-heme
iron complexes are involved in the enzymatic oxygenolysis of cate-
chols (2), and heme complex participates in the catalytic center
of the enzymatic dioxygenation of tryptophan followed by oxidative
cleavage of the heterocyclic ring (3). Similar oxygenolysis of the

241

heterocyclic ring has been observed in the enzymatic dioxygenation
of flavonols, where cupric ion has been shown to be essential for
the catalysis (4). Heme complex is also known to participate in

the reaction center of enzymatic monooxygenation, one of which is
aromatic hydroxylation. Arene oxides are now widely believed as
the key intermediate leading to the characteristic NIH-shift (5)
seen in the enzyme reaction.

It is interesting that heme complex behaves not only as the
"oxygen carrier," but also as catalyst for the dioxygenation and
monooxygenation depending on the nature of the environment where
the heme complex is situated. On the other hand, cobalt(II) complexes coordinated with N-bases have been demonstrated to interact
reversibly with dioxygen displaying a model of the heme oxygen
carrier (6).

Typical cobalt(II) complexes which interact reversibly with
dioxygen are Co(salen) and Co(salpr), cobalt(II)-Schiff's base
complexes. Co(salen) forms a binuclear cobalt-dioxygen complex in

Co(salen)

N,N-dimethylformamide (DMF) or dimethyl sulfoxide (DMSO), whereas
Co(salpr), a five coordination complex, gives a mononuclear cobalt-
dioxygen complex. Activation process for the formation of these
dioxygen complexes is now well understood as an electron transfer
from the Co(II) ion to the dioxygen (7,8). Thus, the complexes
[Co(salen)(DMF)]$_2$O$_2$ and [Co(salpr)]O$_2$ can be regarded as a Co(III)-
peroxide and a Co(III)-superoxide, respectively. Co(salen) can

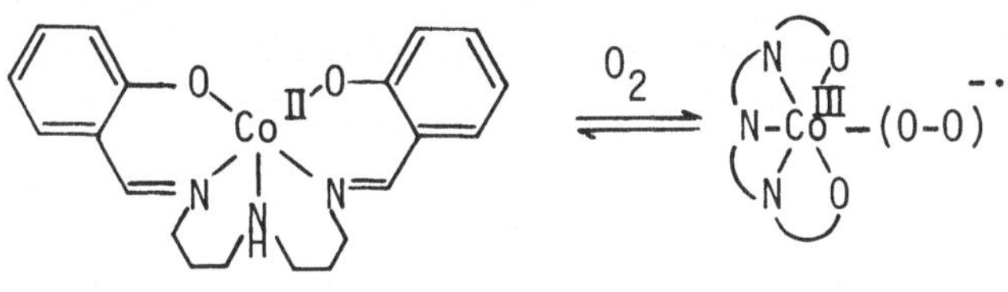

Co(salpr)

also form a mononuclear superoxide complex in pyridine or in the presence of imidazole derivatives. Co(salen) was shown to catalyze the oxidation of phenols and sulfides (9,10). These findings prompted us to investigate whether or not the cobalt-Schiff's base complexes can display catalytic activity for the oxygenation of organic substrates related to biological systems with high selectivity similar to that observed in the biological oxygenations.

We have found that these cobalt complexes catalyze the dioxygen incorporation into some phenols, flavonols, and indoles related to tryptophan with high selectivity. Dioxygen bubbling through the solutions of 4-alkyl-2,6-di-tert-butylphenols (1) in methanol in the presence of catalytic amount of Co(salpr) at room temperature leads to the quantitative formation of p-quinols (2), which are spontaneously converted in a strongly basic medium to the corresponding hydroquinones (3) with migration of the substituent R to the adjacent carbon on the ring. If the substituent is t-Bu or i-Pr, keto forms 4 of the hydroquinones are obtained quantitatively, indicating that the migration of R is initiated by the ketonization of quinolate anion (11,12). Therefore, the oxygenation of 1 in over all reaction may be considered to be a high selective aromatic hydroxylation involving the NIH-shift which does not pass through the arene oxide intermediate.

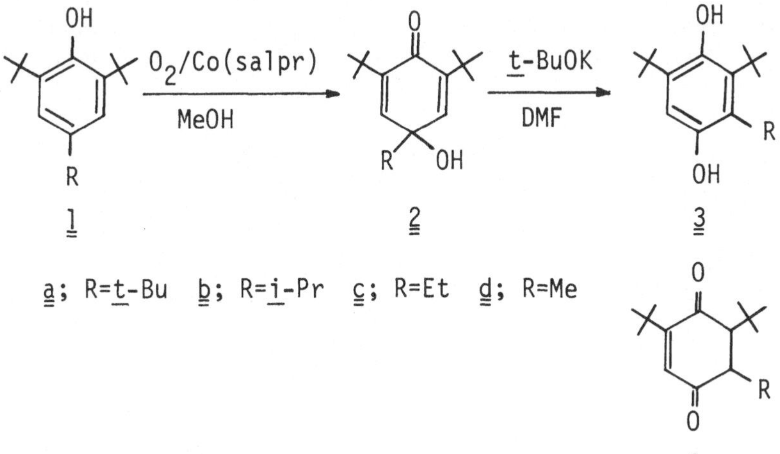

a; R=t-Bu   b; R=i-Pr   c; R=Et   d; R=Me

Flavonols ($\underline{5}$) have been found to be oxygenated by catalysis of Co(salen) in DMF, where the heterocyclic ring undergoes oxidative cleavage to give depsides ($\underline{6}$), carbon monoxide, and carbon dioxide. The oxygenolysis can take place at room temperature by dioxygen bubbling but only in DMF or in DMSO. The highly selective formation of depsides as seen in the enzyme reaction has been observed. Carbon dioxide has been found to result from the catalytic oxygenation of carbon monoxide under the reaction conditions. The oxygenolysis is accelerated when the substituents R_1 and R_2 of $\underline{5}$ are replaced with OH or OMe. The substituent effect on the reactivity has

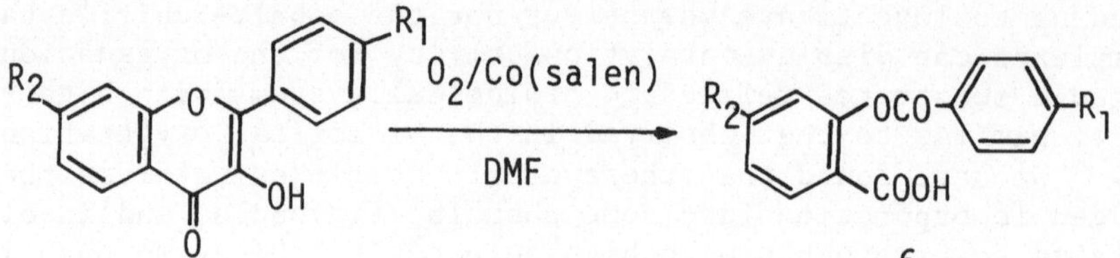

$$R_1, R_2 = H, OH, OMe$$

$$CO \xrightarrow[\text{DMF}]{O_2/Co(salen)} CO_2$$

been found to be in the same order as that observed in the enzyme reaction. Thus, the catalytic oxygenolysis of $\underline{5}$ provides a model reaction for the enzyme (Quercetinase) reaction (13).

3-Substituted indoles ($\underline{7}$) are also found to undergo oxygenolysis of the heterocyclic ring by catalysis of Co(salen). Formylaminoacetophenone derivatives ($\underline{8}$) are selectively formed as seen in the enzyme (Tryptophan-2,3-dioxygenase) reaction. The reactivity

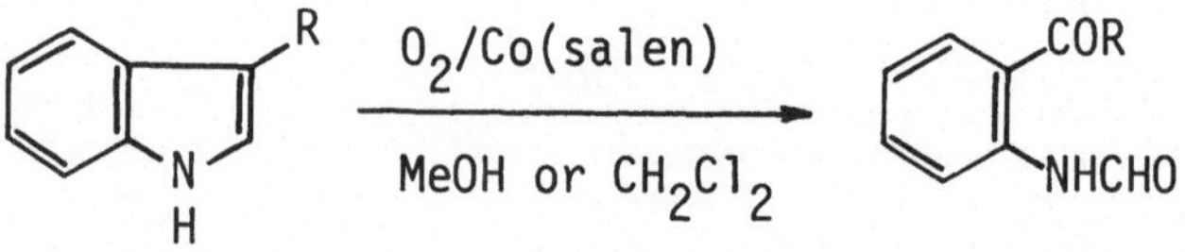

$$R = CH_3, CH_2CH_2CO_2Me, CH_2CH_2NHCOCH_3, CH_2\underset{NHCOCH_3}{CH-CO_2Me}$$

of the indoles is dependent on the size of the substituent R in $\underline{7}$.

The bigger size leads to the less reactivity. The substituent effect has been found to be implicated in the donor ability of $\underline{7}$ for the formation of charge transfer complex with 1,3,5-trinitrobenzene (14).

Since all these substrates for the cobalt complex-catalyzed oxygenation are not susceptible to dioxygen in the absence of the catalyst, it is clear that the activation of dioxygen is essentially required for these catalyses. Question arises how the activation of dioxygen is implicated in the catalytic oxygenation. Some detailed investigation of the oxygenation of 4-substituted 2,6-di-_tert_-butylphenols has provided arguments for a possible mechanism.

The catalytic oxygenation of $\underline{\underline{1a}}$ takes place only in MeOH at room temperature. If the oxygenation of $\underline{1a}$ with Co(salpr) was carried out in CH_2Cl_2 at room temperature or in MeOH at 0°C, the reaction proceeds stoichiometrically to give a peroxy complex ($\underline{9}$), isolated as beutiful crystals, in quantitative yield. A similar stoi-

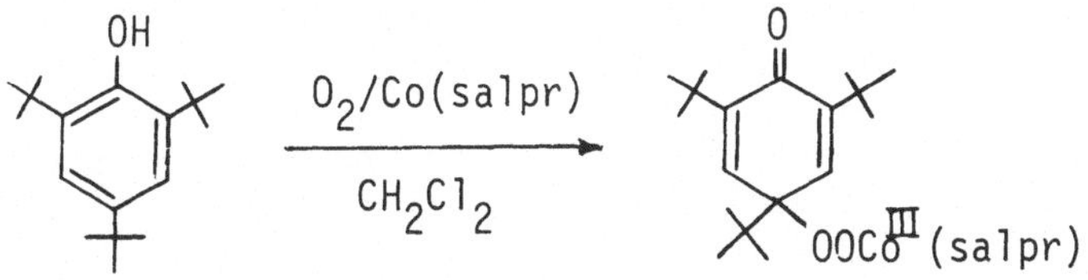

chiometric peroxy complex formation has been observed in the oxygenation of 4-aryl-2,6-di-_tert_-butylphenols ($\underline{\underline{10}}$) with Co(salpr) in CH_2Cl_2. Interestingly, however, in this case dioxygen attacks only the ortho position of $\underline{\underline{10}}$ to give $\underline{o}$-peroxy complexes ($\underline{\underline{11}}$) quantitatively. No $\underline{p}$-peroxy complex of type $\underline{9}$ was detectable in the reaction mixture (15). Similar $\underline{o}$-peroxy complex has been postulated for the dioxygenase reaction of phenolic substrates, so that the peroxy complex $\underline{\underline{11}}$ provides an interesting model peroxy intermediate for the dioxygenase reaction (16).

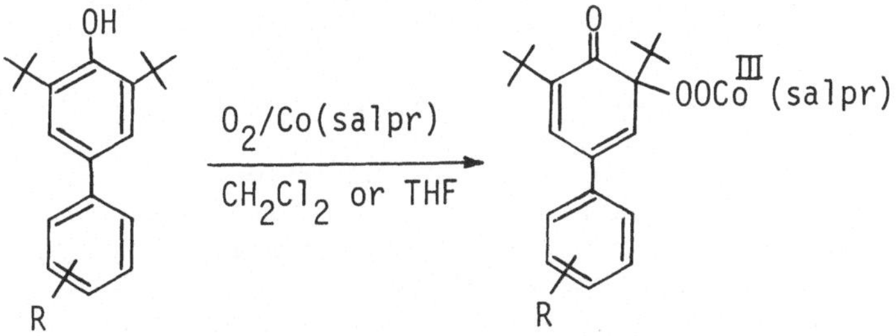

R = 4-OMe, 3-OMe, 2-OMe, 4-Me, 3-Me, 2-Me, 4-Cl,H

R = H or OH

At the initial stage of the catalytic oxygenation of 1a in MeOH at room temperature, only p-quinol 2a and the peroxy complex 9 are detectable in the reaction mixture, suggesting that the per-oxy complex is a primary intermediate in the catalysis. The peroxy complex 9 itself does not give directly the p-quinol 2a but in MeOH in the presence of Co(salpr) 9 gives 2a. The mechanism by which 2a is formed from 9 has been found to involve liberation of hydroper-oxide 12 through an equilibrium between 9 and MeOH with equilibrium constant of 6 X 10^{-8} followed by the reduction of 12 to 2a probably by the Harber-Weis mechanism. Thus, the catalysis is initiated by the secondary formation of 12, and the resulting Co(III) species oxidize the starting phenol to the corresponding phenoxy radical with reproduction of Co(salpr).

12

In order to know the role of dioxygen activation, the stoichi-ometric formation of peroxy complexes 9 and 12 has been kinetically investigated. During the oxygenation, 1.25 mol/mol of oxygen was taken up against the starting phenols and Co(salpr) incorporated into the peroxy complexes. No hydrogen peroxide was detected in the reaction mixtures, which showed neutral pH. The stoichiometry of this reaction is therefore depicted as follows:

$$1a \ (10) \ + \ Co(salpr) \ + \ 5/4 \ O_2 \longrightarrow 9 \ (12) \ + \ 1/2 \ H_2O$$

ESR studies showed the intermediary formation of the corresponding phenoxy radicals in this reaction. A solution of Co(salpr) in CH$_2$Cl$_2$ bubbled with oxygen at room temperature displayed the typical ESR signal for CoO$_2$ (1:1) complex (eight lines, a_{Co} = 14 G)(17). When a small amount of 1a or 10 was added into the mixture under inter-ception of oxygen, the signal was diminished with simultaneous ap-pearance of a new signal assigned as the corresponding phenoxy radical (18). Upon oxygen bubbling through the resulting mixture, the signal of the phenoxy radical disappeared with reappearance of the signal of the CoO$_2$ complex. The vicissitudes of the ESR signals

could be repeated several times, and the peroxy complexes were
obtained from the final solutions.

These results led us to argue possible mechanisms for the
stoichiometric formation of the peroxy complexes $\underline{9}$ and $\underline{\underline{12}}$ as follows:

(i) $Co(salpr) + O_2 \longrightarrow (salpr)Co^{III}(O_2)^{-\cdot}$

(ii) $ArOH + (salpr)Co^{III}(O_2)^{-\cdot} \longrightarrow ArO\cdot + (salpr)Co^{III}(OOH)^{-}$

(iii) $(salpr)Co^{III}(OOH)^{-} \longrightarrow Co(salpr) + 1/2\ H_2O + 3/4\ O_2$

(iv) $ArO\cdot + O_2 \longrightarrow O=ArOO\cdot$

(v) $O=ArOO\cdot + Co(salpr) \longrightarrow O=ArOOCo^{III}(salpr)$

(vi) $ArO\cdot + (salpr)Co^{III}(O_2)^{-\cdot} \longrightarrow O=ArOOCo^{III}(salpr)$

(vii) $ArO\cdot + Co(salpr) \rightleftharpoons ArO^{-}Co^{III}(salpr)$

(viii) $ArO^{-}Co^{III}(salpr) + O_2 \longrightarrow O=ArOOCo^{III}(salpr)$

$ArOH = \underline{\underline{1a}},\ \underline{\underline{10}}$ $O=ArOOCo^{III}(salpr) = \underline{9},\ \underline{\underline{12}}$

The CoO_2 complex formed in the first step abstracts hydrogen from
the phenol to give the corresponding phenoxy radical and hydroperoxy cobalt(III) complex. According to the stoichiometry of this reaction, the hydroperoxy complex should be decomposed following the
step (iii). For the final step leading the phenoxy radical to the
peroxy complex $\underline{9}$ or $\underline{\underline{12}}$, path ways (iv) - (vi) involving complexation of peroxy radical resulting from coupling between the phenoxy
radical and dioxygen or direct coupling between the phenoxy radical
and the CoO_2 complex seemed to be atractive. However, the reaction
of phenoxy radicals $\underline{\underline{13}}$ with a solution of Co(salpr) saturated with
oxygen was found to give predominantly peroxides $\underline{\underline{14}}$ together with
a minor amount of the peroxy complexes $\underline{9}$ and $\underline{\underline{12}}$. The result ruled out

$\underline{\underline{13}}$ $\underline{\underline{14}}$

$\underline{\underline{a}}$; R = $\underline{\underline{t}}$-Bu $\underline{\underline{b}}$; R = 4-MeOPh

the occurence of processes (iv) - (vi). Therefore, we considered an
alternative mechanism in which the formation of the peroxy complex
is preceded by the equilibrium involving electron transfer from
Co(salpr) to the phenoxy radical (equations vii and viii). This
mechanism is strongly supported by the following experimental facts.

The phenoxy radicals 13a and 13b were treated with a solution
of Co(salpr) in CH$_2$Cl$_2$ at 20 °C under nitrogen. The ESR signal of
the phenoxy radicals has been found to decay with time. Degree of
the decay increases with increasing amount of Co(salpr) added,
suggesting that the phenoxy radicals may be reduced with Co(salpr)
in an equilibrium. Aliquotes taken up at intervals were treated
with dilute sulfuric acid, and the mixture was bubbled with oxygen to convert the unreacted phenoxy radical into peroxide 14,
which was then determined by NMR spectroscopy together with the
phenol formed. As shown in Figure 1, similar results were obtained
with both phenoxy radicals 13a and 13b. The formation of the phenol
increased with increasing amount of Co(salpr) added, indicating the
occurence of the equilibrium of equation (vii).

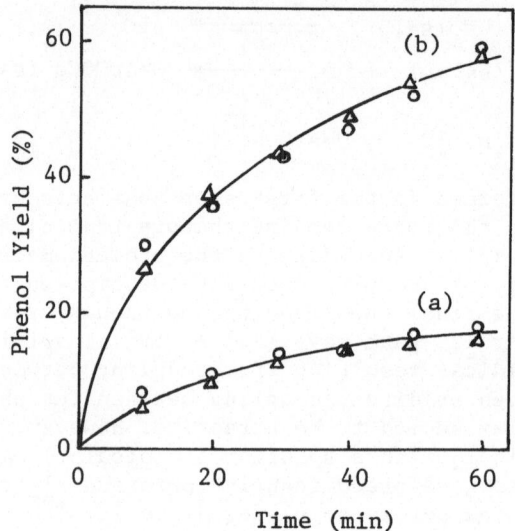

Figure 1. The reaction of phenoxy radicals
13a and 13b with Co(salpr) under nitrogen
to give the corresponding phenols 1a (O)
and 10(R=4-OMe) (Δ). (a) Phenoxy radical
(1 mmol) + Co(salpr) (1 mmol) and (b) phenoxy
radical (1 mmol) + Co(salpr) (5 mmol), in
CH$_2$Cl$_2$ (40 ml) at 20°C.

Effect of the added Co(salpr) on the rate for the oxygenation
of 1a and 10 was examined in comparison with the rate for the oxygenation of the corresponding phenolate anions (the rate for
t-BuOK-catalyzed oxygenation of the phenols). Oxygen was bubbled
after a mixture of the phenoxy radical and Co(salpr) had been
equilibrated for 30 min under nitrogen. As seen from Figure 2, the
oxygenation of the phenoxy radical was accelerated by increasing
amount of the added Co(salpr) in the case of 13b whereas retarded

in the case of 13a. In any case, however, the oxygenation rate approached to that for the oxygenation of the corresponding phenolate anion with increasing amount of the added Co(salpr). The product

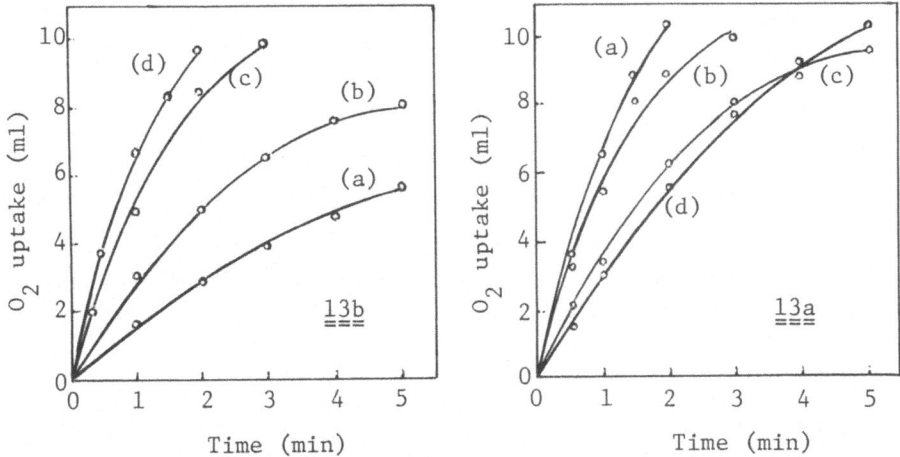

Figure 2. The effect of Co(salpr) on the rate for the oxygenation of the phenoxy radicals 13a and 13b, compared with the rate for the base-catalyzed oxygenation of the parent phenols. (a) Phenoxy radical (1 mmol) alone, (b) phenoxy radical (1 mmol) + Co(salpr) (1 mmol), and (c) phenoxy radical (1 mmol) + Co(salpr) (5 mmol), in CH_2Cl_2 (40 ml) at 20°C. (d) Phenol (1 mmol) + t-BuOK (1 mmol) in DMF (40 ml) for the phenol 1a and in t-BuOH (40 ml) for the phenol 10 (R = 4-OMe).

analysis showed that the yield of the peroxy complex increases with increasing amount of the added Co(salpr) at the expense of the production of the peroxide 14 resulted from coupling between the phenoxy radical and dioxygen. These results indicate that the final step involves the phenolate anion which is reasonably considered to exist as a coordination complex with Co(III)(salpr). Dioxygen can be inserted into   the phenolate complex leading to the quantitative formation of the peroxy complex. This is an analogous reaction to the dioxygen insertion into organocobalt complexes (19).

    The highly regioselective peroxy complex formation depending on the nature of substituent at para position of the phenols is also rationalyzed only by assuming the oxygenation of the phenolate anion, but not of the phenoxy radical 14. Evidently, in the t-BuOK-catalyzed oxygenation of 1a, dioxygen attacks only the para position of 1a to give p-hydroperoxide 12, whereas it attacks exclusively the ortho position of 10 to give the corresponding o-hydroperoxide under the conditions depicted in Figure 2 (20,21).

From all these findings it is concluded that in the present Co(salpr)-catalyzed oxygenation the activation of dioxygen by Co(salpr) through the formation of CoO_2 complex is essential but the activated dioxygen is not incorporated directly to the phenols. The significance of the activation of dioxygen is to activate the substrate. In the present case, hydrogen abstraction is the activation of substrate. The observed high regioselectivity in the oxygenation is the result of coordination of the activated substrate to Co(salpr), which can be stabilized through electron transfer equilibrium, providing a specific position for the dioxygen insertion.

Similar argument may be applied for the significance of the dioxygen activation in the model catalytic oxygenolysis of flavonols 5 and 3-substituted indoles 7.

In the Co(salen)-catalyzed oxygenolysis of flavonols 5 giving depsides 6, DMF as well as DMSO is essential as solvent. No reaction took place in tetrahydrofuran, methylene chloride, or acetic acid. The catalysis is retarded remarkably in pyridine. This tendency corresponds to the dependence of solvent on the formation of dioxygen complex with Co(salen). That is, $(DMF)_2[Co(salen)]_2O_2$ in DMF and $Co(salen)(Py)O_2$ in pyridine are formed and no dioxygen complex is formed in tetrahydrofuran, methylene chloride, or acetic acid (22, 23). Since Co(salen) does not display affinity towards dioxygen without axial donor ligand, the fact that no oxygenation of 5 takes place in methylene chloride indicates that flavonols 5 are not able for coordinating to Co(salen). The oxygenolysis occurs only under conditions where the CoO_2(2:1) complex can be formed. The remarkable lowering of the reactivity in pyridine showed that the CoO_2(2:1) complex is only slightly reactive. This is supported by the fact that Co(salpr) is also less reactive towards the oxygenolysis. These results suggest that CoO_2(1:1) species which is cosidered to be formed transiently through an equilibrium in the solution of the CoO_2(2:1) complex during the oxygenation of Co(salen) in DMF may not be the reactive species in the catalytic oxygenolysis. It is therefore reasonably assumed that the complex $(DMF)_2[Co(salen)]_2O_2$ is the reactive species in the oxygenolysis of 5 in DMF. The activated dioxygen species in this CoO_2(2:1) complex can neither be incorporated directly into the substrate nor have an ability to oxidize the substrate because flavonols 5 are not susceptible in alkaline medium to hydrogen peroxide whose electronic state is just similar to that of the dioxygen species in the CoO_2(2:1) complex. The following mechanism can therefore be reasonably proposed.

The coordinated dioxygen species in the CoO_2(2:1) complex is a base (O_2^{2-}), so that the oxygenolysis is considered to be initiated by an acid-base reaction involving proton transfer from the enolic hydroxy group in 5 to the coordinated dioxygen species of the CoO_2(2:1) complex. This may result in the formation of Co(III)-flavonolate complex intermediate 15. This is an analogous intermediate to that discussed in the Co(salpr)-catalyzed oxygenation of phenols 1a and 10. Ground-state dioxygen can be inserted into the intermediate to give depside 6 probably via a peroxy complex 16.

In fact, oxygen bubbling at room temperature through a solution of 5 in DMF in the presence of t-BuOK (oxygenation of the flavonolate) has been found to give depside 6 and carbon monoxide in excellent yield (24).

The oxygenolysis of 5 is also catalyzed by $CuCl_2$. The active catalyzing species in this case is most probably a Cu(II)-chelate complex with 5, which is readily formed under the reaction conditions, but the catalysis is less effective than that by Co(salen). Catalytic oxygenation of 5 in the presence of other one-electron oxidizing agents gave a complex reaction mixture. For the highly selective oxygenolysis like the enzyme reaction, the intermediary formation of flavonolate complex seems to be of significance.

The above arguments indicate that the significance of dioxygen activation in the Co(salen)-catalyzed oxygenolysis of 5 is again to activate the substrate, but the activation process is different from that in the Co(salpr)-catalyzed oxygenation of phenols 1a and 10.

The Co(salen)-catalyzed oxygenolysis of 3-substituted indoles 7 giving formylaminoacetophenone derivatives 8 also undergoes solvent effect. Contrary to the case of flavonols 5, the fastest reaction rate was observed in methylene chloride, under which conditions the solvent did not react with Co(salen), and the catalysis

was remarkably retarded in DMF. The catalysis also proceeds in
methanol with a slightly slower rate than in methylene chloride.
No reaction takes place in pyridine or acetic acid. Co(salpr) is
little reactive, showing again that the CoO_2(1:1) complex is not
the reactive species in the catalytic oxygenolysis of indoles 7.
The observed solvent effect and the fact that the reactivity of 7
is proportional to its donor ability strongly suggest that the
coordination of 7 to Co(salen) is an important process for the
oxygenolysis. In the absence of dioxygen, no significant change
in the electronic spectrum of Co(salen) in methylene chloride has
been observed in the presence or absence of the substrate at the
same concentration as for the catalytic oxygenolysis. Therefore a
concerted coordination of the substrate and dioxygen to Co(salen)
is reasonably assumed for the catalytic oxygenation of 7 in methylene chloride. The following inner sphere electron transfer process
is reasonably proposed.

In the first step, the transient coordination complex may be considered in analogous with the CoO_2 complex with DMF and bulky
substituents in 7 nulify the electronic advantage for coordination
with Co(salen) owing to the steric hindrance. The subsequent step
involves electron transfer from the substrate to Co(III) center
followed by stabilization of the resulting indolyl radical species
through coordination involving electron transfer. Dioxygen is then
inserted into the resulting complex with the activated substrate.
The highly selective dioxygen addition to 7 in the catalytic oxygenation is rationalized by assuming the stabilization of the activated substrate. In fact, catalytic oxygenation of 7 in the

presence of other one-electron oxidizing agents gave a complex reaction mixture. The retard of the reaction in DMF is rationalized
by assuming that the DMF ligand in $(DMF)_2[Co(salen)]_2O_2$ complex
primarily formed in the system is not easily replaced by the substrate. No reactivity in acetic acid may be explained in such a
way that acetic acid interfere with coordination of the substrate
and dioxygen to Co(salen) owing to solvation through hydrogen bonding of the substrate and of the Schiff's base in the cobalt complex.

Cu(I)-Py complex has also been found to catalyze the oxygen-
olysis of $\underline{7}$ to $\underline{8}$ in good yield, whereas $Co(Ac)_2$, $CoCl_2$, and
Co(acac) were not reactive.

In summary, the oxygenation of organic substrates relating to
biological substances using cobalt(II)-Schiff's base complexes
leads to the highly selective dioxygen insertion into the substrates
as seen in the reaction of dioxygenases. In these Co(II) complex-
catalyzed oxygenation, activation of dioxygen is essential.
Significance of the activation of dioxygen in the oxygenation is
to activate the substrates. Direct insertion of the activated dioxygen into the substrates is unlikely.

Process of the activation of substrates through the dixoygen
activation depends on nature of the substrate and can be summarized
as follows:

1) Substrates are activated through hydrogen abstraction. The
hydrogen abstraction is caused by an outer sphere mechanism, where
the coordinated superoxide species may play the role. This mechanism can operate only for limited substrates having low redox potential.

2) Substrates are activated through coordination. This mechanism
is applied to substrates which do not undergo the hydrogen abstraction by the outer sphere mechanism but coordinate easily to the
square-planer Co(II) complex. The inner sphere electron (or hydrogen) transfer from substrates to dioxygen through the centered
Co(II) ion occurs in such a way that the activated substrates are
stabilized. The selectivity of the oxygenation depends on the stability of interaction between the activated substrates and Co(II)
complex. The oxygenolysis of indoles is classified into this
category.

3) Substrates are activated through proton transfer. This mechanism is applied to oxygenation of substrates which have neither
susceptibility to the hydrogen abstraction by the outer sphere
mechanism nor the ability to coordinate with the square-planer
Co(II) complex but the anionic form of the substrates is subject
to oxygenation. Flavonols are conformable to this mechanism.
The CoO_2 (2:1) complex behaves as a base.

These conclusions must be confirmed by further precise kinetics of the reactions, which are currently investigated.

REFERENCES

1 M. B. Dearden, C. R. E. Jefcoate, J. R. Lindsay-Smith, Adv. Chem. Ser., 77, 260 (1968). G. A. Hamilton, Adv. Enzymol., 32, 55 (1969). V. Ullich, Angew. Chem., 84, 689 (1972).
2 M. Nozaki,"Molecular Mechanism of Oxygen Activation," ed. O. Hayaishi, New York, London, Academic Press, p 135 (1974).
3 Y. Ishimura, M. Nozaki, O. Hayaishi, T. Nakamura, and I. Yamazaki, J. Biol. Chem., 245, 3539 (1970). P. Feigelson and F. O. Brady, "Molecular Mechanism of Oxygen Activation," ed. O. Hayaishi, New York, London, Academic Press, p 87 (1974).
4 T. Oka, F. J. Simpson, and H. G. Krishnamurty, Can. J. Microbiol. 18, 493 (1972).
5 J. Daly, G. Guroff, D. Jerina, S. Udenfriend, and B. Witkop, Adv. Chem. Ser., 77, 279 (1973). Ref. 16
6 L. H. Vogt, Jr., H. M. Faigenbaum, and S. E. Wiberley, Chem. Rev., 63, 269 (1963).
7 R. G. Wilkins, Adv. Chem. Ser., 100, 111 (1971).
8 J. S. Valentine, Chem. Rev., 73, 235 (1973).
9 H. M. van Dort, H. J. Geursen, Recuil., 86, 520 (1967).
10 L. H. Vogt, Jr., J. G. Wirth, and H. L. Finkbeiner, J. Org. Chem., 34, 273 (1969).
11 A. Nishinaga, K. Watanabe, and T. Matsuura, Tetrahedron Lett., 1291 (1974).
12 A. Nishinaga, T. Itahara, T. Matsuura, S. Berger, G. Henes, and A. Rieker, Chem. Ber., 109, 1530 (1976).
13 A. Nishinaga, T. Tojo, and T. Matsuura, Chem. Commun., 896 (1974).
14 A. Nishinaga, Chem. Lett., 273 (1975).
15 A. Nishinaga, K. Nishizawa, H. Tomita, and T. Matsuura, J. Am. Chem. Soc., 99, 1287 (1977).
16 G. A. Hamilton,"Molecular Mechanism of Oxygen Activation," ed. O. Hayaishi, New York, London, Academic Press, p 405,1974.
17 H. Kon and N. L. Sharpless, Spectrosc. Lett., 1, 49 (1968).
18 E. Müller, K. Ley, K. Scheffler, and R. Meyer, Chem. Ber., 91, 2682 (1958).
19 C. Giannotte, C. Fontain, and B. Septe, J. Organometal., 71, 107 (1974). F. R. Jensen and R. C. Kiskis, J. Am. Chem. Soc., 97, 5825 (1975).
20 A. Nishinaga, T. Itahara, and T. Matsuura, Chem. Lett., 667 (1974).
21 A. Nishinaga and A. Rieker, J. Am. Chem. Soc., 98, 4667 (1976).
22 C. Floriani and F. Calderazzo, J. Chem. Soc.(A), 946 (1969).
23 S. Koda, A. Misono, and Y. Uchida, Bull. Chem. Soc. Japan, 43, 3143 (1970).
24 A. Nishinaga and T. Matsuura, Chem. Commun., 9 (1973).

ON THE MECHANISM OF VANADIUM CATALYZED OXIDATION BY t-BUTYL HYDRO-

PEROXIDE

Ruggero Curci[†], Fulvio Di Furia and Giorgio Modena[*]

Centro Meccanismi di Reazioni Organiche del C.N.R.
Istituto di Chimica Organica, Università, Via Marzolo 1,
Padova, Italy and (†) Istituto di Chimica Organica,
Università, Via Amendola 173, Bari, Italy

The wide spread interest in the metal catalyzed epoxidation
of olefins gave rise to a very abundant patent literature[1] but the
reaction mechanism was largely uncovered when we started, some years
ago, this investigation.[2,3]
    Indicator and Brill in their pioneering work in this area[4] had
shown that metal acetylacetonates like those of vanadium, molybdenum,
tungsten and titanium in their higher oxidation states, and possibly
of few others metals, are able to catalyze a clean, polar like,
oxygen transfer from a hydroperoxide (usually t-butyl hydroperoxide)
to an olefin. Other metals, characterized by having close lying
oxidation states, promote normally one-electron transfer processes
either oxidative or reductive and consequently cause homolysis of
the hydroperoxide.[1] Radical oxidation _via_ a chain process often
follows. Such reactions are almost always unselective and little
or none epoxide is formed.

    In this early study, carried out employing the olefin itself
as solvent, the kinetics of the process was not fully explored, however, the authors could show that the epoxidation of 1-octene by t-
butyl-hydroperoxide, as catalyzed either by vanadyl or molybdenyl
acetylacetonates follows a relatively simple rate law:

$$-d(Bu^t O_2 H)/dt = k(Bu^t O_2 H)(catalyst)^{1/2}$$

    A much more detailed mechanistic work was published a few
years later by Gould _et al_.[5] They centered their attention on the
epoxidation of cyclohexene as catalyzed by vanadyl acetylacetonate.

A few other transition metal complexes were also studied subsequently.[6]

The reactions were strongly inhibited by Bu^tOH, a reaction product, and also rapid catalyst deactivation was experienced. Kinetic data were therefore evalued by calculating the initial rates of disappearance of the peroxide, at very low conversions, extrapolated to zero alcohol concentration.

The dependency, both on hydroperoxide and on catalyst, was found to be unitary when the two species were in relatively low concentrations. At higher hydroperoxide concentrations the order in the peroxide itself dropped toward zero as though a saturation process of the catalyst was operating. In fact the authors showed that the rates of the reactions fitted the Michaelis-Menten treatment of enzymes kinetics.[7] Evidence against free radical pathways and favoring heterolytic processes were also offered, so that the transfer of the peroxidic oxygen to the nucleophilic double bond from a vanadium-hydroperoxide complex was proposed as the rate determining step of the reaction.

The Gould's conclusions are rather convincing; however the system studied is quite complex and not suitable for a thorough mechanistic study where the oxidation and coordination state of the metal in the reaction conditions and its catalytic role are investigated.

Our approach was to obtain a drastic simplification of the system by using alcohols as solvents;[2,3] this avoids in fact the great change in the properties of the medium with the progress of the reaction caused by formation of Bu^tOH. The oxidizing ability of the system is, as expected, greatly reduced on passing from hydrocarbons to alcoholic solvents which compete with the peroxide for the coordination to the metal ion.[1c] This difficulty was overcome by employing a much more reactive substrate such as di-n-butyl sulfide. On the other hand we have shown that the electrophilic oxidation of sulfides and the epoxidation of alkenes have identical mechanistic features[8]

We will report hereinafter the results obtained in the study of sulfides and alkenes oxidation by Bu^tO_2H as catalyzed by vanadium derivatives. Bu_2^nS was oxidized in ethanol, according to the stoichiometry to give the parent sulfoxide in a quantitative yield.[2]

$$Bu_2^nS + Bu^tO_2H \xrightarrow{VO(acac)_2} Bu_2^nSO + Bu^tOH \qquad (2)$$

At relatively low hydroperoxide concentrations and with sulfide in large excess, the disappearance of the hydroperoxide, follows a very regular and reproducible pseudo-first order kinetics up to more than 80% conversion.

Under these conditions we also observed first order dependence of the reaction rates on the sulfide concentration in all the range studied (up to 0.45M) and on the catalyst concentration in the range from $1\text{-}20\times10^{-4}$M.

At higher catalyst concentrations however, a strong decrease in catalytic efficiency was observed.

The order in Bu^tO_2H was unitary at relatively low concentrations ($\sim 5\times10^{-2}$M) though deviations were observed at higher concentrations.

Therefore, the rate law in the range of concentrations investigated (see Table 1) is:

$$-d(Bu^tO_2H)/dt = k_{2_{(obs)}}(Bu_2^nS)(Bu^tO_2H) \tag{3}$$

where $k_2(obs) = k_3(catalyst)$ (4)

Addition of nitrobenzene, an efficient radical scavenger, has no effect on the rates which are, on the other hand, affected by the nucleophilicity of the substrates as it is expected for an electrophilic oxidation.[8] Pertinent data are collected in Table 2.

Table 1. Effect of reactants and catalyst concentration on rates of the vanadium catalyzed oxidation of di-n-butyl sulfide by t-butyl hydroperoxide in ethanol at 25.0° under N_2.

$(Bu_2^nS)_o$/M	$10^3(Bu^tO_2H)_o$/M	$10^3(Cat)_o$/M	$k_3^a/1\ mol^{-2}s^{-1}$
0.20	11.0	1.0	1.80
0.30	11.0	1.0	1.63
0.45	11.0	1.0	1.67
0.50	20.2	1.2	2.53
0.50	40.1	1.2	2.68
0.50	60.8	1.2	2.43
0.20	10.5	0.97	2.13
0.20	10.5	1.40	2.00

a From pseudo-first order kinetic runs as k (obs) = k /(Bu S) and k = k (obs)/(Cat)

Table 2. **Relative** rates of oxidation of some organic substrates by
t-butyl hydroperoxide catalyzed by added $VO(acac)_2$ in dry ethanol
at 25.0° under N_2.

Substrate	Relative rates
Bu_2^nS	100
$PhSBu^n$	58
Bu_2^nSO	1.7

This preliminary investigation convinced us on the suitability
of the solvent system adopted and consequently a detailed study was
initiated.[9]

First of all it was deemed worthwhile to confirm the Gould
suggestion[5] that the active species of vanadium is in the (V) oxidation state.

Solutions of $VO(acac)_2$ in ethanol have electronic spectra characterized by maxima in the visible region at 775 nm and at 580
$nm^{2,10-12}$ which may be attributed to d-d transitions. Addition of
excess of $Bu\,O_2H$ causes the rapid disappearance of these two maxima that is consistent with the removal of the d unpaired electron.
The newly formed vanadium species is stable since the spectra do not
show any further appreciable change for relatively large time intervals.

The change in oxidation state has been also observed by esr
spectroscopy: the $VO(acac)_2$ solutions in ethanol present a eight
resonance lines spectrum[11] which disappear upon addition of the
peroxide as expected for the transformation from a paramagnetic
species to a diamagnetic one. For $(VO(acac)_2)_0 = 5 \times 10^{-3}\underline{M}$ and
$(Bu^tO_2H)_0 = 5 \times 10^{-2}\underline{M}$, estimates on the basis of peak height measurements suggest a half-life of $\underline{ca}$ 20 s for the disappearance of V(IV)
resonance lines. The process $V(IV) \rightarrow V(V)$ was also observed by
stopped-flow experiments $(\lambda = 475)$ which show the very fast formation of an intermediate which subsuquently disappears: for
$(VO(acac)_2)_0 = 1.5 \times 10^{-3}\underline{M}$ and $(Bu\,O_2H)_0 = 4.6 \times 10^{-2}\underline{M}$ maximum concentration of the intermediate: $\underline{ca}$ 8 s, total decay of it $\underline{ca}$ 20 s.

We may then conclude that the catalyst in the reaction medium
studied is a V(V) species which is formed in a fast and irreversible
process so that oxidation of V(IV) is complete well before appreciable amount of the hydroperoxide has been consumed in the oxidation
reaction. Further evidence on this point stems out from the combined spectroscopic and kinetic results reported below.

Electronic spectra of $VO(OEt)_3$, and $VO(OPr^i)_3$ in ethanol (as in methanol and 2-propanol) are identical (though different in different solvents) and also equal to the spectra of $VO(acac)_2$ in the presence of excess of Bu^tO_2H provided that the absorbance of the peroxide is taken into account.[9] Since a complete independence of electronic spectra on the nature of the ligands is quite unlikely, the conclusion is that fast ligand exchange of the V(V) species occurs and that in alcoholic solutions acetylacetone is completely removed from coordination shpere of the metal. This is consistent with what has been recently reported in a nmr study on ligand exchange of V(V) alkoxides.[13]

The electronic spectra are strongly acid and base dependent (see Fig. 1) addition of acid gives rise to a new band, centered around 400 nm while the uv region of the spectra is not significantly affected. On the contrary, addition of base causes a depression of the intensity of the band around 250 nm.

The increase in optical density at 400 nm (see Fig. 2) upon addition of acid follows a properly shaped sigmoid curve suggesting the occurence of a protonation process.

This has been confirmed, in ethanol solvent, by carrying out potentiometric experiments which provide the apparent pH values, as measured by a glass electrode, for several acid concentrations both in the absence and in the presence of $VO(OPr^i)_3$ in alcohol solutions.

The experimental technique is analogous to that employed by Rossotti and Rossotti[14] who investigated the acid-base behaviour of solutions of vanadium (V) in water at different pH values. The curve we obtained for ΔE <u>vs</u> log $(MeSO_3H)$ matches very well the analogous spectroscopic one.[9]

The same dependence of electronic spectra upon added acid is shown by solutions of $VO(acac)_2$ when excess of Bu^tO_2H is present and its absorbance is taken into account. This must be considered a further evidence that V(IV) is oxidized to V(V) by the peroxide and that the higher oxidation state is stable under the above stated conditions.

The addition of increasing base concentration to ester solutions (MeO^- in MeOH, EtO^- in EtOH, Pr^iO^- in Pr^iOH) produces, in the uv region of the electronic spectra, a decrease in optical density. See Fig. 1. Also in this case the change of optical density follows a properly sigmoid shaped curve as an indication of an acid-base process. (See Fig. 3)

It is worth of notice that whereas protonation processes occur in very different acid concentration ranges in the three alcohols,

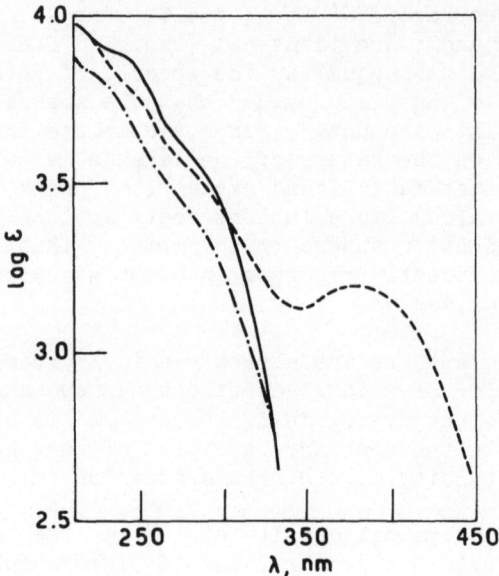

Fig. 1. Electronic spectra of $VO(OPr^i)_3$, 1-10x10^{-4}M, in ethanol, at 25.0° at different medium acidities. (——) "neutral" alcohol; (----) 9.0x10^{-2}M methanesulfonic acid; (-·-·) 5.5x10^{-3}M EtONa.

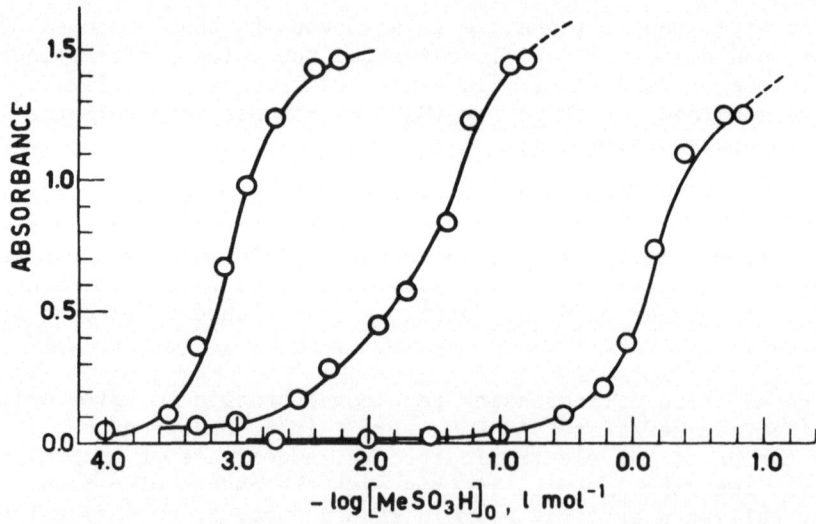

Fig. 2, Effect of increasing concentration of added methanesulfonic acid upon the absorbance of 0.001 M $VO(OPr^i)$, at 25.0° in MeOH, EtOH, and Pr^{i}OH respectively, from left to right.

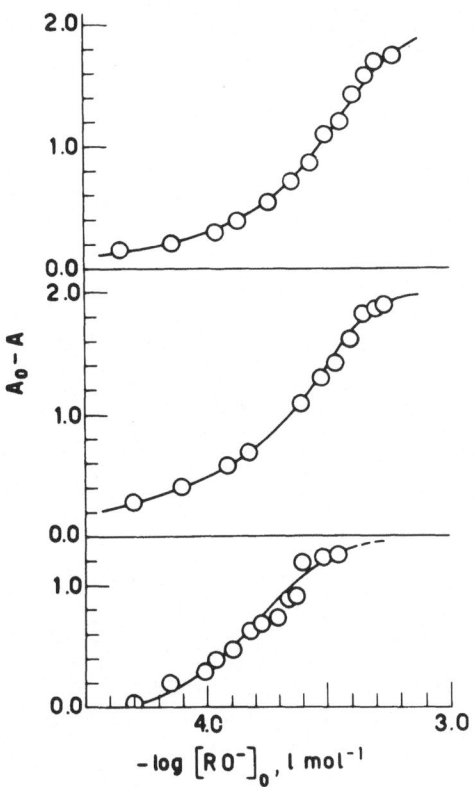

Fig. 3. Variation of difference in absorbance at ∿250 caused by addition of base to solutions of $0.5 \times 10^{-}$ $\underline{M}$ $VO(OPr^i)_3$ at $25.0°$ dissolved in three alcohols, MeOH, EtOH, $Pr^i\underline{OH}$, respectively from the top to the bottom. A_o is the absorbance at zero added base.

anionic species are formed in the same range of base concentration.

Trialkyl esters of V(V) in alcoholic solutions may be therefore considered as weak bases and relatively strong lewis acids.

By consequence, in the "neutral" alcohols, anionic, neutral or cationic species of $VO(OR)_3$ can exist depending on the acid-base properties of the solvent.

An indication of this is also given by the following observation: addition of minute amounts of acid causes an increase in the absorbance in the uv region before any change occurs in the visible range. This increase is well pronounced in ethanol and 2-propanol, and much less in methanol. A reasonable rationale for this phenomenon is that some of the catalyst, in the "initially neutral" solvent, is in its anionic form.

The following processes may therefore explain the spectroscopic features observed:

$$2VO(acac)_2 + Bu^tO_2H + 6ROH \longrightarrow 2VO(OR)_3 + 4acacH + Bu^tOH + H_2O \qquad (5)$$
$$\underset{\underset{\sim}{1}}{}$$

$$VO(OR)_3 + 3R'OH \rightleftharpoons VO(OR')_3 + 3ROH \qquad (6)$$

$$H^+ + VO(OR)_4^- \rightleftharpoons VO(OR)_3 + ROH \rightleftharpoons VO(OR)_3H^+ + RO^- \qquad (7)$$
$$\underset{\underset{\sim}{2}}{} \qquad\qquad\qquad\qquad \underset{\underset{\sim}{3}}{}$$

The spectroscopic behaviour of $VO(acac)_2$ in the presence of Bu^tO_2H, and vanadate esters finds a close correspondence in the kinetic results.[9]

The effect of acid and base on sulfide oxidation rates is summarized in Fig. 4.

Let us first look at the right hand side of Figure 4, i.e. from the maximum of rate constants onwards. In this range of acid concentration the rate constants are independent from the nature of the catalyst added to the reaction mixture what should mean, in agreement with spectroscopic results, that the coordination sphere of vanadium (V) is essentially constituted of the solvent itself which had, in a preliminary fast reactions, substituted the previous ligands. It may be noticed that the rate constants, after having reached a maximum, tend to decrease with increasing acid concentration. Why the rates are depressed by increasing acidity of the medium and why the esters are poorer catalysts than the neutral species is matter of some concern. We will return on this point later on.

One other point to comment on is that the position of the maximum as a function of acid concentration depends on the nature of the alcohol. The maximum occurs at increasing acid concentration following the order methanol, ethanol, 2-propanol. The same sequence, see Fig. 2, is found for the appearance of the band centered at around 400 nm, which we have attributed to the conjugate acid of the ester.

The pattern of the kinetic results in the region between neutral alcohols and the rate maxima seems more complex even though it may be easily rationalized on the basis of the solvolytic equilibria (7). Let us first consider the results in 2-propanol, the most basic of the solvents investigated. The raising in rate con-

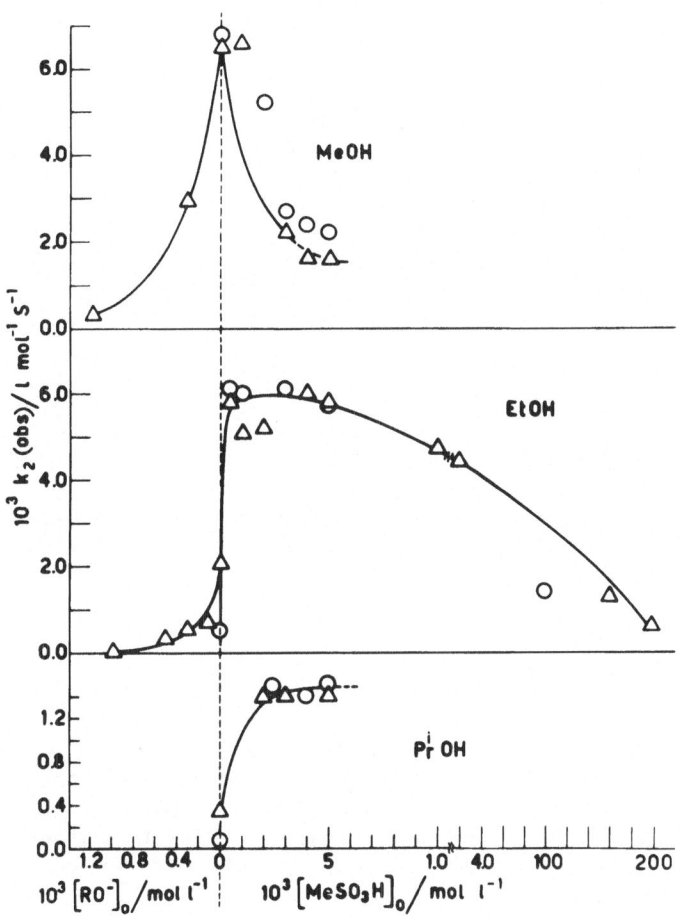

Fig. 4. Effect of added base (left side) and $MeSO_3H$ (right side) upon the rates of oxidation of Bu^n_2S by Bu^tO_2H as catalyzed by added $VO(OEt)_3$ (circles) or by $VO(acac)_2$ (triangles) at $25.0°$. The concentration of vanadium (V) species was $1.0 \times 10^{-3} \underline{M}$.

stants with increasing acid concentration which corresponds to the
increasing absorbance in the uv region we mentioned above has to
be connected with the neutralisation of the anionic form of the
catalyst (see Eq. (7)). The same phenomenon occurs in ethanol
whereas no such increase is observed in methanol, and, indeed, in
this solvent, protonation of the neutral species appears to begin
as soon as acid is added. In this region there are differences in
rate between $VO(acac)_2$ and trialkyl esters, in 2-propanol and ethanol, but not in methanol. As in the former two solvents, the spectroscopic and kinetic results indicate that some of the catalyst
is in its anionic form, these differences are likely related to
the actual acidity of the medium which must also be affected by the
weakly acidic acetylacetone and, possibly, by its oxidation products.

Consistent with the strong depression of reaction rates caused
by the presence of anionic species is the effect of added base
which is shown in the left hand part of the Fig. 4.

Though an accurate analysis has not been carried out nor it
seems worthwhile, it is clear that the decrease in rate cannot be
related to the stoichiometric concentration of anionic vanadium (V)
complexes. In fact the decrease in rate is much greater than expected even if one assumes that anionic V(V) species has none catalytic activity. The rationale seems to be that the anionic form
of the catalyst causes aggregation of the V(V) species that is
fully in agreement with what is known on the properties of V(V)
species.[15-18]

The kinetic orders in the hydroperoxide and in the catalyst,
which were found to be unitary within a narrow range of concentrations, as reported above, were reexamined, in ethanol, in the light
of the observed dependence of the rate constants on the acidity of
the system. No such reinvestigation was carried out for the order
in the substrate which was previously found to be one in a large
concentration interval (see Table 1); moreover, no evidence of
coordination of the sulfide to vanadium catalyst can be observed
spectroscopically in any conditions.

The order in the catalyst, which is unitary at low concentrations, exhibits a fall off with increasing concentration which depends on the acidity of the medium. Figure 5 shows this behaviour
for "neutral" ethanol and for two different acid concentrations
respectively.

It may be observed that in the presence of CH_3SO_3H, but not in
the absence of acid, the fall off may be described by a change of
the kinetic order from 1 to 1/2. An appealing rationale for this
behaviour is that the presence of acid limits the polymerization of
the catalyst to a simple dimerization.

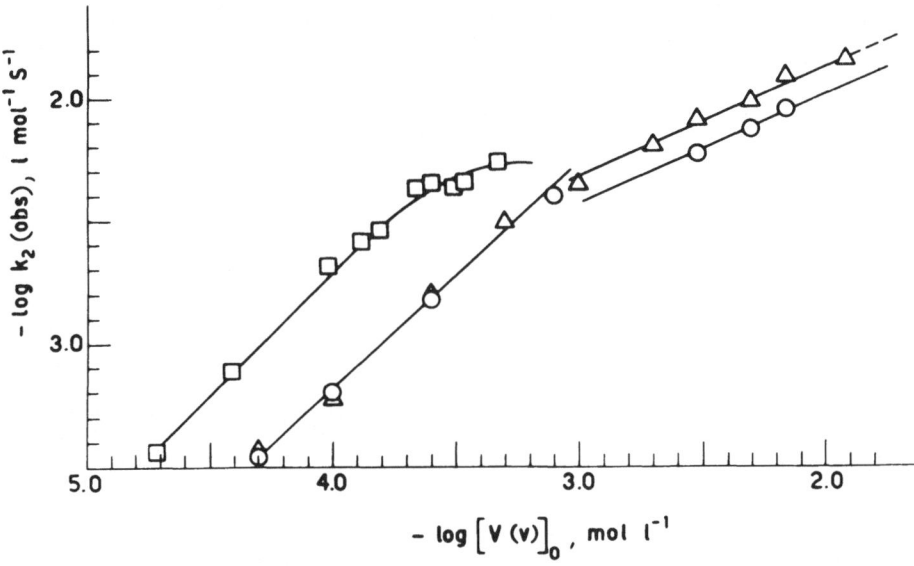

Fig. 5. Effect of catalyst concentration on rates of the vanadium catalyzed oxidation of di-n-butyl sulfide by t-butylhydroperoxide in ethanol, at 25.0°, at different medium acidities.
Squares: "nuetral alcohol"; circles: $(MeSO_3H) = 5.0x10^{-4}M$; tri-angles: $(MeSO_3H) = 20x10^{-3}M$

The effect of t-butyl hydroperoxide concentration on the rate constants was also reexamined. As reported in Table 3 still a difference may be observed between neutral and acidic media.

In particular, in the presence of methanesulfonic acid, deviations from first-order dependence occurs only at very high peroxide concentration and perhaps they may be due to trivial medium effect. On the contrary in "neutral" alcohol the decrease in k_3 values is sizeable and may be connected with saturation of the catalyst. Indeed, this data can be treated by the Lineweaver-Burk equation.[2,19]

In conclusion, the relevant points of this investigation may be summarized as follows:

The most active catalyst in ROH solutions is the neutral species $VO(OR)_3$ which is formed by metathesis or by oxidation and metathesis from trialkyl esters or $VO(acac)_2$ respectively. In fact when anionic species of the catalyst are present and possibly oligomerization occurs rates are considerably depressed. Protonation of the ester also results in a decrease of its catalytic activity.

Table 3.   Effect of t-butyl hydroperoxide concentration on rates of the vanadium catalyzed oxidation of di-n-butyl sulfide at different medium acidity, in ethanol, at 25.0° under N_2.

$10^3(Bu^tO_2H)_o/\underline{M}$	$k_3^a/$ l mole^{-2}s^{-1}	
	"Neutral" alcohol	$(MeSO_3H) = 20x10^{-3}M$
5.00	–	4.3
10.00	–	4.5
20.2	1.27	–
57.0	–	4.7
60.8	1.21	–
80.4	1.14	–
108.0	–	5.4
147.0	–	5.1
160.0	1.02	–
195.0	–	4.8
314.0	0.85	–

$\underline{a}$ See note $\underline{a}$, Table 1.

Reaction rates in the three alcohols, measured in the acid concentration range where the catalyst is the neutral ester, are very similar as it should be for solvents of very similar nature. The alrge differences observed in rates in neutral alcohols derive mainly from different acid-base properties of the three solvents which affect the solvolytic equilibria of the catalyst.

As far as the detailed picture of the rate determining step of the reaction is concerned, one must obviously consider the involvement of the vanadate ester of the alcohol present in the system.

It may either coordinate the Bu^tO_2H giving a pentacoordinate species, as in Scheme A, or give a mixed vanadate ester, as in Scheme B.

SCHEME A

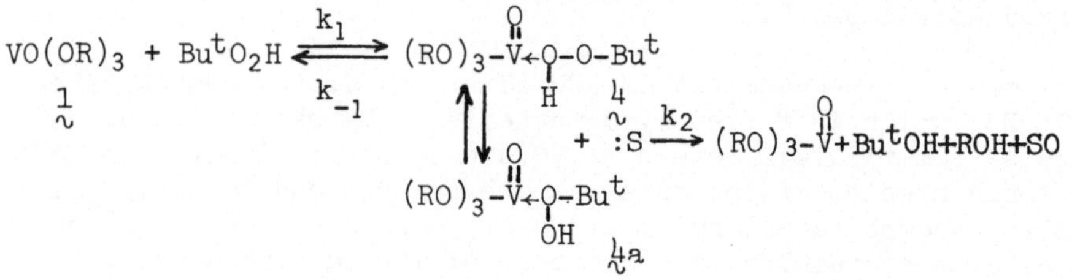

$$VO(OR)_3 + Bu^tO_2H \underset{k_{-1}}{\overset{k_1}{\rightleftharpoons}} (RO)_3\text{-}\overset{\overset{O}{\|}}{V}\text{←}O\text{-}O\text{-}Bu^t$$

$$(RO)_3\text{-}\overset{\overset{O}{\|}}{V}\text{←}O\text{-}Bu^t + \underset{\sim}{:S} \overset{k_2}{\longrightarrow} (RO)_3\text{-}\overset{\overset{O}{\|}}{V} + Bu^tOH + ROH + SO \quad (8)$$

$$(RO)_3\text{-}\overset{\overset{O}{\|}}{V}\text{←}O\text{-}Bu^t$$

SCHEME B

$$VO(OR)_3 + Bu^t O_2H \xrightleftharpoons[k'_{-1}]{k'_1} (RO)_2\text{-}\overset{O}{\underset{5}{V}}\text{-}O\text{-}O\text{-}Bu^t + ROH \qquad (9)$$

$$\underset{5}{5} + :S + ROH \xrightarrow{k'_2} (RO)_3\text{-}\overset{O}{V} + Bu^tOH + SO \qquad (10)$$

Both of these hypothesis have been previously advanced:[1c,2,5] they differ in assuming that the catalytic active species is penta-coordinate (A) or tetracoordinate (B). All the information so far available for vanadium (V) indicates that the tetracoordination is the preferred[15,16] one and the experiments above discussed as well as other author's results point to a very easy ligand exchange for vanadate esters.[13,17] This, on one hand, requires that species like 5 must be present in the system and on the other hand makes the suggestion of a pentacoordinate vanadium (V) species a not necessary hypothesis to be demonstrated. No direct evidence favouring such species are in fact available and therefore we are keen to consider that a species like 5 is the catalytic active intermediate and prefer to discuss the results from this point of view.

Scheme B requires that in the transition state of the rate determining step the oxygen transferred to the substrate must be that linked to vanadium and hence a 1,2-vanadium shift must operate as in 6:

$$(RO)_3\text{-}\overset{\overset{\displaystyle O}{\|}}{\underset{\underset{\displaystyle O\text{-}Bu^t}{|}}{V}}\text{-}O \quad \cdots :S \qquad\qquad (RO)_3\text{-}\overset{\overset{\displaystyle O}{\|}}{\underset{\underset{\displaystyle OBu^t}{|}}{V}}\text{-}O \quad \cdots :S$$

$$\qquad\qquad\qquad\qquad\qquad\qquad\qquad\qquad R\text{-}O\text{--}H$$

$$\underset{6}{6} \qquad\qquad\qquad\qquad\qquad\qquad \underset{6a}{6a}$$

This resemble to the formal 1,2-proton shift in the uncatalyzed oxidation by hydroperoxides.[20] It is well known that, in the latter example, the proton transfer occurs with the participation of protic solvent molecules; we cannot exclude that in the vanadium catalysis too, a more complex vanadium transfer, with the aid of solvent molecules, may occur: (6a).

Following Scheme B the effect of acid on rates may either be due to a depression of k'_2 or to a shift to the left of equilibrium (9). The former hypothesis is rather unlikely as just the opposite would be expected for a reaction of the kind here considered. Equilibrium (9) is formally independent on acid concentration as charges are neither formed nor destroyed. However vanadate esters are basic species, as discussed previously, and consequently they will be solvated via hydrogen bonding by protic solvents. Furthermore esters 5 must be weaker bases than the homoesters 4 since the peroxy group is more electron withdrawing than the alkoxy ligand. It seems rather likely that these differences in basicity parallel differences, in the same direction, in solvation energies. As increasing acidity causes increasing strength of hydrogen bonds, and hence in solvation energies, it may be predicted a shift of equilibrium (9) to the left; i.e. the species more solvated are favoured by increasing solvating ability of the medium.

In conclusion the effect of acid might be a simple medium effect acting on equilibrium (9) even before true protonation occurs, although the same arguments would be easily applied to the equilibrium between the conjugate acids of 4 and 5.

Though this suggestion is highly speculative, an indirect evidence in favour of it is that the catalyst saturation phenomenon observed in "neutral" alcohol is almost suppressed in the presence of acid[2,9] what is consistent with a decrease in the association constant of eq. (9).

<u>Epoxidation of alkenes</u>. The oxidizing system so far discussed even though epoxidation of cyclohexene has been achieved,[2] is not certainly very efficient as far as olefins epoxidation is concerned.

Indeed the high temperatures, 60°, and the long reaction times required make the always present decomposition of the peroxide competitive with the oxidation of the substrate.[1c]

On the other hand vanadium catalyzed epoxidations in hydrocarbon solvents are well known at least at a qualitative synthetic level.[1]

We thought it worthwhile to investigate the mechanistic features of these reactions in benzene with small amount of alcohol added so that vanadate esters may be formed.[21]

A similar technique was formerly adopted by Gould.[5] For comparison purpose with our previous work we measured the rates of VO(acac)$_2$ catalyzed oxidation of di-n-butyl sulfide and p-chlorophenyl methyl sulfide by t-butylhydroperoxide both in the absence and in the presence of added methanesulphonic acid.[21] The data, collected in Table 4 show that continuous decrease on rates of sulfide oxidation in benzene is observed upon addition of increasing alcohol concentration. In benzene-methanol mixtures rates of Bu$_2^n$S oxidation are faster than those in benzene-2-propanol. The same order of reactivity was previously found in the two pure alcohols.[9] Addition of increasing amount of CH$_3$SO$_3$H accelerates the rate of oxidation of p-Cl-C$_6$H$_4$SCH$_3$ in benzene-ethanol. Similar effect was

Table 4.   Effect of added alcohols and methanesulfonic acid on rates of the vanadium catalyzed oxidation of di-n-butyl sulfide and p-chlorophenyl methyl sulfide by t-butyl hydroperoxide, in benzene, at 25.0°, under N_2.

Sulfide	Alcohol		10^4 $MeSO_3H/M$	$k_3^a/lM^{-2}s^{-1}$
	Type	Molarity		
Bu_2^nS	–	–	–	591.
	MeOH	0.12	–	594.
		0.25	–	612.
		0.62	–	436.
		1.25	–	371.
		1.87	–	114.
		2.50	–	74.
		6.20	–	27.
		24.7[b]	–	8.3
	Pr^iOH	0.08	–	53.
		0.16	–	27.
		0.32	–	10.
		0.64	–	3.1
		3.20	–	2.9
		13.1[b]	–	0.2
$p-ClC_6H_4-S-CH_3$	EtOH	0.13	–	30.
		0.13	1.0	70.
		0.13	4.0	173.
		0.17	–	22.
		0.43	–	8.6
		0.43	1.0	13.0
		0.43	2.0	15.0
		0.43	4.0	18.0
		0.85	–	4.3
		1.71	–	2.4
		1.71	1.0	3.9
		1.71	4.0	4.2

[a] See note a, Table 1; [b] Molarity corresponding to 100% alcohol

observed in ethanol solvent and the explanation given should still be the same. The enhancement on rates by added acid is larger when alcohol is in low concentration, as shown in Table 4.

On the ground of the above observations we may assume that oxidation of sulfides in benzene-alcohol solvent proceeds by the same general mechanistic scheme previously established. Furthermore, for added alcohol-hydroperoxide ratios of ca 10, with alcohol ca 0.1 M, pseudo first order kinetics are still regular, i.e. autoinhibition is not observed.

The oxidation rates, in these conditions, are still fairly high and comparable with those observed in benzene solvent alone which are ca two orders of magnitude larger than in alcohol. This mixed solvent system was then adopted to study from a mechanistic point of view the oxidation of the much less reactive olefins.

In the conditions above reported for the oxidation of sulfides, but at 60° we observed that epoxidation of cyclohexene, when the substrate is in large excess in respect of the peroxide, follows a regular pseudo-first order kinetic law. Moreover stoichiometric experiments showed that formation of cyclohexene oxide is nearly quantitative.[21]

Preliminary experiments have also been carried out on some other olefins and data are collected in Table 5.

As it may be observed the rates of epoxidation follows the expected order as far as the substitution pattern around the double bond is concerned. They in fact parallel fairly well the rates of oxidation by peracids.[22] Even though the mechanistic features of alkene epoxidation with $Bu^tO_2H/V(V)$ have not yet been fully explored, every evidence, including the similarity with peracid epoxidation, calls for a reaction mechanism very similar, if not identical, for sulphide oxidation in alcoholic solvents.

In apparent contradiction with what above stated, is the dramatic differences between $Bu^tO_2H/V(V)$ system and peracids in

Table 5. Relative rates of epoxidation of some representative al-kenes by t-butyl hydroperoxide catalyzed by added $VO(acac)_2$, 0.001 M, in benzene −0.12 M ethanol, at 60°, under N_2, by peracetic acid (PA) in acetic acid, at 25.8° and by perbenzoic acid (PBA) in benzene, at 30°.

Relative rates

	$Bu^tO_2H/VO(acac)_2$	(PA)	(PBA)
1-octene	1	1	1
2-trans-octene	7	24	−
cyclohexene	4	27	7
4-tert-butyl cyclohexene	6	26	−
geraniol	17,000	−	−

a Second order rate constants in the conditions above stated: 0.03×10^{-3} l $M^{-1}s^{-1}$, for VO(acac) ; 2.15×10^{-3} l $M^{-1}s^{-1}$ for PA and 30×10^{-3} l $M^{-1}s^{-1}$ for PBA. Data from Refs 21 and 22.

the oxidation of geraniol, an allylic alcohol, which by the metal catalyzed reaction is oxidized at the allylic double bond about 10^3 times faster than cyclohexene (see Table 5). It is known that peracids attack the isolated double bond at a rate comparable with that of a not otherwise functionalized dialkylsubstituted olefin.[22]

This peculiar feature of metal catalyzed epoxidation of allylic alcohols had been observed by Sharphess et al.,[23] who reported that geraniol gives in 98% yield the previously unknown 2,3-epoxide. Both the exceptionally high regioselectivity and reactivity of t-butyl hydroperoxide in the presence of V(V) catalysts were explained by invoking complexation of the alcohol to the metal ion. On the basis of the results previously reported on the nature of the actual V(V) catalyst in the presence of alcohols, the formation of alkyl esters of V(V) in which alkoxy group(s) derive from allylic alcohols seems a very likely process.

Further support for this suggestion is found in the results of the experiments on the asymmetric oxidation[24] which are discussed below.

Asymmetric oxidations. The catalytic species in alcoholic solvent has been shown to be the vanadate ester $VO(OR)_3$. Moreover the same species is likely to exist also in hydrocarbon solvents provided that enough alcohol is added. The concentration of alcohol needed is in fact quite low so that oxidation rates are still acceptable. All these results suggested to carry out oxidation of prochiral substrates by $Bu^t_2O_2H/V(V)$ in benzene-chiral alcohol mixtures with the aim to obtain optically active oxidation products.

The purposes of these experiments were essentially two:
a) to collect one more evidence that vanadate esters are the catalysts also in mixed hydrocarbon alcohol solvents.
b) to establish a way of direct asymmetric oxidation alternative to the only one previously known: i.e. the use of chiral peracids.[25]

The substrates initially employed were the two prochiral sulfides p-tolyl methyl sulfide and phenyl-methyl sulfide.

The results are reported in Table 6.

It may be observed that sizeable optical yields are obtained in the oxidation of the two sulfides using a variety of chiral alcohols as cosolvents. In particular, menthol seems to be the most effective in obtaining asymmetric synthesis. Indeed the efficiency of the oxidizing system when menthol is employed competes effectively with chiral peracid oxidation of sulfides.[25]

In benzene-menthol solvent very recently we performed also the asymmetric epoxidation of prochiral alkenes. Only preliminary re-

Table 6. Asymmetric induction in the oxidation of prochiral sulfides by Bu^tO_2H in benzene (toluene)-chiral alcohols mixtures.

Substrate	$10^4(VO(acac)_2)_o/M$	t^o	Alcohol		Sulphoxide [a]	
			Type	(%)	$(\alpha)_D^{20}$	%e.e.
p-CH_3-C_6H_4-S-CH_3	25	0°	(-)-2-octanol	(15)	-0.3	0.2
	17.5	0°	(-)-borneol	(6)	+1.7	1.2
	25.0	0°	(-)-TADE[b]	(15)	-3.2	2.2
	13.5	0°	(-)-menthol	(5)	-7.0	4.8
	25.0	0°	"	(15)	-12.9	8.9
	25.0	0°	"	(30)	-14.4	9.8
	24.5	-10°	"	(5)	9.2	6.3
C_6H_5-S-CH_3	25.	0°	"	(30)	6.7	4.7

[a] See Ref. 24; [b] Tartaric Acid Diethyl Ester

sults are so far available but they must be considered encouraging.[26] Cis and trans 2-octene have been oxidized to give the parent optically active epoxides with low but significant enantiomeric excess. Also, 4-t-butyl cyclohexene gives upon epoxidation a mixture of optically active cis and trans 4-t-butylcyclohexene oxides.[26]

In the same benzene-menthol mixtures employed for sulfides and non functionalized alkenes, we found that no asymmetric induction is obtainable for the epoxidation of geraniol.[26]

This finding, though surprising at first sight, can be rationalized by and is consistent with the general mechanistic picture so far discussed. Indeed, geraniol was, in our experiments, in concentration comparable with menthol and if the former is, as it is likely, a better ligand than the latter, the vanadium is present as allyl vanadate ester only. It follows that the catalyst is not any longer chiral and therefore asymmetric induction is not possible. Furthermore this rationale offers an explanation of the exceptional reactivity of allylic alcohols since in this case, and only in this one, both peroxide and substrate are coordinated to vanadium.

Acknowledgement. One of us (r.c.) wishes to thank NATO Sicentific Committee for partial support.

REFERENCES

1.   (a) R. Hiatt in "Oxidation", Vol. 2, R. L. Augustine and D. J. Trecker, Eds., Marcel Dekker, Inc., New York, N.Y., 1971, Chap. 3; (b) G. Sosnovsky and D. J. Rawlinson in "Organic Peroxides", Vol. 2, D. Swern, Ed., Wiley-Interscience, New York, N.Y., 1971, Chap. 2; (c) R. A. Sheldon and J. K. Kochi, Advances in Catalysis, 25, 271 (1976).

2.   R. Curci, F. Di Furia, R. Testi, and G. Modena, J. C. S. Perkin II, 752 (1974).

3.   R. Curci, F. Di Furia, and G. Modena, J. C. S. Perkin II, 576 (1977).

4.   N. Indictor and W. F. Brill, J. Org. Chem., 30, 2074 (1965).

5.   E. S. Gould, R. R. Hiatt and K. C. Irwin, J. Amer. Chem. Soc., 90, 4573 (1968).

6.   C.-C. Su, J. W. Reed and E. S. Gould, Inorg. Chem., 12, 337 (1973).

7.   M. Orhanovic and R. G. Wilkins, J. Amer. Chem. Soc., 89, 278 (1967).

8.   R. Curci, R. A. Di Prete, J. O. Edwards and G. Modena, J. Org. Chem., 35, 740 (1970).

9.   S. Cenci, R. Curci, F. Di Furia, J. O. Edwards, and G. Modena, J. C. S. Perkin II, in press.

10.   C. J. Ballhausen and H. B. Gray, Inorg. Chem., 1, 111 (1962).

11.   I. Bernal and P. H. Rieger, Inorg. Chem., 2, 256 (1963).

12.   (a) J. Selbin, R. T. Ortolano, and F. J. Smith, Inorg. Chem., 2, 1315 (1963); (b) T. R. Ortolano, J. Selbin, and S. P. McGlynn, J. Chem. Phys., 41, 262 (1964).

13.   P. J. White, M. J. Kaus, J. O. Edwards, and P. H. Rieger, J. C. S. Chem. Comm., 429 (1976).

14.   F. J. C. Rossotti and H. Rossotti, Acta Chem. Scand., 10, 957 (1956).

15.   J. A. Connor and E. A. Ebsworth, Advances in Inorganic Chemistry and Radiochemistry, 6, 279 (1964).

16.   F. Cartan and C. N. Caughlan, J. Phys. Chem., 64, 1756 (1960).

17. (a) R. K. Mittal and R. C. Mehrotra, Z. Anorg. Chem., 327, 311
     (1964); (b) ibid., 355, 328 (1967).

18. A. Lachowitz, W. Höbold, and K.-H. Thiele, Z. Anorg. Chem.,
     418, 65 (1975).

19. K. J. Laidler, "Chemical Kinetics", McGraw-Hill, London-New
     York, 1965, 2nd Edn., pp. 474ff.

20. R. Curci and J. O. Edwards in "Organic Peroxides", Vol. 1,
     D. Swern, Ed., Wiley-Interscience, New York, N.Y., 1970,
     Chap. 4.

21. R. Curci, F. Di Furia and G. Modena, unpublished results.

22. D. Swern in "Organic Peroxides", Vol. 2, D. Swern, Ed., Wiley-
     Interscience, New York, N.Y., 1971, Chap. 5.

23. K. B. Sharpless and R. C. Michaelson, J. Amer. Chem. Soc., 95,
     6136 (1973).

24. F. Di Furia, G. Modena, and R. Curci, Tetrahedron Lett., 50,
     4637 (1976).

25. J. D. Morrison and H. S. Mosher, "Asymmetric Organic Reactions",
     Prentice-Hall, Englewood Cliffs, N.J., U.S.A., 1971, Chap. 8.

26. R. Curci, F. Di Furia, and G. Modena, unpublished results.

SYNTHETIC REACTIONS BY COPPER COMPLEX CATALYSTS

Takeo Saegusa and Yoshihiko Ito

Department of Synthetic Chemistry, Faculty of

Engineering, Kyoto University, Kyoto, Japan

Organic Syntheses utilizing copper compounds have been remarkably developed. This paper describes several synthetic reactions caused by copper catalysts, which were found in this laboratory.

1. Synthesis of 1,4-Diketones by Oxidative Dimerization of Ketones.

Transition-metal induced oxidative dimerization of carbanions has constituted a convenient method for the carbon-carbon bond formation in organic synthesis. Copper-induced oxidative dimerizations of carbanions, which are stabilized by sulfonyl, phosphory, imidoyl and alkoxycarbonyl groups, have been hitherto known. Therefore, the most straightforward approach to 1,4-diketones seemed to be the oxidative dimerization of ketone enolates. However, no successful dimerization of ketone enolates by transition-metal salts had been realized prior to our studies which disclosed that 1,4-diketones were readily prepared by the following two reactions.

1.1 The Reaction of Ketone Enolates with $CuCl_2$.[1]

The $CuCl_2$-induced dimerization of ketone enolates was simply performed in one flask by treating lithium enolate, which was generated in situ at $-78°C$ from ketone and lithium diisopropylamide in THF, with $CuCl_2$ in dimethylformamide. The dimerization of ketones bearing one kind of enolizable hydrogen gave a single 1,4-diketone according to eq 1. The use of dimethylformamide as a cosolvent was crucial in the oxidative dimerization of ketone enolates.

$$2 \; R-\underset{\underset{OLi}{|}}{C}=C\overset{R'}{\underset{R''}{\diagdown}} + \; 2 \; CuCl_2 \longrightarrow$$

$$\underset{\underset{O \; R''R''O}{\parallel \; | \; | \; \parallel}}{R-\underset{\overset{|}{R'}R'}{C-C-C-C-R}} + \quad \begin{matrix} 2 \; CuCl \\ \\ 2 \; LiCl \end{matrix} \qquad (1)$$

Methyl ketones (RCOCH$_3$) were dimerized to the corresponding 1,4-diketones (RCOCH$_2$CH$_2$COR) in excellent to moderate yields. The product yields depend on alkyl substitution at the coupling site. Increasing alkyl substitution at the coupling site resulted in a remarkable reduction in the yield of 1,4-diketone. This trend in the ketone dimerization was also observed in the 1,4-diketone synthesis by the oxidative coupling of silyl enol ether with Ag$_2$O, which will be described later.

The dimerization of ketones having two different enolizable hydrogens could, in principle, give a mixture of three possible isomers of 1,4-diketones, as shown in eq 2.

$$RR'CHCOCHR''R''' \xrightarrow[\text{2. } CuCl_2 \text{ in DMF}]{\text{1. } LiN(i-Pr)_2}$$

$$\left\{ \begin{array}{l} RR'CH\underset{\parallel}{C}C(R'')(R''')C(R'')(R''')\underset{\parallel}{C}CHRR' \\ \quad\quad\;\; O \quad\quad\quad\quad\quad\quad\quad\quad\quad\;\; O \\ \\ RR'CH\underset{\parallel}{C}C(R'')(R''')C(R)(R')\underset{\parallel}{C}CHR''R'' \qquad (2) \\ \quad\quad\;\; O \quad\quad\quad\quad\quad\quad\quad\;\; O \\ \\ R''R'''CH\underset{\parallel}{C}C(R)(R')C(R)(R')\underset{\parallel}{C}CHR''R''' \\ \quad\quad\quad\;\; O \quad\quad\quad\quad\quad\quad\quad\;\; O \end{array} \right.$$

Actually, the least crowded 1,4-diketone was, as expected, produced predominantly together with the more crowded 1,4-diketone as a minor product, both of which were separated and isolated by preparative GLC or TLC. The most crowded 1,4-diketone was produced in a negligible amount under the reaction conditions employed. Some results are summarized in Table 1.

As expected, the cross coupling of two different ketones, each of which has two enolizable hydrogens, yielded a complex mixture of all possible 1,4-diketones including unsymmetrical 1,4-diketones as cross coupling products; but, it was found that,

Table 1. Synthesis of 1,4-Diketones by the Reaction of Ketone Enolates with $CuCl_2$

Starting Ketone	1,4-Diketones (yield, %)	
$(CH_3)_3CCOCH_3$	$(CH_3)_3CCOCH_2CH_2COC(CH_3)_3$	(95)
$(CH_3)_2CHCOCH_3$	$(CH_3)_2CHCOCH_2CH_2COCH(CH_3)_2$	(89)
	$(CH_3)_2CHCOCH_2C(CH_3)_2COCH_3$	(3)
$C_2H_5COC_2H_5$	$C_2H_5COCH(CH_3)CH(CH_3)COC_2H_5$	(32)
$CH_3COC_2H_5$	$C_2H_5COCH_2CH_2COC_2H_5$	(58)
	$C_2H_5COCH_2CH(CH_3)COCH_3$	(12)
$CH_3COC_6H_{13}$	$C_6H_{13}COCH_2CH_2COC_6H_{13}$	(62)
	$C_6H_{13}COCH_2CH(C_5H_{11})COCH_3$	(18)
CH_3COPh	$PhCOCH_2CH_2COPh$	(95)
CH_3CH_2COPh	$PhCOCH(CH_3)CH(CH_3)COPh$	(28)
$(CH_3)_2CHCOPh$	$PhCOC(CH_3)_2C(CH_3)_2COPh$	(2)
$CH_3COC(CH_3)_2CO_2C_2H_5$	$H_5C_2O_2CC(CH_3)_2COCH_2CH_2COC(CH_3)_2CO_2C_2H_5$	(64)

(78)

in the cross coupling reaction of two different methyl ketones
(CH_3COR and CH_3COR'), the use of a large excess of one ketone
enolate over another led to the formation of a specific
unsymmetrical 1,4-diketone, $RCOCH_2CH_2COR'$, in a synthetically
useful yield and selectivity; e.g., (Z)-8-Undecene-2,5-dione, a
precursor of cis-jasmone, was prepared in a satisfactory yield
and selectivity by the cross coupling reaction of 3 mol of
acetone enolate and 1 mol of (Z)-5-octen-2-one enolate.

$$3 \; CH_3COCH_3 \quad + \quad 1 \; (Z)\text{-}CH_3CO(CH_2)_2CH{=}CHC_2H_5$$

$$\xrightarrow[\text{2. 4.5 equiv } CuCl_2]{\text{1. 4.5 equiv } LiN(i\text{-}Pr)_2}$$

$$\begin{cases} (Z)\text{-}CH_3COCH_2CH_2CO(CH_2)_2CH{=}CHC_2H_5 \\ \qquad\qquad\qquad\qquad\qquad\qquad\qquad (68\%) \\ (Z)\text{-}CH_3COCH_2CH(COCH_3)CH_2CH{=}CHC_2H_5 \\ \qquad\qquad\qquad\qquad\qquad\qquad\qquad (1\%) \\ CH_3CO(CH_2)_2COCH_3 \end{cases}$$

$$(3)$$

1.2   The Reaction of Silyl Enol Ethers with Ag_2O [2]

On heating a mixture of silyl enol ether with Ag_2O in
dimethyl sulfoxide, 1,4-diketone was produced in moderate yield,
according to eq 4. This reaction provides a convenient method
for the oxidative dimerization of ketones leading to 1,4-
diketones, since silyl enol ethers can be readily prepared from
the corresponding ketones.

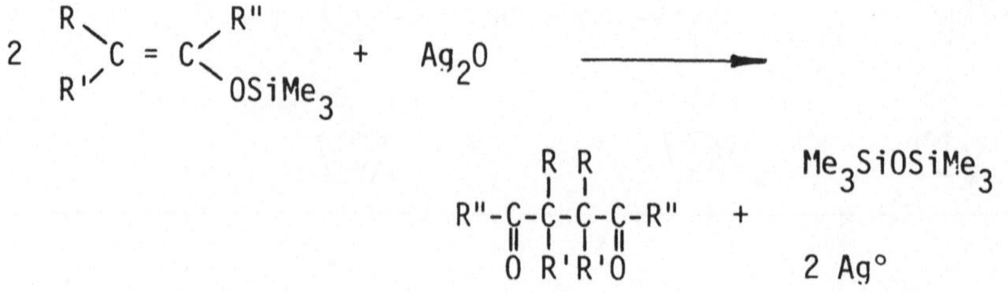

$$2 \quad \begin{array}{c} R \\ R' \end{array}\!\!\!\diagdown\!\!C = C\!\!\diagup\!\!\!\begin{array}{c} R'' \\ OSiMe_3 \end{array} \quad + \quad Ag_2O \quad \longrightarrow$$

$$\begin{array}{c} \quad R \;\; R \\ R''\text{-}C\text{-}C\text{-}C\text{-}C\text{-}R'' \\ \quad \| \; | \; | \; \| \\ \quad O \; R'R'O \end{array} \quad + \quad \begin{array}{c} Me_3SiOSiMe_3 \\ \\ 2 \; Ag° \end{array} \quad (4)$$

The solvent of dimethyl sulfoxide was very important in the coupling reaction. Similar results were obtained in aprotic polar solvents such as hexamethylphosphoric triamide, dimethylformamide and acetonitrile. However, no reaction occurred in toluene and diglyme. An important feature of the oxidative dimerization of ketones is regiospecific formation of 1,4-diketone. For instance, 2-trimethylsilyloxy-3-methyl-1-butene was reacted with Ag_2O in dimethylsulfoxide at 65° for 2 hr to produce 2,7-dimethyl-3,6-octanedione in 81% yield uncontaminated with its isomers. Some results are summarized in Table 2.

Table 2. Synthesis of 1,4-Diketones by the Reaction of Silyl Enol Ethers with Ag_2O

Silyl Enol Ether	Reaction Conditions	1,4-Diketone (%)
$CH_2=C(Ph)OSiMe_3$	65°, 2 hr	$PhCOCH_2CH_2COPh$ (73)
$CH_2=C(i-C_3H_7)OSiMe_3$	65°, 2 hr	$(CH_3)_2CHCOCH_2CH_2COCH(CH_3)_2$ (81)
$(CH_3)_2C=C(CH_3)OSiMe_3$	100°, 5 hr	$CH_3COC(CH_3)_2C(CH_3)_2COCH_3$ (38)
$CH_2=C(C_2H_5)OSiMe_3$	65°, 2 hr	$C_2H_5COCH_2CH_2COC_2H_5$ (76)
$CH_3CH=C(CH_3)OSiMe_3$	80°, 3 hr	$CH_3COCH(CH_3)CH(CH_3)COCH_3$ (23)
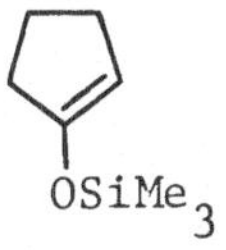	100°, 2 hr	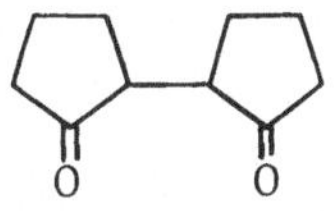 (61)
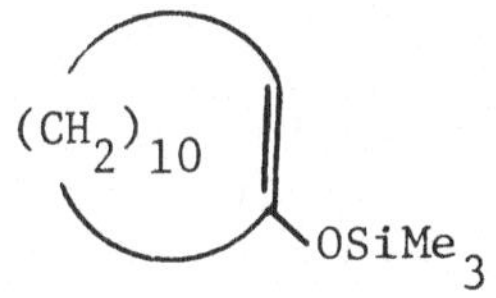	85°, 3 hr	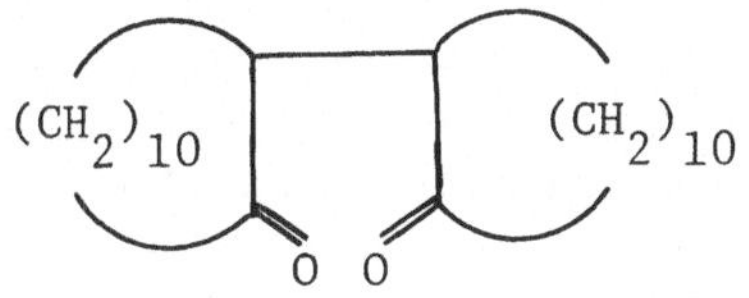 (45)

2. Heterocycle Syntheses by Means of Group IB Metal Compounds

2. 1.   Synthesis of 2-Oxazoline, 5,6-Dihydro-4H-1,3-oxazine,
        2-Imidazoline, and 2-Thiazoline. [3]

Several years ago, we found that primary and secondary
amines reacted with isonitriles in the presence of IB and IIB
metal salts of the periodic table such as copper(I), silver(I),
gold(I), zinc(II), to produce N,N'-substituted formamidines in
almost quantitative yields.

$$RR'NH \quad + \quad R''NC \quad \xrightarrow{\text{CuCl, etc.}} \quad RR'N-CH=NR'' \qquad (5)$$

Recently Balch, et. al. reported a carbene-coordinated gold(I)
complex in the gold(III)-catalyzed reaction of isonitrile with
amine, which is considered to be a key intermediate of the
"formimidoylation reaction".
    Based on the "formimidoylation reaction", we have found a
new and versatile method for heterocycle synthesis, i.e., the
reactions of isonitriles with diamines, with aminoalcohols, and
with aminothiols in the presence of copper(I) and silver (I) salts
to produce the heterocycles consisting of -N=CH-Y-unit (Y=NH, O,
S) such as 2-imidazoline, 2-oxazoline, 5,6-dihydro-4H-1,3-oxazine,
and thiazoline in moderate yields, according to eq 6.

$$R-NC \quad + \quad H_2N-(CH_2)_n-YH \quad \xrightarrow{\text{Cu(I), Ag(I)}}$$

$$Y = NH, O, S \qquad \begin{pmatrix} (CH_2)_n \\ N=C\diagdown Y \\ \quad\ H \end{pmatrix} \quad + \quad RNH_2 \qquad (6)$$

Y = NH, O, S
n = 2, 3

Moreover, palladium(II) chloride was found to be an excellent
catalyst for heterocycle syntheses. The versatility of
palladium(II) chloride as a catalyst is demonstrated in the
reaction of aminoalcohols and aminothiols having a polar
carboxylate substituent with isonitriles, e.g., methyl threonate
and methyl cysteinate react with t-butyl isocyanide in the
presence of palladium(II) chloride to produce 4-methoxycarbonyl-5-
methyl-2-oxazoline and 4-methoxycarbonyl-2-thiazoline, respectively. Copper(I) chloride and silver(I) chloride are inactive toward

such reactions. Some results of the heterocycle syntheses are summarized in Table 3.

Of particular interest is that the carbene-coordinated palladium (II) complexes, which are the key intermediates in the heterocycle syntheses, can be isolated from the reactions of the palladium(II) chloride complex with tert-butyl isocyanide with β-aminobutanol, methyl threonate, and o-aminophenol. These

Table 3. Preparation of Oxazolines, Oxazines, Imidazolines, and Thiazolines

Substrate	Catalyst	Product (%)
$H_2N-CH_2CH_2-OH$	AgCl	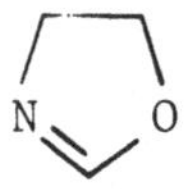 oxazoline ring (N, O) (67)
$H_2N-CH_2-CH(CH_3)-OH$	AgCl	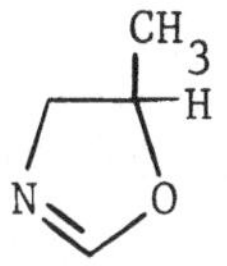 methyl oxazoline ring (N, O, CH₃, H) (72)
$H_2N-CH_2CH_2CH_2-OH$	AgCl	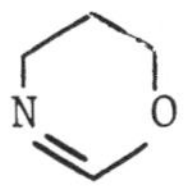 oxazine ring (N, O) (66)
$H_2N-CH_2CH_2-NH_2$	AgCl	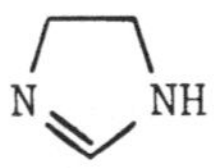 imidazoline ring (N, NH) (80)
$H_2N-CH_2CH_2-SH$	AgCl	 thiazoline ring (N, S) (88)
$H_2N-CH(COOCH_3)-CH(CH_3)-OH$	PdCl$_2$	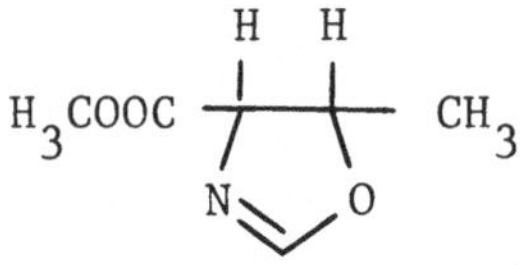 H_3COOC—oxazoline ring (H, H, N, O)—CH_3 (98)
$H_2N-CH(COOCH_3)-CH_2-SH$	PdCl$_2$	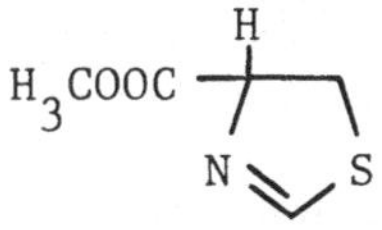 H_3COOC—thiazoline ring (H, N, S) (78)

complexes give the corresponding oxazolines on the pyrolysis at
200° or on the treatment with isonitrile ligand at about 80°.

$$
\underset{\underset{X\ \ Y}{|\ \ |}}{HOCHCHNH_2} \quad + \quad Pd(II)Cl_2(t\text{-}C_4H_9NC)_2
$$

$$
\longrightarrow \quad
\begin{array}{c}
Cl \diagdown \\
\qquad Pd \\
Cl \diagup
\end{array}
\begin{array}{l}
\cdots CNC_4H_9\text{-}t \\
 NHC_4H_9\text{-}t \\
\cdots :C \diagup \\
 \diagdown NHCHCHOH \\
 \underset{X\ \ Y}{|\ \ |}
\end{array}
\qquad (7)
$$

2.2.   Synthesis of Indole Derivatives [4]

In the continuation of our studies on synthetic reactions
caused by Cu_2O-isonitrile complex, it has been found that the
so-called active methylene compounds such as malonate, acetoacetate and cyclopentadiene were reacted with Cu_2O in the
presence of isonitrile to produce organocopper(I) isonitrile
complexes, which undergo an isonitrile insertion into their
copper-carbon bonds.   On the basis of the isonitrile insertion
of organocopper(I) isonitrile complexes, we found a new
synthetic method of indole derivatives by the Cu_2O-catalyzed
reactions of o-cyanomethylphenyl isocyanide and o-methoxy-
carbonylmethylphenyl isocyanide, in which intermediacy of the
corresponding organocopper(I) isonitrile complexes may be assumed.

$$
(8)
$$

a : X=CN
b : X=CO_2CH_3

o-(α-Cyanoalkyl)phenyl isocyanide and o-(α-methoxycarbonyl-
alkyl)phenyl isocyanide, which were readily prepared by the base
catalyzed alkylation of o-cyanomethylphenyl isocyanide and o-
methoxycarbonylmethylphenyl isocyanide, respectively, were also
cyclized by Cu_2O catalyst producing 3-alkyl-3-cyano-3H-indole
and 3-alkyl-3-methoxycarbonyl-3H-indole in moderate yields.

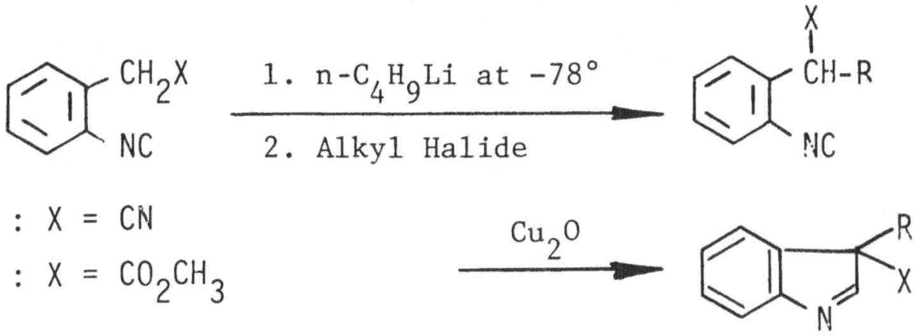

$$a : X = CN$$
$$b : X = CO_2CH_3$$

(9)

Some results of the indole synthesis are summerized in Table 4.

Table 4.   Indole Synthesis by Cu_2O Catalyzed Cyclizations of
o-(α-Cyanoalkyl)phenyl Isocyanides

o-(α-Cyanoalkyl)phenyl Isocyanide and o-(α-Methoxycarbonylalkyl)-phenyl Isocyanide	Reaction[a] °C/hr	Indole Derivatives (%)
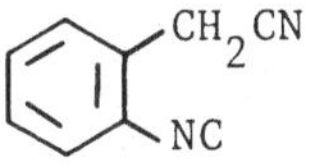	55/6	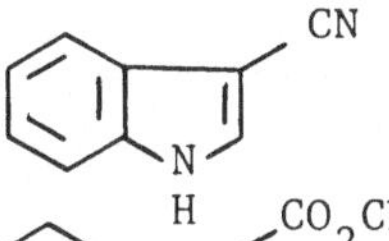 (86)
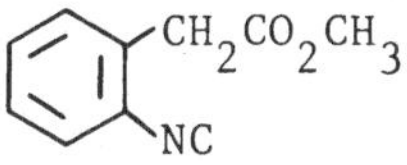	55/6	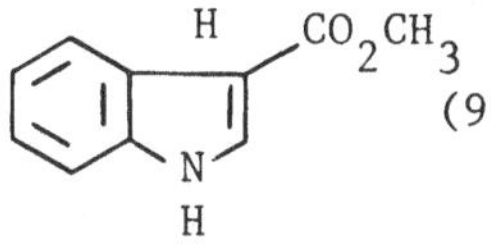 (92)
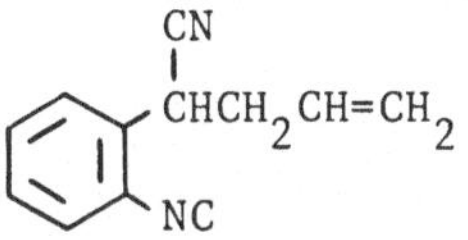	80/13	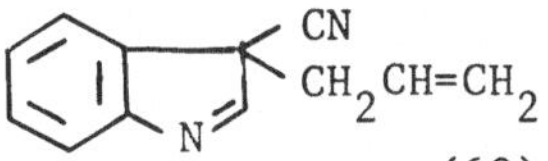 (60)
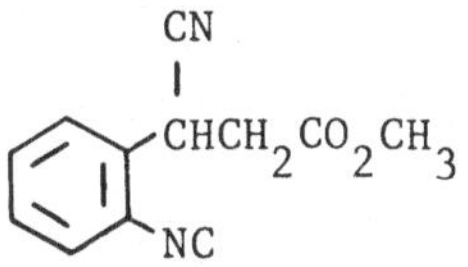	80/6	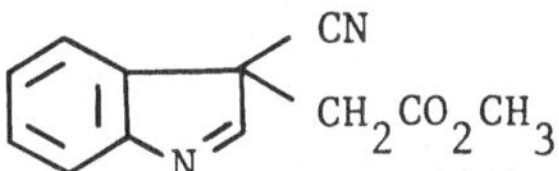 (43)
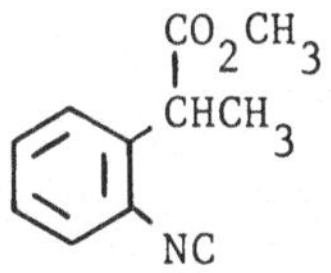	80/30	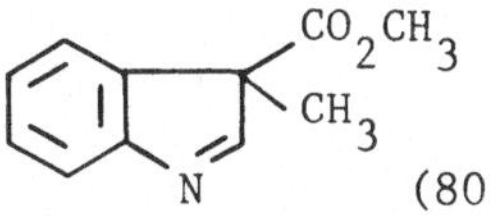 (80)

a) Cyclizations were performed by heating a mixture of
o-(α-cyanoalkyl)phenyl isocyanide or o-(α-methoxycarbonyl-
alkyl)phenyl isocyanide (2 mmol) and Cu_2O (0.025 mmol) in 2
ml of benzene.

284 T. SAEGUSA AND Y. ITO

References

1. Y. Ito, T. Konoike, T. Harada and T. Saegusa, J. Am. Chem. Soc., 99, 1487 (1977).
2. Y. Ito, T. Konoike and T. Saegusa, J. Am. Chem. Soc.,97, 649 (1975).
3. a) Y. Ito, Y. Inubushi, M. Zenbayashi, S. Tomita and T. Saegusa, J. Am. Chem. Soc.,95, 4447 (1973).
   b) Y. Ito, T. Hirao and T. Saegusa, J. Organometal. Chem.,131, 121 (1977).
4. Y. Ito, Y. Inubushi, T. Sugaya, K. Kobayashi and T. Saegusa, Bull. Chem. Soc. Japan, in press.

SYNTHETIC ANALOGUES OF THE ACTIVE SITE IN BLUE COPPER PROTEINS

Tobin J. Marks

Department of Chemistry and the Materials Research Center
Northwestern University
Evanston, Illinois 60201

A number of copper-containing proteins, which are involved in electron transfer and/or oxidase activity, contain an unusual form of copper. The "blue" or "type 1" binding sites[1] of such metalloproteins as azurin, plastocyanin, stellacyanin, laccase, ceruloplasmin, cytochrome oxidase, and ascorbate oxidase exhibit chemical and spectral characteristics which are unique among copper complexes. These characteristics include generally high redox potentials (i.e., the Cu(I) form is unusually stable), and for the oxidized, Cu(II) form of the proteins, intense optical absorption at ca. 600 nm (ϵ = 1000-5000 $M^{-1}cm^{-1}$) as well as unusually small copper hyperfine interaction in the electron paramagnetic resonance (EPR) spectra ($A_\parallel$ = 3.3-9.0 mK).[1] The structure of the blue active site has been probed by a wide variety of techniques. On the basis of chemical,[1] metal substitution (cobalt[2] nickel[3]). sequence,[4] resonance Raman,[5] CW and pulsed EPR,[6] carbon[7] and proton[8] nuclear magnetic resonance (NMR), electronic spectral[2,9] circular dichroism, and magnetic circular dichroism,[2,9] infrared spectral,[10] EXAFS,[11] and X-ray photoelectron[12] spectral studies, two basic active site configurations have been proposed (A and B). Both structural models emphasize that, in most cases, two

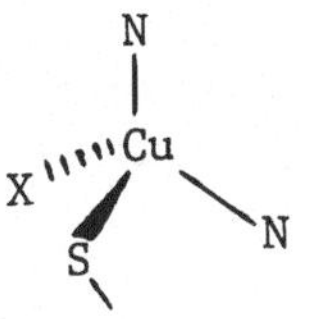

A 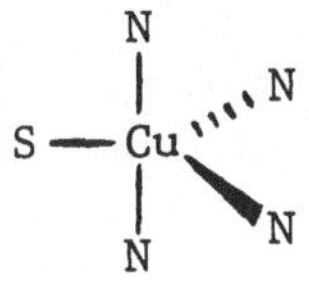   B

285

of the coordinated nitrogen atoms shown are probably from histidine
imidazoles and that one sulfur atom is a cysteinyl mercaptide.
The proponents of the four-coordinate models (A) have proposed
that X may be an amide nitrogen atom from a deprotonated amide,
an oxygen atom from a tyrosine phenolate, a thioether sulfur atom
from a methionine, or another cysteinyl sulfur atom. The chemical
and physicochemical differences among the various blue proteins
stongly suggest that there may actually be three subclasses of
cupric binding environment: the azurin or plastocyanin type, the
stellacyanin type, and the laccase type.[1,13,14] Variation of the
identity of X would be one plausible explanation for differences
among these proteins. Recent single crystal X-ray diffraction
results[14] on a poplar (Populus nigra var. italica) plastocyanin
are consistent with structure A in which X is a methionine sulfur
atom and both N's are imidazole nitrogen atoms. This result is
illustrated in Figure 1. The coordination geometry of the Cu(II)
ion appears to be, at the present stage of refinement, distorted
tetrahedral.[14]

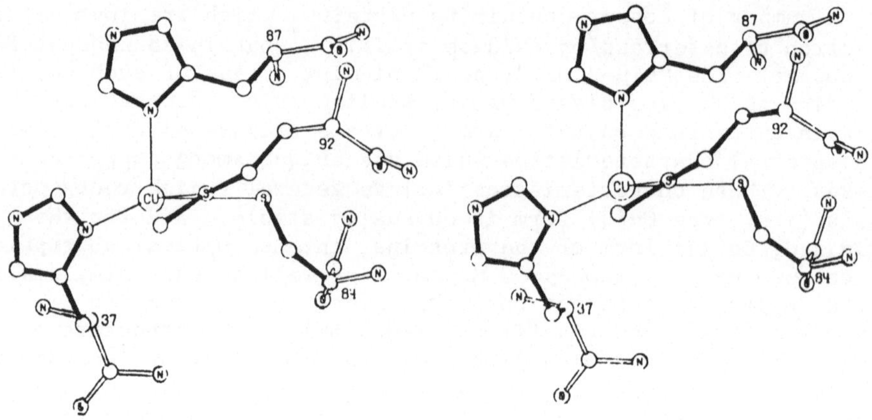

Figure 1. Stereoscopic view of the copper binding environment in
poplar plastocyanin from reference 14.

      Although the aforementioned results have considerably ex-
panded our chemical and physical understanding of the properties
of the blue copper proteins, many questions remain to be answered
and a number of new ones have arisen. Very little is actually
known from an inorganic chemical standpoint about the properties
of copper ions in binding sites such as A. Such coordination
compounds of copper are completely unknown in the classical literature and no precedent exists for the characteristics of $CuN_2X(SR)$
species. How the spectral and chemical properties of a copper ion

in such an unusual environment respond to modifications in the
ligation sphere and coordination geometry are a mystery. These
fundamental inorganic chemical issues have a direct bearing on the
biochemistry. The spectral, chemical, and structure/function
relationships among the various type 1 proteins are poorly understood. The proof by X-ray diffraction that the copper in poplar
plastocyanin is coordinated to a methionine sulfur actually provides support for the idea that X in structure A can have vari-
ability. Stellacyanin does not possess a methionine residue,[1]
and clearly structural differences must exist among the blue
binding sites. Furthermore, the discussion to this point has
focused on the oxidized, Cu(II) form of the blue proteins. Almost
no structural information is available for the reduced form of the
various blue proteins since these Cu(I) species have few distinc-
tive spectral properties. Clearly, an understanding of the
electron transfer activity must be founded on details concerning
both halves of the redox couple.

Valuable information on a number of the above questions can
be provided by studies of simple, low molecular weight analogue
compounds which embody the essential features of the blue copper
core, but which allow lines of experimentation and modification
which are impossible with a molecule as complex as a metallo-
protein. Activity in the area of synthetic analogues of the blue
copper proteins has resulted in much interesting new chemistry,[15]
however, the compounds produced have generally not been well defined in terms of composition and structure, or, in several cases,
possess physiologically implausible ligands and/or polymetallic
structures. The purpose of this article is to summarize recent
studies at Northwestern University on both oxidized and reduced
type 1 active site analogues.[16] It has proven possible to synthesize coordination compounds of structure A where X is a heterocyclic nitrogen atom. These compounds constitute the only well
defined, mononuclear species produced to date which approximate
any of the proposed type 1 core structures and which, in addition,
exhibit many of the characteristic spectral properties of the blue
copper proteins. Furthermore, an analogous synthetic methodology
has yielded cobalt complexes which are analogues of the cobalt-
substituted blue proteins.[25]

SYNTHETIC CONSIDERATIONS

We have previously employed the anionic hydrotris(1-pyrazolyl)
borato (C) and hydrotris(3,5-dimethyl-1-pyrazolyl)borato (D) ligands to simulate polyimidazole copper binding sites.[17] To
assemble complexes of structure E, it was necessary to react the
pyrazolylborate anions with appropriate copper mercaptide reagents.
To avoid aggregated products, the more sterically encumbered,
methylated ligand D was utilized. Two factors were of importance

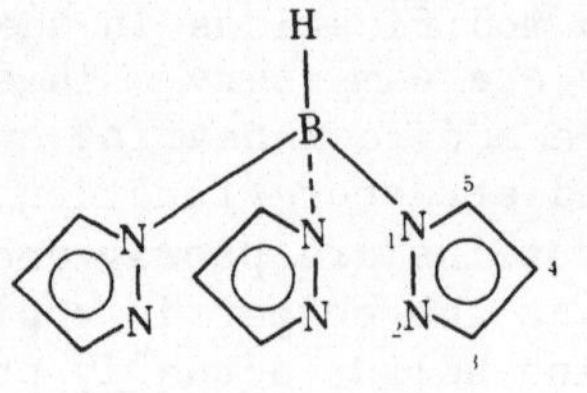

C

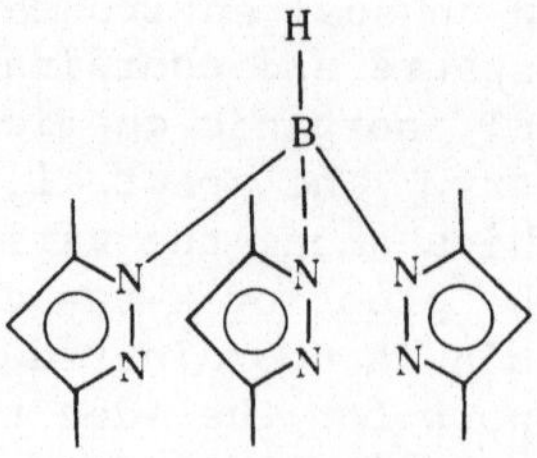

D

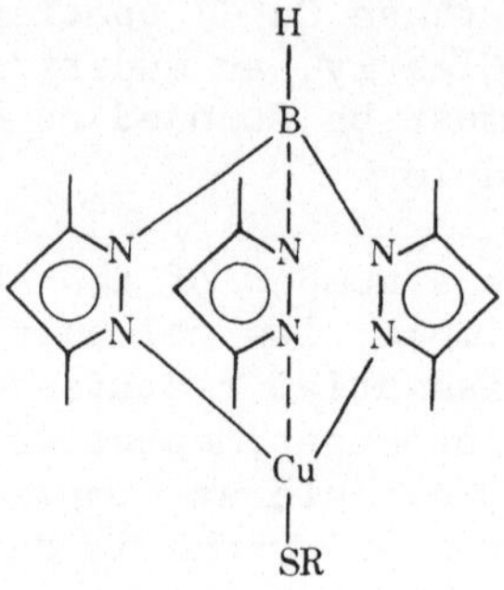

E

in confronting the well known tendency of cupric ion to oxidize
mercaptides (equation (1)) [1,18]

$$2Cu^{+2} + 2RS^- \longrightarrow 2Cu^{+1} + RSSR \qquad (1)$$

First, trispyrazolylborato ligands are known to place a large
amount of electron density on copper ions as judged by the relatively low energies of the C-O stretching frequencies in
$HB(R_2pz)_3CuCO$ derivatives. [17,19] This factor is expected to lend
stability to a cupric mercaptide by diminishing the oxidizing
power of the Cu(II) ion, and may also be a factor in stabilizing
the blue site in the native protein, since the pK_a of imidazole
(7.1)[20] suggests even greater electron donating power than
pyrazole (2.5).[20] The possibility of varying the mercaptide
ligand was also recognized and the first experiments were conducted with less readily oxidized mercaptides such as p-nitro-
phenylthiolate. [21]

The synthetic approach to the monovalent copper derivatives
is rather straightforward and is shown in equations (2) and (3).

$$\text{CuCl} + \text{K(SR)} \xrightarrow{\text{ethanol}} \text{Cu(SR)} + \text{KCl} \tag{2}$$

$$\text{Cu(SR)} + \text{K[HB}(3,5\text{-Me}_2\text{pz})_3] \xrightarrow[\text{room temperature}]{\text{tetrahydrofuran}}$$

$$\text{K[Cu(HB}(3,5\text{-Me}_2\text{pz})_3)(\text{SR})] \tag{3}$$

1a SR = p-NO$_2$C$_6$H$_4$S

1b SR = O-ethylcysteinate

The new compounds were characterized by elemental analysis as well as by infrared, laser Raman, and proton NMR spectroscopy. The O-ethylcysteinate complex is colorless (as are the reduced type 1 proteins) while the violet color of the p-nitrophenylthiolate complex (λ = 550 nm, ϵ = 1600 m^{-1}cm^{-1}) arises from the mercaptide moiety.[17] The energy and intensity of the p-nitrophenylthiolate optical absorption is sensitive to the oxidation state of coordinated metal ion.[22] Single crystals of the nitrophenylthiolate derivative can be grown from acetone/ether/hexane solutions at -10°C. The results of a single crystal X-ray diffraction study of K[Cu(HB(3,5-Me$_2$pz)$_3$)(p-NO$_2$C$_6$H$_4$S)]·2 acetone are presented in Figure 2. The crystal structure consists of anionic CuN$_3$S moieties (shown) and weakly associated K(acetone)$_2^+$ cations. The coordination geometry about the copper ion is significantly distorted from tetrahedral with N-Cu-S angles of 115.7(7)°, 130.3(7)°, and 131(1)°. The N-Cu-N angles were found to be 86.8(9)°, 89.3(9)°, and 90.7(9)°. The Cu-S distance of 2.19(1)Å and the Cu-N distances of 2.00(2), 2.06(2), and 2.19(1)Å compare favorably to metrical parameters in other Cu(I) complexes.[17,23]

The route to the synthetic representations of the oxidized, Cu(II) forms of the type 1 core is shown in equations (4) and (5). Great precaution must be exercized in the second step since

$$\text{Cu(ClO}_4)_2 + \text{KSR} \xrightarrow{\text{ethanol}} \text{Cu(SR)(ClO}_4) + \text{KClO}_4 \tag{4}$$

$$\text{Cu(SR)(ClO}_4) + \text{K[HB}(3,5\text{-Me}_2\text{pz})_3] \xrightarrow[-78°]{\text{tetrahydrofuran}}$$

$$\text{Cu(HB}(3,5\text{-Me}_2\text{pz})_3)(\text{SR}) + \text{KClO}_4 \tag{5}$$

2a SR = p-NO$_2$C$_6$H$_5$S
2b SR = O-ethylcysteinate

the deep blue solutions of the Cu(II) mercaptide derivatives, 2a and 2b, decompose above ca. -30°C. Among the thermal decomposition products are the monovalent complexes 1a and 1b.

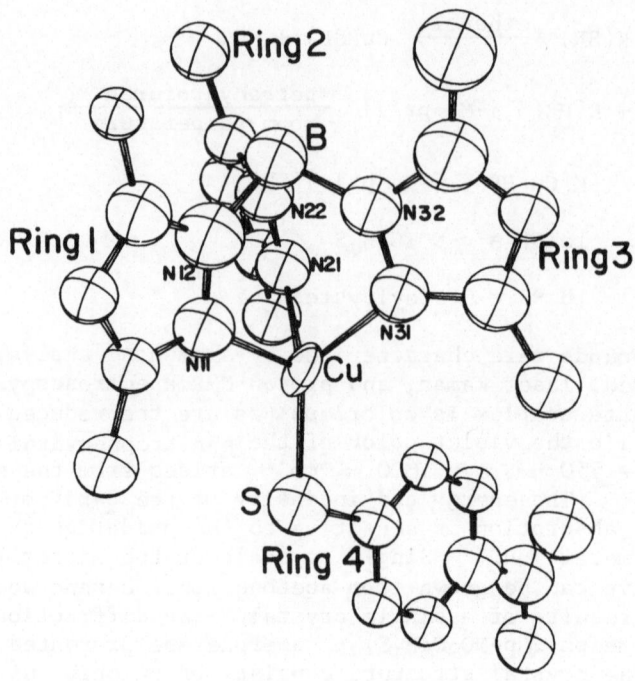

Figure 2. The molecular structure of the copper-containing portion of $K[Cu(HB(3,5-Me_2pz)_3)(\underline{p}-NO_2C_6H_4S)]\cdot2$ acetone from reference 16a.

Blue-black solids can be isolated by filtering the reaction mixtures of equation (5) at low temperature and then removing the solvent under high vacuum at ca. -40°C. The solids were characterized as 2a and 2b by elemental analysis and infrared spectroscopy. Based upon a variety of chemical and spectral data (vide infra) we formulate the structures of 2a and 2b as E. Chemical connection between species 1 and 2 is demonstrated by the fact that solutions of 2 can be reduced with potassium metal at -78°C (equation (6)) to regenerate the cuprous derivatives (1) in essentially quantitative yield. Additional support for

$$Cu(HB(3,5-Me_2pz)_3)(SR) + K \longrightarrow K[(Cu(HB(3,5-Me_2pz)_3)(SR)] \qquad (6)$$

the proposed constitution of 2a and 2b is derived from the synthesis of the analogous N,N-dimethyldithiocarbamate derivative 3 (equation (7)).[16c] The dithiocarbamate

$$Cu[S_2CN(CH_3)_2]Cl + K[HB(3,5\text{-}Me_2pz)_3] \xrightarrow{\text{tetrahydrofuran}}$$

$$\longrightarrow Cu(HB(3,5\text{-}Me_2pz)_3)(S_2CN(CH_3)_2) + KCl \qquad (7)$$

$$\underline{3}$$

ligand is known to be compatible with high metal oxidation states[24]
and consistent with this, the dark green, crystalline compound $\underline{3}$
is stable indefinitely at room temperature; it can be characterized
by elemental analysis, crysoscopic molecular weight in benzene,
EPR, etc. [16c] We propose structure $\underline{F}$ for this cupric dithiocarbamate complex.

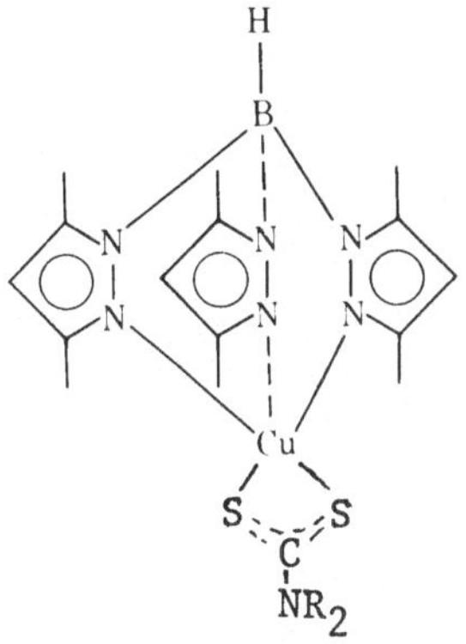

$$\underline{F}$$

Synthetic analogues of the cobalt-substituted type 1 site, $\underline{4}$,
can be prepared by the reaction sequence shown in equations (8)
and (9). [25] These air-sensitive

$$Co(ClO_4)_2 + KSR \xrightarrow{\text{ethanol}} Cu(SR)(ClO_4) + KClO_4 \qquad (8)$$

$$Co(SR)(ClO_4) + K[HB(3,5\text{-}Me_2pz)_3] \xrightarrow[78°]{\text{tetrahydrofuran}}$$

$$\longrightarrow Co(HB(3,5\text{-}Me_2pz)_3)(SR) + KClO_4 \qquad (9)$$

$\underline{4}$a　SR = $\underline{p}$ - $NO_2C_6H_4S$
$\underline{4}$b　SR = $\overline{O}$ - ethylcysteinate
$\underline{4}$c　SR = C_6F_5S

new complexes are stable at room temperature and have been characterized by elemental analysis, cryoscopic molecular weight, infrared spectroscopy, and proton NMR spectroscopy (relatively narrow,
isotropically shifted resonances are observed). [25]

SPECTRAL STUDIES
Electronic Spectra

As already noted, a characteristic feature of the type 1 site is an intense optical absorption at 600-625 nm ($\epsilon = 1000 - 5000$ M^{-1} cm^{-1})[1]. Additional, weaker absorptions are usually observed at 450 nm ($\epsilon \approx 300 - 1000$ $M^{-1}cm^{-1}$) and 800 nm ($\epsilon \approx 400$ $M^{-1}cm^{-1}$)[1]. A representative spectrum of an azurin is presented in Figure 3.

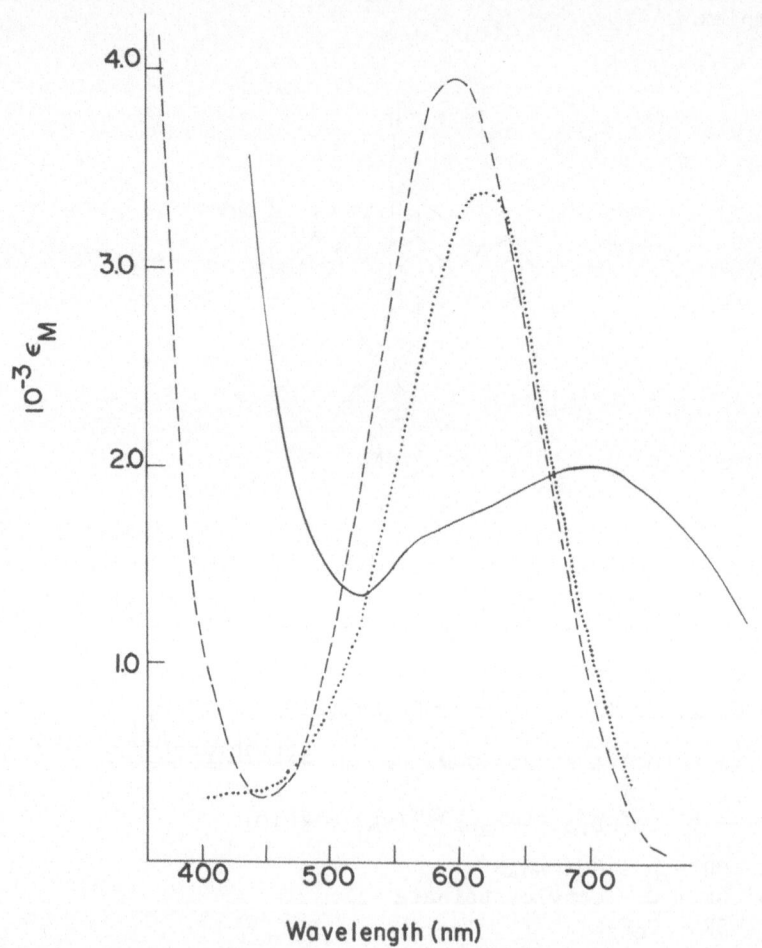

Figure 3. Electronic spectra of: ($\cdots$) <u>Pseudomonas aeruginosa</u> azurin from reference 26; (----) Cu(HB(3,5-Me$_2$pz)$_3$)(p-NO$_2$C$_6$H$_4$S) at 78°C from reference 16a; (——)Cu(HB(3,5-Me$_2$pz)$_3$)(0-ethylcysteinate) at -78°C from reference 16a.

The 600 nm band in the blue proteins has been suggested to arise from sulfur (σS) $\longrightarrow$ metal charge transfer.[1,9,15] Reference to Figure 3 reveals close agreement between the azurin optical spectrum and that of 2a (λmax = 599 nm, $\varepsilon \approx$ 3900 M^{-1} cm^{-1}) or 2b (λmax = 680 nm, $\varepsilon \approx$ 2000 M^{-1} cm^{-1}). That the 600 nm transition is indeed ligand-to-metal charge-transfer in character is confirmed by three lines of reasoning. First, the extinction coefficient is far too large for a d-d transition. Second, the shift of the 600 nm transition to higher energy as the mercaptide becomes more electron-withdrawing (2b $\longrightarrow$ 2a) is exactly what is predicted for a ligand-to-metal charge transfer transition.[27] Importantly, this type of observation can only be made with carefully selected synthetic analogues. Thirdly, the S $\rightarrow$ Cu(II) transition energy is nicely predicted by the empirical, optical electronegativity approach of Jørgensen.[27,28] As expressed in equation (9), the charge transfer frequency (ν_{CT}) is

$$\nu_{CT} = 30,000(\chi_{ligand} - \chi_{metal})cm^{-1} \qquad (9)$$

proportional to the difference in optical electronegativities of the ligand (χ_{ligand}) and the metal ion (χ_{metal}). Employing in equation (9) a tabulated[28] $\chi_{optical}$ for distorted tetrahedral Cu(II) of 2.3 and a tabulated[28] χ_{ligand} of 2.8(diethylthiophosphate) or 2.9(diethylsulfide) yields a predicted charge-transfer transition at 560-670 nm, which is in excellent agreement with experiment. The alkoxide complex, Cu(HB(3,5-Me$_2$pz)$_3$)(OC$_6$H$_5$),[16b] lacks the 600 nm transition, as expected.

It can also be seen that there is close correspondence between the optical spectra of the Co(II) - substituted blue proteins and those of 4a, 4b, and 4c. The cobalt proteins exhibit what is proposed to be an intense σS $\rightarrow$ Co(II) charge-transfer transition at approximately 340 nm.[25] The synthetic analogues exhibit maxima at 388 nm (R = cysteinyl), 318 nm (R = p-NO$_2$C$_4$H$_6$), and < 310 nm (R = C$_6$F$_5$). The ordering of the transition energies with respect to the electron-withdrawing properties of the mercaptides is again in accord with ligand-to-metal charge transfer. Furthermore, using the Cu(II) protein and analogue optical data, it is possible to solve equation (9) for the χ_{ligand} values. Substituting these data and the tabulated[28] tetrahedral χ_{cobalt} = 1.8 into equation (9) yields predicted charge-transfer energies of ca. 330 nm for the protein and ca. 310-340 nm for the synthetic analogues. These results are in excellent agreement with experiment and lend strong support to the contention that the 600 nm type 1 transition is S $\rightarrow$ Cu(II) charge-transfer in nature.

The Co(II) - substituted proteins exhibit weak absorptions in the 600 nm region which have been assigned to ligand field transitions.[2] The multiplicity is proposed to arise from the splitting (due to lowered symmetry) of the $^4T_1$ states which are degenerate in a tetrahedral ligand field. Thus cobalt substituted plastocyanin exhibits transitions at 600 nm (ϵ = 650 M^{-1} cm^{-1}) and 500 nm (ϵ = 540 M^{-1} cm^{-1}). Closely analogous to the protein result is the optical spectrum of 4a which exhibits bands at 625 nm (ϵ = 670 $M^{-1}cm^{-1}$) and 586 nm (ϵ = 550 $M^{-1}cm^{-1}$) as well as a shoulder on the low energy absorption at 675 nm (ϵ = 250 M^{-1} cm^{-1}).[25] Clearly, the ligand field in the cobaltous pyrazolylborate mercaptide complexes resembles that in the protein.

Resonance Raman Investigations

The type 1 proteins exhibit characteristic resonance-enhanced Raman scattering[29] when excited within the 600 nm optical transition.[5] Vibrations in the range 350-470 cm^{-1} have been assigned to Cu-N stretching modes and a single band at 260 cm^{-1} has been associated with Cu-S stretching. Though these assignments appear to be reasonable,[30] they are by no means secure.[31] Representative laser Raman spectra (spinning sample configuration) of 1 and 2 complexes are shown in Figure 4. The spectrum of 2a is remarkably similar to that of the blue proteins. The polarized bands at 395, 360, and 339 cm^{-1} are consistent with the three Cu-N stretching modes expected for the low molecular symmetry observed in 1a (Figure 1) and anticipated for E. The polarized emission at 276 cm^{-1} is assigned to the Cu-S stretching mode. Support for this assignment is provided by Raman studies on Cu(HB(3,5-Me_2pz)$_3$)-(OC_6H_5)[16b] in which this transition is absent. Figure 4 also presents Raman spectra (not resonance enhanced) of the cuprous analogue complexes. These provide the only conjecture available as to the nature of the reduced protein spectrum. Since there is not resonance enhancement, core vibrational spectra of the reduced proteins have not been observed. It is especially noteworthy that the Raman spectra of the Cu(I) and Cu(II) analogue complexes are so similar. Similarity in the structures and metal-ligand force constants should minimize the inner-sphere reorganization energy[32a,b] associated with electron transfer in both protein and analogues. Clearly the type 1 core has been designed for optimum Franck-Condon overlap, though the protein redox characteristic reflect many other fascinating subtleties as well.[32b,c,d]

Electron Paramagnetic Resonance Studies

The blue copper proteins display characteristic spin Hamiltonian parameters of $g_{||} \approx 2.19$-2.30 and $g_{\perp} \approx 2.03$-2.07 with an unusually small copper hyperfine coupling constant ($A_{||}$) of 3.3-9.0 millikaysers (mK) and $A_{\perp} \approx 0$.[1,6] The reason for the small magnitude of $A_{||}$ has evoked considerable discussion and explanations based upon the nature of the copper-mercaptide bond and/or the geometry of the copper site have been put forward. The synthetic analogues provide an opportunity to test some of these

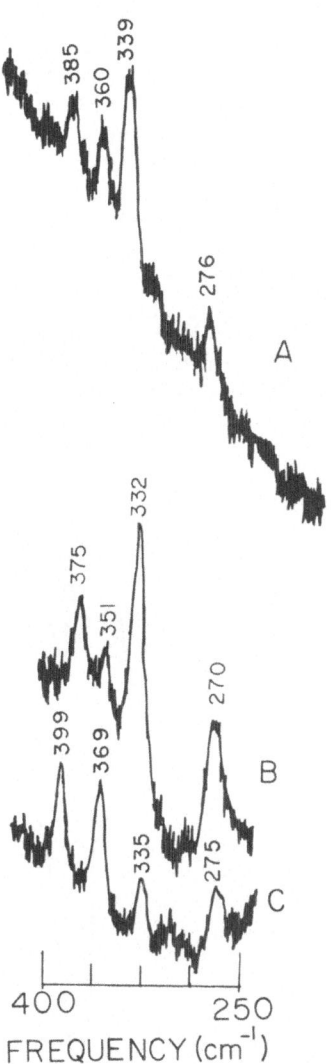

FREQUENCY (cm⁻¹)

Figure 4. Laser Raman spectra of: A: Cu(HB(3,5-Me₂pz)₃(p-NO₂-
C₆H₄S) at -80° in tetrahydrofuran solution with Kr⁺(6471Å) excitation.   B.   K[Cu(HB(3,5-Me₃pz)₄(p-NO₂C₆H₄S)] at room temperature
as a polycrystalline solid with Ar⁺ (5145 Å) excitation.   C.
K[Cu(HB(3,5-Me₂pz)₃(O-ethylcysteinate)] at room temperature as a
polycrystalline solid with Ar⁺(5145Å) excitation. From ref. 16a.

ideas.   The EPR spectrum of 2a in a tetrahydrofuran glass at -196°
C is presented in Figure 5.   Spin quantization by standard techniques[33] shows that this spectrum represents $\gtrsim 98\%$ of the copper in
the sample.   There is no evidence of molecular association which
might be indicated by $\Delta M_s = 2$, half-field transitions.[34]   For 2a
it is found that $g_{\parallel} = 2.286$ and $g_{\perp} = 2.067$.   These parameters are
close to typical type 1 values as exemplified by $g_{\parallel} = 2.273$ and
$g_{\perp} = 2.049$ in Bordetella pertussis azurin[1b] or be $g_{\parallel} = 2.287$ and
$g_{\perp} = 2.051$ in Rhus vernicifera stellacyanin.[1b] For 2b it is found
that $g_{\parallel} = 2.203$ and $g_{\perp} = 2.040$, which can be favorably compared
with $g_{\parallel} = 2.19$ and $g_{\perp} = 2.03$ for the type 1 site in Polyporus
versicolor laccase.[16] Clearly, the g values of the N_3CuS synthetic analogues are comparable to those of the blue copper proteins.
Several empirical studies[6,35] have shown that $g_{\parallel}$ values in Cu(II)
complexes reflect, in order of decreasing importance, ligands,
coordination geometry, and overall charge on the complex. Reference
to tabulated data[6,35] reveals that the g values of the Cu(HB(3,5-
Me$_2$pz)$_3$)(SR) complexes are in the range expected for CuN_3S coordination.

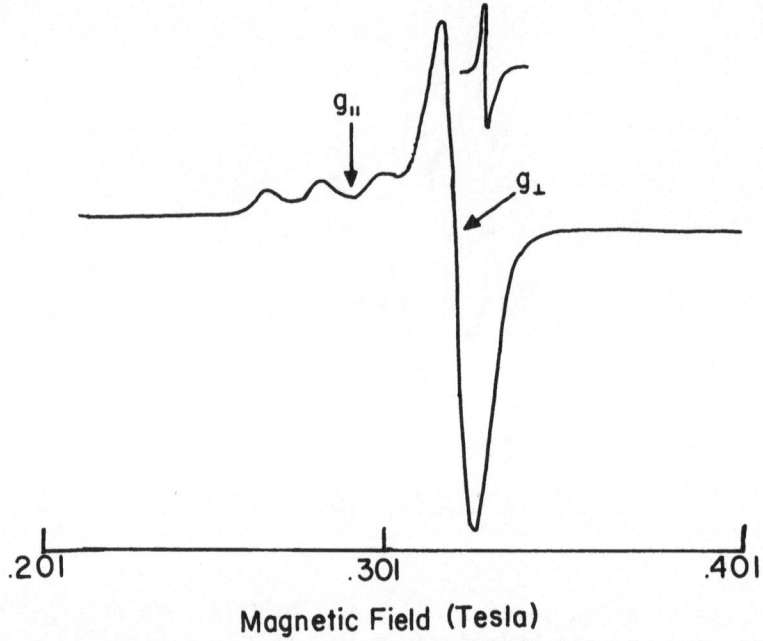

Magnetic Field (Tesla)

Figure 5.   Electron paramagnetic spectrum of Cu(HB(3,5-Me$_2$pz)$_3$)-
(p-NO$_2$C$_6$H$_4$S)(9.225 GHz) in a tetrahydrofuran glass at -196°C.
The inset resonance is the signal of the calibrant, strong pitch.
From reference 16a.

Hyperfine and superhyperfine information derived from the EPR spectra of 2a and 2b is also informative. The ^{63}Cu and ^{65}Cu (I = 3/2) hyperfine splitting in the $g_{||}$ parallel direction is considerably larger than for the type 1 sites. For 2a, $A_{||}$ = 17.1 mK and for 2b, $A_{||}$ = 17.0 mK. As for the $g_{||}$ parameters, empirical studies[6,35] reveal the $A_{||}$ values in Cu(II)complexes are dependent on ligands, coordination geometry, and overall charge. The relationship between $A_{||}$ and these factors is not as clear-cut as for $g_{||}$. However, it is certain from the literature data that the neutral charge of 2a and 2b is not, in itself, sufficient to explain the increased magnitudes of $A_{||}$. It also follows from this work that the presence of a cupric mercaptide functionality does not automatically guarantee a low value of $A_{||}$, as many investigators previously believed.[1] The spin Hamiltonian parameters for the complex Cu(HB(3,5-Me$_2$pz)$_3$)(OC$_6$H$_5$) further substantiate this: $g_{||}$ = 2.308, $g_{\perp}$ = 2.07, and $A_{||}$ = 16.7 mK. The most reasonable explanation for the variance of $A_{||}$ for 2a and 2b from the protein values is that the synthetic analogues are distorted from tetrahedral coordination by elongation of the ligation polyhedron along a three-fold axis, while flattening of the tetrahedron appears, from existing data,[6,35] to be more compatible with protein-like copper hyperfine interaction. Scattered wave Xα calculations, now in progress, will hopefully shed more light on this issue.[36]

The EPR spectra of 2a and 2b also provide further confirmation of structure E. Superhyperfine splitting (a seven line pattern) is observed in the $g_{||}$ direction and is consistent with the coordination of three magnetically equivalent nitrogen atoms (for ^{14}N, I = 1). The magnitude of A_N, ca. 0.8 mK, is in the range determined by ENDOR for the coordinated nitrogen atoms of Rhus vernicifera stellacyanin, viz. 1.2-1.6 mK.[37]

CONCLUSIONS

This work demonstrates that, with the appropriate synthetic strategy, it is possible to assemble low molecular weight approximations to the active sites of the blue copper proteins. The cuprous and cupric CuN$_3$(SR) redox pairs reported here represent the closest approach yet achieved to a proposed active site configuration. Considering the variation that appears to exist in the structure and properties of the native copper binding environments, and the lack of precedence for any structures of this type, studies of synthetic analogues offer the possibility of developing the spectroscopic and chemical criteria necessary to understand type 1 structure/function relationships. Further efforts will obviously be necessary to refine better the congruence between synthetic and native systems.

ACKNOWLEDGMENTS

This work was supported by generous grants from the National Science Foundation (CHE84-10341A02), the Northwestern Materials Research Center (NSF DMR72-0319A06), and the Camille and Henry Dreyfus Foundation. It would not have been possible without the valuable collaboration of Prof. J. A. Ibers, Dr. J. S. Thompson, and Mr. R. B. Osborne.

REFERENCES

1. a. H. Beinert, Coord. Chem. Rev., 15, 119(1977).
   b. J. A. Fee, Struct. Bond., 23, 1(1975).
   c. A. S Brill, "Transition Metals in Biochemistry," Springer-Verlag, New York, N. Y., 1977, Chapt. 3.
   d. R. Malkin in "Inorganic Biochemistry," G. L. Eichorn, Ed., Elsevier, Amsterdam, 1973, Vol. 2, Chapt. 21.

2. a. D. R McMillin, R. Rosenberg, and H. B. Gray, Proc. Nat. Acad. USA, 71, 4760(1974).
   b. D. R. McMillin, R. A. Holwerda, and H. B. Gray, Proc. Nat. Acad. Sci. USA, 71, 1339(1974).
   c. E. I. Solomon, J. Rawlings. D. R. McMillin, P. J. Stephens, and H. B Gray. J. Amer. Chem. Soc., 98, 8046(1976).

3. D. L. Tennent, B. L. Haufenstein, Jr., and D. R. McMillin, Abstracts, 175th National Meeting of the American Chemical Society, Anaheim, California, March,1978, No. INOR 56.

4. a. L. Ryden and J.-O. Lundgren, Nature, 261, 344(1976).
   b. G. McLendon and A. E. Martell, J. Inorg. Nucl. Chem., 39, 191(1977).

5. a. O. Siiman, N. M. Young and P. R. Carey, J. Amer. Chem. Soc., 98, 744(1976).
   b. V. Miskowski, S.-P. Tang, T. G Spiro, and T. H. Moss, Biochem., 14, 1244(1975).
   c. L. Tosi, A. Garnier, M. Hervé, and M. Steinbuch, Biochem. Biophys. Res. Comm., 65, 100(1975).

6. a. T. Vänngard in "Biological Applications of Electron Spin Resonance," H. M. Swartz, J. R. Bolton, and D. C. Borg, Eds., Wiley, New York, N. Y., 1972, Chapt. 9.
   b. H. Yokoi and A. W. Addison, Inorg. Chem., 16, 1341(1977), and references therein.
   c. U. Sakaguchi and A. W. Addison, J. Amer. Chem. Soc., 99, 5189(1977).
   d. W. B. Mims and J. Peisach, Biochem. 15, 3863(1976).
   e. J. Peisach and W. B. Mims. Proc. Nat. Acad. Sci. USA, in press. We thank Prof. Peisach for a preprint.

7. a. J. L. Markley, E. L. Ulrich, and D. W. Krogmann, Biochem. Biophys. Res. Comm., 78, 106(1977).
   b. K. Ugurbil, R. S. Norton, A. Allerhand, and R. Bersohn, Biochem., 16, 886(1977).

8. a. K. Ugurbil and R. Bersohn, Biochem., 16, 3016 (1977).
   b. H. A. O. Hill, J. C. Leer, B. E. Smith, and C. B. Storm, Biochem. Biophys. Res. Comm., 70, 331 (1976).
   c. J. L. Markley, E. L. Ulrich, S. P. Berg. D. W. Krogmann, Biochem., 14, 4428 (1975).
9. a. E. I. Solomon, J. W. Hare, and H. B. Gray, Proc. Nat. Acad. Sci. USA, 73, 1389 (1976).
   b. A. R. Amundsen, J. Whelan, and B. Bosnich, J. Amer. Chem. Soc., 99, 6730 (1977).
10. J. W. Hare, E. I. Solomon, and H. B. Gray, J. Amer. Chem. Soc., 98, 3205 (1976).
11. T. D. Tullius, P. Frank, and K. O. Hodgson, Proc. Nat. Acad. Sci. USA, in press. We thank Prof. Hodgson for a preprint.
12. S. Larsson, J. Amer. Chem. Soc., 99, 7708 (1977), and references therein.
13. H. B. Gray, Award Address, 175th National Meeting of the American Chemical Society, Anaheim, California, March, 1978.
14. P. M. Colman, H. C. Freeman, J. M. Guss, M. Murata, V. A. Norris, J. A. M. Ramshaw, and M. P. Venkatappa, Nature, 272, 319 (1978).
15. a. Y. Sugiura and Y. Hirayama, J. Amer. Chem. Soc., 99, 1581 (1977).
    b. Y. Henry and A Dobry-Duclaux, J Chim. Phys. Phys. Chim. Biol., 73, 1068 (1976).
    c. T. E. Jones, D. B. Rorabacher, and L. A. Ochrymowycz, J. Amer. Chem. Soc., 97, 7485 (1975).
    d. V. M. Miskowski, J. A. Thick, R. Solomon, and H. J. Schugar, J. Amer. Chem. Soc., 98, 9344 (1976).
    e. R. D. Bereman, F. T. Wang, J. Najdzionek, and D. M. Braitsch. J. Amer. Chem. Soc., 99, 8266 (1976).
    f. Y. Agnus, R. Louis, M. Schappacher, A. Mitschler, J. Fischer, and R. Weiss, J. Amer. Chem. Soc., in press. We thank Prof. Weiss for a preprint.
16. a. J. S. Thompson, T. J. Marks, and J. A. Ibers, Proc. Nat. Acad. Sci. USA, 74, 3114 (1977).
    b. J. S. Thompson, T. J. Marks, and J. A. Ibers, submitted for publication.
    c. R. B. Osborne and T. J. Marks, unpublished results.
    d. P. L. Dedert and T. J. Marks, unpublished results.
17. C. Mealli, C. S. Arcus, J. L. Wilkinson, T. J. Marks, and J. A. Ibers, J. Amer. Chem. Soc., 98, 711 (1976).
18. P. J. M. W. L. Beilser and H. C. Freeman, J. Amer. Chem. Soc., 99, 6890 (1977) and references therein.
19. M. I. Bruce and A. P. P. Ostazewski, J. Chem. Soc. Dalton Trans., 2433 (1973).
20. "Handbook of Biochemistry," 2nd edition, CRC Press, Cleveland, 1973, p. J-211.
21. S C. Tang, S. Koch, G. C. Papaefthymiou, S. Foner, R. B. Frankel, J. A Ibers, and R. H. Holm, J. Amer. Chem. Soc., 98, 2414 (1976).
22. J. S. Thompson, unpublished results.

23. F.H. Jardine, <u>Adv</u>. <u>Inorg</u>. <u>Chem</u>. <u>Radiochem</u>., <u>17</u>, 115(1975).

24. J. Willemse, J.A. Cras, J.J. Steggerda, and C.P. Keijzers, <u>Struct</u>. <u>Bond</u>., <u>28</u>, 93(1976).

25. J.S. Thompson, T.J. Marks, and J.A. Ibers, manuscript in preparation.

26. A.S. Brill, G.F Bryce, and H.J. Maria, <u>Biochim</u>. <u>Biophys</u>. <u>Acta</u>. <u>154</u>, 342(1968).

27. A.B.P. Lever, "Inorganic Electronic Spectroscopy," Elsevier, Amsterdam, 1968, Chapt. 8.

28. C.K. Jørgensen, <u>Prog</u>. <u>Inorg</u>. <u>Chem</u>., <u>12</u>, 101(1970).

29. a.   T.G. Spiro and P. Stein, <u>Annu</u>. <u>Rev</u>. <u>Phys</u>. <u>Chem</u>., <u>28</u>, 501 (1977).

    b.   B.B. Johnson and W.L. Peticolas, <u>Annu</u>. <u>Rev</u>. <u>Phys</u>. <u>Chem</u>., <u>27</u>, 465(1976).

30. a.   A. Müller, E.J. Baran, and R.O. Carter, <u>Struct</u>. <u>Bond</u>., <u>26</u>,

    b.   B.C. Cornilsen and K. Nakamoto, <u>J</u>. <u>Inorg</u>. <u>Nucl</u> <u>Chem</u>., <u>36</u>, 2467(1974).

    c.   K. Nakamoto, <u>Angew</u>. <u>Chem</u>. <u>Int</u>. <u>Ed</u>., <u>11</u>, 666(1972).

31. L. Tosi and A. Garnier, <u>J</u>. <u>Chem</u>. <u>Soc</u>., <u>Dalton</u> <u>Trans</u>., 53(1978).

32. a.   H. Taube, <u>Adv</u>. <u>Chem</u>. <u>Ser</u>., <u>162</u>, 127(1977).

    b.   S. Wherland and H.B. Gray in "Biological Aspects of Inorganic Chemistry," A.W. Addison, W.R. Cullen, D. Dolphin, and B.R. James, Eds., Wiley-Interscience, New York, N.Y., 1977, Chapt. 10.

    c.   M.G. Segal and A.G. Sykes, <u>J</u>. <u>Chem</u>. <u>Soc</u>. <u>Chem</u>. <u>Comm</u>., 764 (1977).

    d.   A.G. Sykes, private communication.

33. R. Aasa and T. Vanngard, <u>J</u>. <u>Mag</u> <u>Resonan</u>., <u>19</u>, 308(1975).

34. a.   D.N. Hendrickson and D.M. Duggan, <u>Amer</u>. <u>Chem</u>. <u>Soc</u>. <u>Symp</u>. <u>Ser</u>., <u>5</u>, 76(1974).

    b.   W.E. Hatfield, <u>Amer</u>. <u>Chem</u>. <u>Soc</u>. <u>Symp</u>. <u>Ser</u>., <u>5</u>, 108(1974).

    c.   S.G.N. Roundhill, D.M. Roundhill, D.R. Bloomquist, C. Lander, R.D. Willett, D.M. Dooley, and H.B. Gray, submitted for publication. We thank Prof. Roundhill for a preprint.

35. a.   J. Peisach and W.E. Blumberg, <u>Archives</u> <u>Biochem</u>. <u>Biophys</u>. 165, 691(1974).

    b.   Y. Murakami, Y. Matsuda, and K. Sakata, <u>Inorg</u>. <u>Chem</u>., <u>10</u> 1734(1971).

36. J.G. Norman, private communication.

37. G.H. Rist, J.S. Hyde, and T. Vänngard, <u>Proc</u>. <u>Nat</u>. <u>Acad</u>. <u>Sci</u>. <u>USA</u>, <u>67</u>, 79(1970).